VERTEBRATE PALAEONTOLOGY

VERTEBRATE PALAEONTOLOGY
Biology and evolution

MICHAEL J. BENTON
University of Bristol

London
UNWIN HYMAN
Boston   Sydney   Wellington

Published by the Academic Division of
Unwin Hyman Ltd
15/17 Broadwick Street, London W1V 1FP, UK

Unwin Hyman Inc.,
955 Massachusetts Avenue, Cambridge, Mass. 02139, USA

Allen & Unwin (Australia) Ltd,
8 Napier Street, North Sydney, NSW 2060, Australia

Allen & Unwin (New Zealand) Ltd in association with
the Port Nicholson Press Ltd,
Compusales Building, 75 Ghuznee Street, Wellington 1,
New Zealand

First published in 1990

British Library Cataloguing in Publication Data
Benton, Michael, *1956–*
    Vertebrate palaeontology.
1. Vertebrate palaeontology, history
I. Title
566.09

ISBN 0–04–566001–8
ISBN 0–04–566002–6

Library of Congress Cataloging-in-Publication Data
Benton, M. J. (Michael J.)
    Vertebrate palaeontology/Michael J. Benton
        p.   cm.
Includes bibliographical references.
ISBN 0–04–566001–8   ISBN 0–04–566002–6 (pbk.)
1. Vertebrates, Fossil.   I. Title
QE841.B44   1990
566—dc20   89–70704
                                             CIP

Typeset in 10 on 12 point Palatino by Columns of Reading,
and printed in Great Britain by Cambridge University Press

Preface

The story of the evolution of the vertebrates, the animals with backbones, is fascinating. There is currently an explosion of new research ideas in the field – the closest fossil relatives of the vertebrates, dramatic new fish specimens unlike anything now living, the adaptations required for the move on to land, the relationships of the early amphibians and reptiles, the biology of the mammal-like reptiles, the origins and biology of the dinosaurs, the role of mass extinction in vertebrate evolution, the earliest mammals, ecology and mammalian diversification, the origins and evolution of human beings.

The evolution of the vertebrates is of interest to us since we are vertebrates, and the subject of our origins has occupied the minds of philosophers, biologists, and palaeontologists for centuries. There is also a fascination for many people in exploring past worlds in detail. The age of the dinosaurs, 230–66 million years ago, or the age of armoured fishes, 520–360 million years ago, are so different from the world today that they seem like different universes. There is no need to read imaginative science fiction stories about imaginary Martians when we have fossil evidence for even more unusual and bizarre vertebrates on the Earth!

There are very few books currently available that describe the past diversity and the evolution of the vertebrates, and those that are do not fully represent the current lines of research. The best reference work is Carroll's *Vertebrate Paleontology and Evolution* (1987) which contains a summary of all extinct vertebrates and their basic anatomy.

My aims in writing this book were twofold. First, I wanted to present a readable account of the history of the vertebrates which is accessible to any interested person, whether having a professional or an amateur interest in the subject. The book follows the time-sequence of major events in the sea and on land, so that it can be read as a continuous narrative, or individual chapters may be read on their own. I have tried to show the adaptations of all major extinct groups, both in words and in pictures. My second aim was to show how the palaeobiological information is obtained by the research scientists, vertebrate palaeontologists, who devote their lives to particular groups or time units. To do this, I have mixed thematic sections with the straightforward narrative account. Detailed accounts are given throughout the book of specific research topics that are proving particularly controversial or novel at present: the biology of a helmeted fish, closest relatives of the tetrapods, a rich fossil deposit of early amphibians, jaw action and diet of

vii

mammal-like reptiles, ecology of dinosaur origins, thermal physiology of the dinosaurs, biology of a giant sauropod dinosaur, dinosaur extinction, the unusual mammals of South America, molecular information on mammalian phylogeny, climatic change and human origins, and so on.

I am indebted to many people who assisted in the production of this book, and who corrected many of the errors of fact or interpretation which existed in early drafts of the text. I thank Roger Jones of Unwin Hyman for suggesting the idea of this book, and Clem Earle and Andy Oppenheimer for editorial work on it. Andrew Milner read the whole text and made many useful suggestions. I also thank the following experts who gave me invaluable help with particular chapters or sections of the book: Peter Andrews, Bob Carroll, Jenny Clack, Joel Cracraft, Susan Evans, Jens Franzen, Nick Fraser, Brian Gardiner, Alan Gentry, Lance Grande, Bev Halstead, Philippe Janvier, Dick Jefferies, Zofia Kielan-Jaworowska, Gillian King, Alec Panchen, Bob Savage, Pascal Tassy, Mike Taylor, Nigel Trewin, and Cyril Walker. I thank Patrick Spencer for producing the Index. My main thanks go to Libby Mulqueeny who drew all of the diagrams for the book in a frenzy of work in the last months of 1988.

Michael J. Benton
August 1989

Contents

CONTENTS

CHAPTER ONE
Vertebrate origins

The oldest presumed fossil vertebrates are rather unimpressive specimens: small black phosphatic fragments of bone from the Late Cambrian (520–505 Myr ago) of Wyoming (Repetski 1978). Under the microscope these fragments can be seen to have a knobbly surface ornament that could represent scales of some kind, but they give no idea of the overall body outline of the animal (Fig. 1.1). They have been named *Anatolepis*, and it is assumed that they belonged to a jawless fish. The important point is that these fossils are composed of **apatite** (calcium phosphate) which is the mineralized constituent of bone, a characteristic of most vertebrates.

Figure 1.1 The oldest probable vertebrate, *Anatolepis* from the Late Cambrian of Crook County, Wyoming, USA. Fragment showing the scaly outer surface of the body armour, and a broken cross-section. Photographs, ×70 (left), ×200 (right). (Courtesy of Dr J. E. Repetski.)

From these small beginnings arose the great diversity of the backboned animals – the armoured fishes of the Silurian and Devonian, the ponderous amphibians of the Carboniferous and Permian, the dinosaurs of the Triassic, Jurassic, and Cretaceous, the bizarre horned mammals of the Tertiary, and human beings (see Appendix, *Geological time*).

The purpose of this chapter is to explore the various forms of evidence that can be used to reconstruct the story of the origin of the vertebrates. The evidence comes from the study of modern animals that are similar in certain ways to the vertebrates, and from fossils. There are two main views on the evolution of the vertebrates from their invertebrate ancestors, one of which is supported by a great deal of evidence, and the other more popular one based surprisingly on less evidence.

SEA SQUIRTS AND THE LANCELET

The vertebrates are members of a larger grouping, termed the phylum Chordata. Chordates, as the name suggests, are primarily characterized by the possession of a **notochord**, a flexible tough rod that runs along the length of the body near the back. The notochord in living chordates is generally made from an outer sheath of collagen, a tough fibrous connective tissue which encloses turgid fluid-filled spaces. There are two basal groups of living chordates, the sea squirts and the cephalochordates.

The sea squirts, or tunicates, subphylum Urochordata, could hardly seem to be less likely relatives of the vertebrates! A typical sea squirt is *Ciona* (Fig. 1.2a), which lives attached to rocks almost worldwide. It is a 100–150 mm tall bag-shaped organism with a translucent outer skin (the tunic) and two openings, or siphons, at the top. The body is firmly fixed to a hard substrate.

The internal structure is fairly complex (Fig. 1.2b). A large **pharynx** fills most of the internal space, and its walls are perforated by hundreds of gill slits, each of which bears a fringe of **cilia**, fine hair-like vibratile structures. Seawater is pumped through the inhalant siphon into the pharynx by beating movements of the cilia, and is then passed through a surrounding cavity, the **atrium**, and ejected through the exhalant siphon. The pharynx serves mainly to capture food particles from the stream of seawater which flows through it. The seawater is drawn into a filter bag of mucus which is produced inside the pharynx by a gland known as the **endostyle**. During feeding, this bag is continuously secreted and drawn into the **oesophagus** together with the food particles which it has filtered from the seawater.

Why is *Ciona* identified as a chordate? The pharynx and other structures are in fact very like those of the cephalochordates and lamprey larvae, but further evidence is to be found in the **larval** stage, when the sea squirt is a tiny free-swimming tadpole-shaped animal with a head and

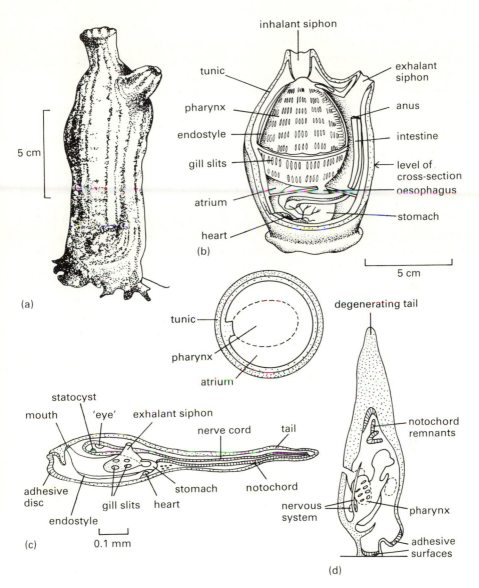

Figure 1.2 The sea squirts: (a) *Ciona*, external view; (b) internal anatomy and cross-section of an adult; (c) swimming larva; (d) metamorphosing form. (Figs (a) & (b) after Storer & Usinger 1965; (c) & (d) modified from Brien 1948.)

a tail. The larval sea squirt (Fig. 1.2c) has a notochord which runs along the tail, and this identifies it as a chordate. There are muscles on either side of the notochord which contract alternately, causing the tail to beat from side to side, and this drives the animal forward in the water. The larva has a dorsal nerve cord, running along the tail just above the notochord, and this expands at the front into a very simple brain which includes a light sensor (an 'eye') and a tilt detector.

The larva eventually settles on a suitable surface. It up-ends onto the tip of its 'snout' and attaches itself by means of adhesive suckers (Fig. 1.2d). The notochord and tail portion wither away, and the pharynx and gut expand to fill up the body cavity. This complete metamorphosis occurs rapidly in order to enable the adult to start feeding.

Another chordate generally reckoned to be related closely to the vertebrates is the amphioxus or lancelet, *Branchiostoma*, a representative of the subphylum Cephalochordata (or Acraniata). The adult amphioxus is convincingly chordate-like, being a 5 cm long cigar-shaped animal which looks like a young lamprey or eel. Amphioxus can swim freely by beating its tail from side to side, but it also burrows in the sediment on the sea-floor. It feeds by filtering food particles out of the seawater. Water carrying food particles is pumped into the mouth and through the pharynx by cilia or the gill slits, and food particles are caught up in a bag of mucus produced by the endostyle, the feeding system seen also in tunicates and in the larvae of the lamprey. The mucus with its contained food particles is pulled into the gut for digestion, while the seawater passes through the gill slits into the atrium. Oxygen is also extracted, and the waste water then exits through the **atriopore**.

The anatomy of amphioxus, with its pharynx, notochord, dorsal nerve cord, V-shaped muscle blocks (**myotomes**), and endostyle (Fig. 1.3) is typically chordate. Swimming and burrowing are by means of lateral contractions of the myotomes acting against the stiff rod-like notochord.

PTEROBRANCHS, ACORN WORMS AND CHORDATE ORIGINS

An unusual group of marine animals may offer further clues to the origin of the chordates. These are the hemichordates, a phylum that includes two superficially very different kinds of marine animals. The first, the pterobranchs such as *Cephalodiscus* (Fig. 1.4a & b), are small animals that live on the sea-bed in the southern hemisphere in loose 'colonies'. *Cephalodiscus* has a plate-like head shield, a collar with five to nine pairs of feeding arms, a sac-like trunk perforated by a pair of gill slits and containing the gut and gonads, and ends in a contractile stalk. Cilia on the arms produce a feeding current, and food particles are captured by mucus on the arms, while water passes out of the pharynx through the

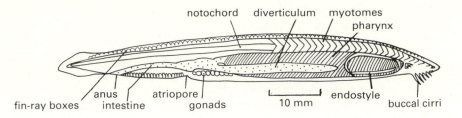

Figure 1.3 Internal anatomy of amphioxus, a cephalochordate. (Modified from Pough *et al.* 1989.)

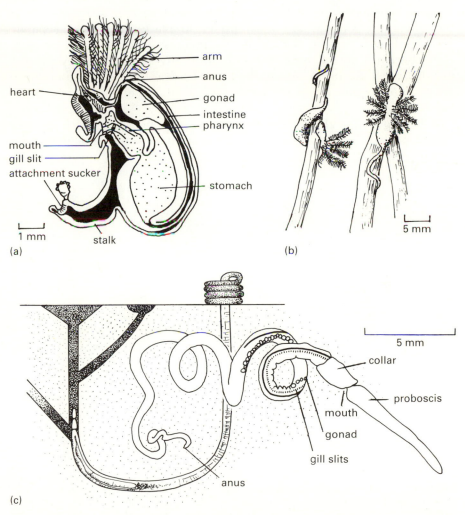

Figure 1.4 Typical hemichordates: (a) *Cephalodiscus*, internal anatomy and (b) mode of life; (c) *Saccoglossus*, mode of life and external anatomy. (Figs (a) & (c), after Barrington 1965; (b), after Andersson 1907.)

gill slits. The animal lives in or around a group of horny tubes which the colony has constructed, and it attaches itself inside these tubes by means of a sucker on the end of the stalk.

The second hemichordate group, the acorn worms such as *Saccoglossus*, are worm-like animals varying in length from 20 mm to 1.8 m. They live in burrows low on the shore in Europe and elsewhere. *Saccoglossus* (Fig. 1.4c) has a long muscular proboscis which fits into a fleshy ring or collar behind (hence the comparison with an acorn). The mouth is placed beneath this collar, and seawater and sand are pumped through the gut and expelled through an anus at the posterior end of the body. The long worm-like body is pierced by small holes at the front end, probably equivalent to the gill slits of *Cephalodiscus*, sea squirts, and amphioxus.

Some biologists have included the pterobranchs and acorn worms in the phylum Chordata, but most prefer to place them in a separate phylum Hemichordata, which is accepted as being closely related. Hemichordates do not have a notochord at any stage, but they possess gill slits and giant nerve cells in the nerve cord of the collar region which are probably equivalent to similar nerve cells in amphioxus and primitive vertebrates.

If the question of chordate relationships is widened in a search for their closest relatives among *major* invertebrate groups, some difficulties arise when only adult forms are examined. It is hard to find any reason for

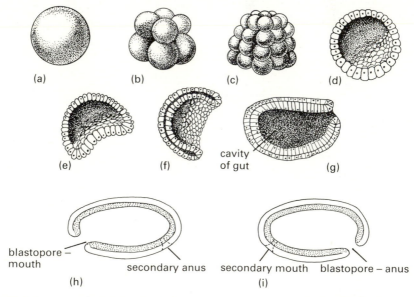

Figure 1.5 (a)–(g) Sequence of cell division in Amphioxus, from the single-cell stage (a), through the blastula stage (d), to the gastrula stage (g). (h) Fate of the blastophore in protostomes, and (i) in deuterostomes (Figs (a)–(g), after Hildebrand 1974, copyright © 1974 John Wiley & Sons, New York; (h) & (i), after Jefferies 1986.)

pairing adults of the phylum Chordata with any particular group of worms, molluscs, arthropods, or other potential relatives. The answer seems to come from **embryology**, the study of early phases of development in, and just out of, the egg.

In early development each animal starts as a single cell. Soon this cell begins to divide, first into two cells, then four, then eight, sixteen, and so on (Fig. 1.5a–c). Eventually a hollow ball of cells is produced, called the **blastula stage** (Fig. 1.5d). A pocket of cells then moves inwards, forming the precursor of the gut and other internal structures. You can imagine pushing in the walls of a hollow rubber squash ball with your thumb to produce a model of this embryonic pattern, known as the **gastrula stage** (Fig. 1.5e–g). Embryologists noticed some time ago that animals fall into two large groups depending on how the gastrula develops. In most invertebrates (the **protostomes**), the opening of the infolded gut structure becomes the mouth (Fig. 1.5h), while in others (the **deuterostomes**), including the chordates, this opening becomes the anus (Fig. 1.5i), and the mouth is a secondary perforation. Such a dramatic turnaround, a switch from mouth to anus, seems incredible, and its evolution is a mystery of course since there are no intermediate stages! However, this peculiarity of embryological development appears to solve the question of the broader relationships of chordates. The deuterostomes are the phyla Chordata, Hemichordata, and Echinodermata.

The closest major group of living relatives of the chordates and hemichordates are thus the echinoderms – sea urchins, star fish, sea lilies, and sea cucumbers.

CLADISTICS

The principal analytical technique used in establishing the relationships of vertebrates and their relatives is called **cladistics** (see Patterson 1982 and Ax 1985 for summaries of the techniques and applications).

The result of a cladistic analysis is a **cladogram**, such as that in Figure 1.6. A cladogram is not an evolutionary tree since there is no absolute time-scale (we don't know when all of the changes took place, but can postulate their relative order). The cladogram shows the closeness of relationship, or recency of common descent, by the arrangement of the groups – the closer they are to each other, and the shorter the linking lines, the closer is the postulated relationship. The cladogram is based on an assessment of **characters**, and shared derived characters (**synapomorphies**) are distinguished from primitive characters, like the possession of a gut, which is widespread outside this group. Derived characters are found only in the members of a particular group which is then said to be **monophyletic**, that is, arisen from a single ancestor and including all descendants of that ancestor (Fig. 1.7a). Most familiar named groups of animals are monophyletic groups (also termed **clades**): examples are the phylum Chordata, the sub-

---

7

Box – continued

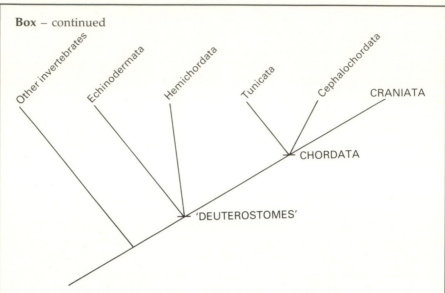

Figure 1.6 Cladogram showing the relationships of the main deuterostome groups.

phylum Vertebrata, the family Canidae (dogs), and so on (see Appendix).

Standard classifications of vertebrates and other groups often include non-monophyletic groups. The commonest examples are **paraphyletic** groups which include only the most primitive descendants of a common ancestor, but exclude some advanced descendants (Fig. 1.7b). A well-known paraphyletic group is the class 'Reptilia' which we can assume arose from a single ancestor, but which excludes some descendants, the birds and the mammals. The upper bounds of the class 'Reptilia' are defined only by the absence of characters (e.g. feathers and hair), and the group cannot be said to be natural.

The other kind of non-monophyletic groups are **polyphyletic**, those whose latest common ancestor is not included in the group, and which are characterized by a convergent feature (Fig. 1.7c). It is hard to think of a good example of vertebrates, but an imaginary class 'Natantia' which would include fishes, seals, and whales, could suffice.

The basis of cladistics is an analysis of characters in which the **polarity** is assessed, the direction of evolution from the primitive to the derived

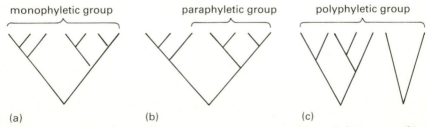

Figure 1.7 Hypothetical cladogram patterns showing (a) a monophyletic group, (b) a paraphyletic group, and (c) a polyphyletic group.

character state. There are often problems in distinguishing just what are derived characters, and what are not: the classic evolutionary dilemma of separating **homologies** from **analogies**. An homology is a feature seen in different organisms that is the same in each – it is anatomically and functionally equivalent, and shows evidence of derivation from a single source – while an analogy is a feature that may look or act in broadly similar ways in different organisms, but which gives evidence of separate origins. An example of an homology is the wing of a robin and the wing of an ostrich. Although the ostrich wing is not used in flight, its location in the body and its detailed structure show that it is a direct equivalent to the robin wing, and the latest common ancestor of robins and ostriches would have had such a wing. The wing of a robin and the wing of a fly are analogies since all the evidence suggests that they arose quite independently, even though they perform similar functions. Homologies, or shared derived characters (synapomorphies), are the clues to the pattern of evolution (the **phylogeny**) since they indicate common ancestry.

CHORDATES AND THE TALE OF THE TAIL

The animal groups we have considered so far may be arranged in a cladogram (Fig. 1.6) which indicates the postulated pattern of relationships based upon a cladistic analysis of the living forms (see Box). The echinoderms and hemichordates are associated with the chordates as their closest relatives (close **outgroups**) since all three phyla are deuterostomes (see p. 7). However, there is some controversy over relationships within the group, which is why there are two unresolved three-way branching points.

Maisey (1986) and others would place the hemichordates as the closest outgroup (the **sister group**) of the Chordata since they both share ciliated gill slits and giant nerve cells which are not seen in echinoderms. Jefferies (1986), on the other hand, considers that the echinoderms are closer relatives of the chordates because of certain fossil evidence (see p. 10).

The chordates all share several derived characters such as the notochord, the dorsal hollow nerve cord, the endostyle organ (equivalent to the thyroid gland of vertebrates), and possibly also the tail. It is generally reckoned that only chordates have true tails. A tail technically may be defined as a distinct somatic region extending behind the visceral cavity, and in particular located entirely behind the anus. Non-chordates, such as insects, worms, molluscs, jellyfish, and sea urchins, do not have tails.

Further problems arise within the Chordata, in trying to sort out the relationships of the tunicates, cephalochordates, and vertebrates. Most authors (e.g. Janvier 1981, Maisey 1986) regard amphioxus as the sister group of the Vertebrata on the basis of up to 15 postulated synapomorphies, but these are matched by some postulated synapomorphies of tunicates and vertebrates found in the fossils (Jefferies 1986). Can these opposing viewpoints be resolved?

Calcichordates – fossil chordates?

The calcichordates are a group of about 60 species of asymmetrical organisms which had a **calcitic** (calcium carbonate) outer skeleton. They date from the Middle Cambrian to Middle Devonian (540–370 Myr). They consist of two parts (Fig. 1.8a), a compact body portion and a long segmented appendage. Most authors have interpreted the calcichordates as aberrant echinoderms, but Jefferies (1986) has argued strongly that they are chordates, placing the various species on the cladogram between echinoderms, cephalochordates, sea squirts, and vertebrates. He interprets the appendage as a tail, while his critics call it a locomotory stem or feeding arm. A major opening in the body is called a mouth by Jefferies, and an anus by others. He interprets a series of openings as pharyngeal gill slits, while others call them inhalant respiratory pores. The argument

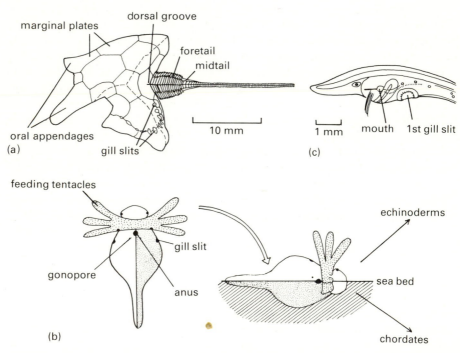

Figure 1.8 The calcichordates: (a) reconstruction of *Ceratocystis*, a Middle Cambrian cornute; (b) dexiothetism, in which the right-hand side of a bilaterally symmetrical hemichordate (left) rotates through 90°, and is largely lost (right); (c) early embryo of Amphioxus, with asymmetrical gill slits and mouth. (Fig. (a) after Jefferies 1969; (b) modified from Jefferies 1986; (c) after Bone 1958.)

hinges largely on interpretations of the fossils, but these do show striking similarities with pterobranchs, tunicates, and cephalochordates, and the model helps to explain some otherwise curious coincidences in the anatomy of the living groups.

The crux of Jefferies' (1986) ideas is **dexiothetism**, the theory that echinoderms and chordates are equivalent to hemichordates turned upright from lying on their right-hand sides (Fig. 1.8b). The water vascular, respiratory, reproductive, and digestive systems of echinoderms all show a strange asymmetry during development. Early larval echinoderms are symmetrical, but then many of the organs in the right-hand side atrophy, or rotate through 90°. Consequently, in a crinoid for example, the coelom of the larval left side comes to face upwards, while the coelom of the right side faces towards the sea-bed.

The development of amphioxus gives support to Jefferies' views: the early embryonic larva is distinctly asymmetrical and seems to show a parallel of its presumed dexiothetic ancestry. The mouth and the gill slits both arise on the left side of the body (Fig. 1.8c). At metamorphosis, the asymmetrical larva becomes nearly symmetrical (Fig. 1.3): the mouth moves to the midline position, and new series of right-hand gill slits appear. Jefferies (1986) argues that this shows how the chordates had a dexiothetic origin from formerly asymmetrical, one-sided beasts. Our present symmetry, he argues, is secondary, a subsequent modification of the left-hand side of our distant ancestors.

Jerreries' (1986) theory of vertebrate origins is supported by a large amount of fossil and embryological evidence. However, it has failed so far to gain wide acceptance partly because of the interpretive nature of much of the fossil evidence: how can we be sure that the soft anatomy of calcichordates has been correctly interpreted? Also, the theory involves a number of major character losses. If hemichordates and calcichordates have true tails, then echinoderms must have lost the tail. Further, the calcite skeleton of the calcichordates and echinoderms was apparently lost three times, on the lines to the cephalochordates, tunicates, and vertebrates (Fig. 1.6). Thirdly, the opposing theory for vertebrate origins, although supported by very little evidence, is seemingly more elegant: this is that we arose ultimately from the sea squirt tadpole.

The 'juvenile adult' theory of craniate origins

Walter Garstang noted the similarities between the larval sea squirt (Fig. 1.2c), adult amphioxus (Fig. 1.3) and vertebrates. The sea squirt tail seemed to Garstang to be a transient appendage that evolved as an outgrowth from the body to ensure wide dispersal of the larvae before they settled. Garstang proposed in 1928 that the evolutionary link between the sea squirts and all subsequent chordates in the evolutionary tree is a process termed **paedomorphosis**, the full development of the gonads and reproductive abilities in an essentially juvenile body. A sea squirt larva of 540 Myr ago, or more, apparently failed to metamorphose and became adult (i.e. reproductively mature) while retaining the tail, notochord, and dorsal nerve cord, and lost the sedentary adult sea squirt stage altogether.

This paedomorphic theory can only be surmized on the basis of the few living representatives of non-craniate chordates. However, it has been accepted by most biologists and palaeontologists because of its elegance, and because of the common occurrence of paedomorphosis elsewhere in vertebrate evolution.

Resolution

It is equally possible to interpret the larval sea squirt, not as a 'forerunner' of adult vertebrates, but rather as a **recapitulation** or 'throw back' to adult calcichordates. In this case, the sea squirt tadpole is a soft-bodied calcichordate, and sea squirt development has then been extended in length in comparison with the calcichordates and vertebrates, a process which is the opposite to paedomorphosis. Only further study of the details of calcichordate anatomy and the early phase of embryological development of living chordates will strengthen Jefferies' (1986) theory, or turn up some evidence for Garstang's (1928) paedomorphic model.

CRANIATES AND THE HEAD

The craniates, or vertebrates, the major group of chordates, form the subject of this book. They are more properly termed craniates since all forms, including the hagfishes and lampreys, have specialized head features (the **cranium** is the core of the skull). The term 'vertebrate' is less appropriate since some of the basal living craniates, the hagfishes, do not have vertebrae, while neither hagfishes nor lampreys have bone. At present it is hard to establish the relationships of these forms and their fossil relatives, but the term 'vertebrate' might better be reserved for a position higher in the cladogram (Janvier 1981). However, the word 'craniate' will be used here interchangeably with the better-known 'vertebrate' in the traditional broad sense.

The basic craniate body plan (Fig. 1.9) shows all of the chordate characters so far described – notochord, dorsal nerve cord, pharyngeal 'gill' slits, postanal tail, myotomes, and so on. The special vertebrate characteristics include a range of features that make up a true head: well-defined sensory organs (nose, eye, ear) with the necessary nervous connections (the **cranial nerves**) and the olfactory, optic, and auditory (otic) regions which make up a true brain. Larval sea squirts and amphioxus have an expansion of the nerve cord at the front end and some sensory organs, as we have seen, but these are not developed to the

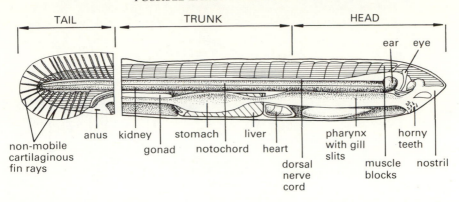

Figure 1.9 The hypothetical 'basic' craniate body plan, shown in longitudinal section. (After Jefferies 1986.)

same level as in craniates. It can be said then that craniates are characterized by their heads, just as chordates, or possibly all deuterostomes, are by their tails.

POSSIBLE EARLY FOSSIL CHORDATES

Over the years, palaeontologists have described a number of fossils as possible early chordates. Regrettably, many of the fossils are poorly preserved in certain crucial parts and it is hard to be sure what they are. However, there are three main categories of possible early chordates: calcichordates (described above), possible cephalochordates, and conodonts.

Pikaia from the Middle Cambrian of Canada is a small fish-like animal, preserved as a flattened carbonized film on the rock. It has a longitudinal bar (? notochord) and divided segments (? myotomes) along its side, and its overall body outline (Fig. 1.10a) is reminiscent of amphioxus.

The conodonts have long been suspected of having chordate affinities because they are composed of apatite. Conodonts are small tooth-like fossils, mainly of Palaeozoic (570–245 Myr) age, and are so abundant in places that they are used for dating the rocks. After many false alerts, the 'conodont animal' was actually found in 1982 (Briggs *et al.* 1983). It was 40 mm long, worm-shaped, and it had a conodont apparatus (an interlocking set of conodont elements) in its head region (Fig. 1.10b). This confirmed that the conodonts were feeding structures. The conodont animal was interpreted as a chordate because it had traces of V-shaped myotomes in the trunk, and a tail fin. Conodont animals are now known from the Carboniferous and the Silurian (Aldridge *et al.* 1986). These latter authors have in fact interpreted the conodont animals not only as chordates, but also as craniates. The conodont apparatus is comparable in

13

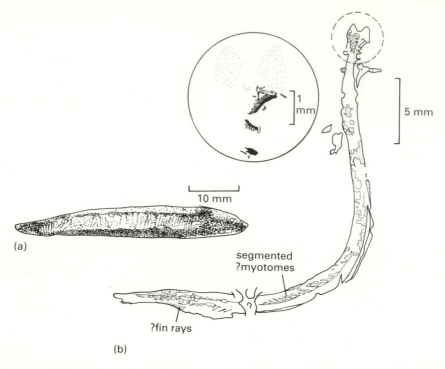

1 mm

5 mm

10 mm

(a)

segmented
?myotomes

?fin rays

(b)

Figure 1.10 Possible early fossil chordates: (a) *Pikaia*, tracing from a photograph; (b) *Clydag-nathus*, the conodont animal, drawing of one specimen, and enlargement of the conodont elements in the head region. (Fig. (a) after Jefferies 1986; (b) after Briggs *et al.* 1983.)

some respects to the tooth arrangement of the jawless hagfishes (see p. 25), and the absence of a bony skeleton in the conodont animal is not seen as an impediment to such a relationship since living hagfishes and lampreys also lack such hard tissues.

CHAPTER TWO

Early fishes

The earliest remains of fishes from the Cambrian do not show much of their overall shape or anatomy, and very little can be deduced about their modes of life. However, fish fossils from Ordovician times (505–438 Myr) onwards are often preserved complete, and with a great deal of fine anatomical detail. It is possible to identify four or five major fish lineages in the Ordovician, Silurian, and Devonian, and to trace their evolution in some detail. Although some of these fishes would look familiar to us, most of them were very different from modern forms. The key episodes of early fish evolution seem to have taken place during the Devonian period (408–360 Myr) when, firstly, the early dominance of the seas and freshwaters by heavily armoured forms gave way to the modern sharks and bony fishes and, secondly, the fishes gave rise to the first land vertebrates, the amphibians.

DEVONIAN ENVIRONMENTS

The Ordovician, Silurian, and Devonian world was very different from today, largely because of an entirely different continental layout. Not only have areas that were once under the sea now been uplifted to form land, and vice versa, but the continents and oceans themselves have moved bodily. The theory of **continental drift** is now familiar, and it is accepted by virtually all geologists.

A wealth of geological and palaeontological evidence confirms that the crust of the Earth is divided into twenty or so major **plates**, each of which moves horizontally relative to the others. By projecting present plate movements back into the past, and by examining other evidence such as the geographic distribution of fossils, geologists have attempted reconstructions of the Earth's surface in the past. Of course, these exercises become more and more uncertain the further back in time one goes. Hence, **palaeogeographic** maps of the Silurian and Devonian worlds are controversial in many respects.

It is possible to distinguish **faunal provinces** among early fishes. For

15

example, there was a Scotto–Norwegian fauna of thelodonts (small jawless fishes) in the Silurian which differed in many respects from the Acadian–Anglo–Welsh fauna (Fig. 2.1). In other words, Silurian fish fossils from Wales and southern England are more like those from the eastern parts of North America and Greenland (Acadia) than those from the central parts of Scotland or from Norway. It is assumed that there were barriers to mixing among these thelodont faunas, such as major land masses or wide oceans. These particular barriers seem to have disappeared largely in the Early Devonian, because a single thelodont fauna occurs nearly worldwide. Most other Silurian and Devonian fish groups show similar patterns of distribution.

The Silurian and Devonian seas and freshwaters are assumed to have been warm, and fish fossil localities are clustered in the equatorial and tropical belt (Fig. 2.1). Important environmental changes took place on land during the Silurian and Devonian, and these affected vertebrate evolution. The first land plants appeared in the Late Silurian. They were small and reed-like, and probably grew around ponds and lakes with their tuberous roots partly in the water. Early Devonian terrestrial rocks very rarely contain fossils of land plants or animals, but by Middle and Late Devonian times, large horsetails and scale trees (lycopods) became quite common. The first land animals may have been scorpions which could live in water and on land. They first appear in the Late Silurian. In

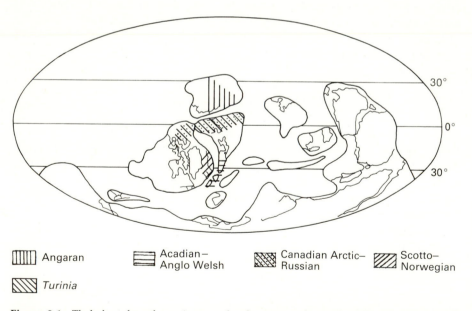

Figure 2.1 Thelodont faunal provinces and palaeogeography in the Silurian. Continental outlines for those times are shown with heavy lines, and modern continental margins are shown with fine lines. The five thelodont provinces are indicated by shading. (After Halstead 1985.)

the Early Devonian, fossils of spiders, mites, and wingless insects have been found, and the diversity of insects increased in the Late Devonian. These new plant and animal groups that spread on to the land in the Late Silurian and Devonian provided new sources of food for animals that could exploit the shallow waters of the lakes and the land around the edges.

THE EXTINCT JAWLESS FISHES

The earliest fishes are included in the paraphyletic class Agnatha (literally, 'no jaw'), and they achieved a great diversity of forms and sizes in the Ordovician to Devonian periods. The main fossil agnathan groups are the early Arandaspida and Astraspida, the heavily armoured Heterostraci, the poorly armoured Thelodonti, the armoured Osteostraci, the Galeaspida, the Anaspida, and the living Petromyzontiformes (lampreys) and Myxinoidea (hagfishes).

Ordovician jawless fishes

The oldest-known vertebrate fossils (Fig. 1.1) are usually classed as heterostracans, but are now placed in the separate orders Arandaspida and Astraspida. They have a mobile tail covered with small protruding pointed plates, and a massive bony head shield made from several large plates that cover the head and front part of the body. In some early forms, this shell extends back to the anus.

Arandaspis from the Middle Ordovician of Australia (Ritchie & Gilbert-Tomlinson 1977) has a head shield (Fig. 2.2a) made from a large **dorsal** (upper) plate that rose to a slight ridge in the midline, and a deep curved **ventral** (lower) plate. Narrow **branchial plates** link these two along the sides, and cover the gill area. Small scales covered the rest of the body behind the head shield. The eyes are very far forward and set close together, the mouth is armed with very thin plates, and there may be two nostrils. The fossils show clear evidence of a sensory structure that is peculiar to all fishes (except hagfishes) – the **lateral line** system. This is a line of open pores within each of which are open nerve endings that can detect slight movements in the water, produced for example by predators. The arrangement of these organs in regular lines allows the fish to detect the direction from which the disturbance is coming and its distance. The outer surface of the plates is ornamented with small lozenge-shaped tubercles in regular patterns (Fig. 2.2b).

Astraspis from the Ordovician of North America (Elliott 1987) also has a series of separate gill openings along the side of an extensive head shield behind the eye (Fig. 2.2c). The body is oval in cross-section, and covered with broad overlapping scales, and the tail terminates as a symmetrical flat-sided structure.

17

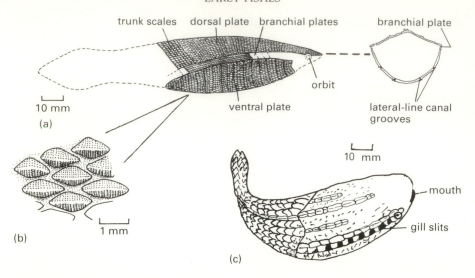

trunk scales   dorsal plate   branchial plates   branchial plate

orbit

10 mm

(a)

ventral plate   lateral-line canal
                       grooves

10 mm

mouth

gill slits

1 mm

(b)

(c)

Figure 2.2 Ordovician jawless fishes: (a) *Arandaspis*, from the Middle Ordovician of Australia, in side view (left), in cross-section (right), and (b) detail of dermal tubercles; (c) reconstruction of *Astraspis* from the Ordovician of North America. (Figs (a) & (b) after Ritchie & Gilbert-Tomlinson 1977; (c) based on Elliott 1987.)

Order Heterostraci

The early agnathans have several gill slits, which distinguishes them from the heterostracans which have one common external gill opening on each side. The heterostracans began to radiate extensively in the mid to Late Silurian and the Early Devonian. Their head shields varied in shape tremendously between different species. They all had in common a broad ornamented plate in the middle of the top of the head, one or more plates on either side of this, and several large elements covering the underside (Halstead 1973). Four main groups emerged, the cyathaspids, amphiaspids, pteraspids, and psammosteids. The cyathaspids (Fig. 2.3a) are completely encased in bony plates and scales. The dorsal, ventral, and branchial plates (or shields) are broadly similar in shape to those of *Arandaspis* but they are not covered with an ornament of tubercles, merely short narrow ridges of dentine. The mouth appears to have operated by movements of small ventral plates, but there are no jaws, so that feeding must have occurred by suction. The body is deep and covered with large bony scales which overlap backwards like the slates on the roof of a house. There is no sign of fins or paddles in *Anglaspis*, so it must have swum by moving its body from side to side, a rather clumsy mode of locomotion because of the weight of the armour and the inability to adjust the direction of movement.

The amphiaspids show complete fusion of the head shield into a single all-round carapace, and the eyes are reduced or lost. It has been suggested that the amphiaspids lived partially buried in the mud where sight was not required: some forms, such as *Eglonaspis* (Fig. 2.3b), have a long tube at the front of the carapace with the mouth opening at the end, possibly used as a kind of 'snorkel' when burrowing.

The pteraspids of the Early and Middle Devonian are much better known, with more than 25 genera, which show considerable variation in the shape of the head shield (Blieck 1984). In *Pteraspis* (Fig. 2.3c) there are large dorsal and ventral plates, the linking branchial plate, as well as a **cornual plate** at the side, an **orbital plate** around the eye, a **rostral plate** forming a pointed 'snout', several small plates around the mouth, and a dorsal spine pointing backwards. The rest of the body is covered with small scales that look more like modern fish scales than the bony plates of

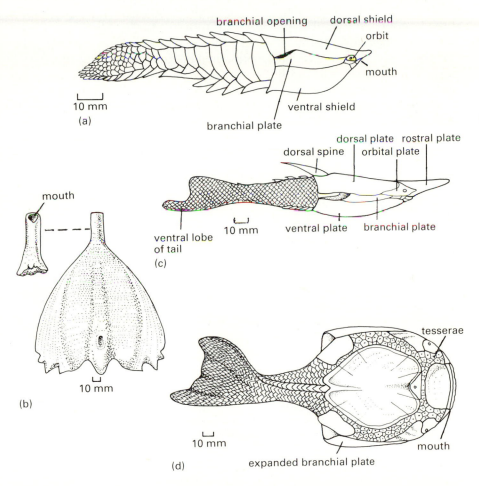

Figure 2.3 Heterostracans: (a) *Anglaspis*; (b) *Eglonaspis*, dorsal view of head shield and underside of mouth tube; (c) *Pteraspis*; and, (d) *Drepanaspis*. (After Moy-Thomas & Miles 1971.)

earlier forms. The tail has a long ventral lobe which would have assisted in control during swimming.

The psammosteids are much flatter than the others (Fig. 2.3d) and have several rows of small scale-plates called **tesserae** lying between the main shields. The flattening of the body has pushed the eyes well apart and turned the mouth upwards.

Order Thelodonti

The thelodonts are poorly known since intact specimens are rare, although isolated scales are common enough in Late Silurian and Early Devonian rocks of various parts of the world. *Phlebolepis* is well-enough preserved for a reconstruction to be attempted (Fig. 2.4a & b; Ritchie 1968; Märss 1986). The 70 mm long body is slightly flattened in shape, with a broad snout, an eye at each side and a wide mouth. There are lateral flaps, a dorsal and ventral 'fin' near the back, and a long lower tail fin. The body is completely covered with small pointed scales, and there is no sign of bone shields in the head region at all. There are eight small gill openings beneath the lateral fins.

The scales (Fig. 2.4c) are lozenge-shaped, hollow beneath, and rising to a point above. The exact shape of the scales and the arrangement of the spines and nodules are used to identify thelodont species from isolated specimens. In cross-section (Fig. 2.4d), the scale is seen to be made from dentine (as in a tooth) around an open pulp cavity (again, as in a tooth).

Order Osteostraci

The Osteostraci radiated in the Late Silurian and Early Devonian. They are heavily armoured in the head region, like the heterostracans, and

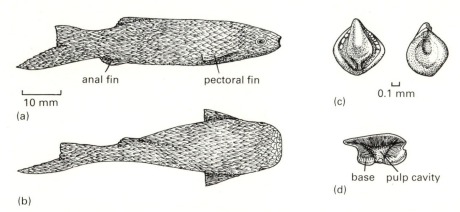

anal fin   pectoral fin

10 mm

(a)

(b)

0.1 mm

(c)

base   pulp cavity

(d)

Figure 2.4 Thelodonts: (a) whole-body restoration of *Phlebolepis*, in lateral and (b) dorsal views; (c) scales of *Logania* in dorsal (left) and ventral (right) views; and (d) scales of *Thelodus* in cross-section. (Figs (a) & (b) after Ritchie 1968; (c) based on Turner 1973; (d) based on Moy-Thomas & Miles 1971.)

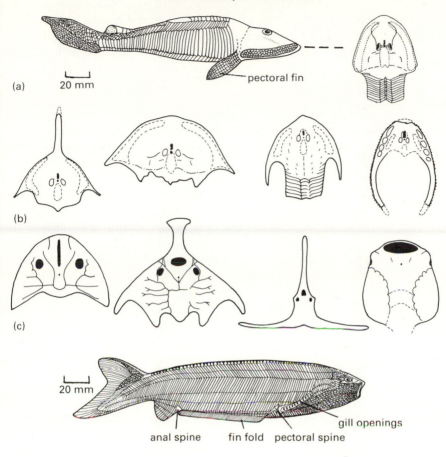

(a)

20 mm

pectoral fin

(b)

(c)

20 mm

gill openings

anal spine   fin fold   pectoral spine

(d)

Figure 2.5 Diverse agnathans of the Silurian and Devonian: (a) & (b) osteostracans; (c) galeaspids; and (d) anaspid. (a) *Hemicyclaspis* in lateral view, and dorsal view of head shield; (b) osteostracan head shield diversity; from left to right, *Boreaspis*, *Benneviaspis*, *Thyestes*, and *Sclerodus*; (c) galeaspid head shield diversity; from left to right, *Eugaleaspis*, *Sanchaspis*, *Lungmenshanaspis*, and *Latirostraspis*; (d) *Pharyngolepis*. (Figs (a) & (d) after Moy-Thomas & Miles 1971; (b) after Stahl 1974, based on Gregory 1951; (c) after Janvier 1984.)

most have a characteristic flattened curved semicircular head shield shaped rather like the toe of a boot. *Hemicyclaspis* from the Early Devonian of Europe has a solid **carapace** made from a single bony plate that enclosed the head region (Fig. 2.5a). Behind the head plate are a pair of **pectoral fins** covered with small scales, and these could presumably have been used in swimming. The body and tail are covered with broad scales on the side and beneath, and narrower ones on top which form a dorsal ridge and a dorsal fin.

In the course of their evolution, the head shield of osteostracans adopted a variety of forms, ranging from an elongate bullet shape in some early examples, to rectangular and hexagonal forms, some with backward-pointing spines, or cornua, and one even with a long rostral spine in front (Fig. 2.5b). Osteostracan biology is detailed below (see pp. 22–24).

Order Galeaspida

The galeaspids are remarkable fossils from the Early Devonian of China (Janvier 1984). Many have broad head shields rather like those of the osteostracans, but others show the development of an impressive array of processes (Fig. 2.5c): curved cornua, multiple cornua pointing backwards, a 'hammer-head' rostral spine, and pointed snout spines that are longer than the head shield itself. Some forms also have very long lateral spines that may have acted like the wings of a glider during swimming, in achieving a very stable body position. They lack paired fins.

Galeaspids have their mouth just beneath the head shield, and they have an additional dorsal opening at the tip of the snout which may be a transverse slit, a broad oval, a heart shape, or a longitudinal slit. Its function has been debated, but it is probably an inhalant opening.

Order Anaspida

The anaspids, like the thelodonts, do not have a heavy head shield. They have a patchy fossil record in the Silurian and Devonian. *Pharyngolepis* (Fig. 2.5d) is a 200 mm long cigar-shaped animal with a terminal mouth, small eyes, and a covering of irregular scales and plates in the head region. The body scales are long and regular, and arranged in several rows. There is a pectoral spine and two pectoral fins beneath, and a tail fin on top of the downwardly bent tail.

BIOLOGY OF THE CEPHALASPIDS

Certain of the cephalaspid heterostracans, such as *Hemicyclaspis* and *Cephalaspis* (Figs 2.5, 2.6) and others, are extremely well preserved, and it has been possible to extract a great deal of anatomical and biological information from the specimens.

The upper surface of the head shield (Fig. 2.6a) shows two oval openings for the eyes, the orbits, and a narrow keyhole-like slit in front of them in the midline, the **nasohypophysial opening**. Behind it, and still in the midline, is a tiny **pineal opening**, associated with the pineal gland in the brain which might have been light-sensitive.

There are three specialized areas on the head shield marked by small scales set in slight depressions which might have had an additional sensory function: the dorsal field in the midline behind the orbits, and the

22

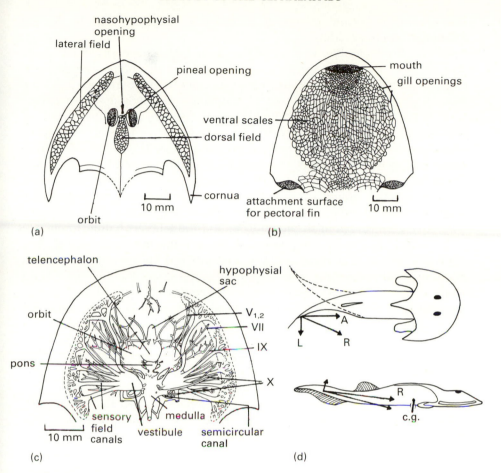

Figure 2.6 Cephalaspid anatomy and function: (a) head shield of *Cephalaspis* in dorsal view, showing sensory fields; (b) head shield of *Hemicyclaspis* in ventral view; (c) internal structure of the head shield of *Kiaeraspis*, showing the brain and related sense organs and nerves; (d) locomotion of *Alaspis* in dorsal (left) and lateral (right) views. Abbreviations in (d): A, anterior component of force produced by tail; L, lateral component; R, resultant of A and L; c.g., centre of gravity. (Figs (a) & (b) after Moy-Thomas & Miles 1971; (c) after Stensiö 1963; (d) after Belles-Isles 1987.)

two lateral fields (Fig. 2.6a). These areas are connected to the auditory region of the brain by large canals in the bone that may have transmitted nerves or contained fluid. The fields may have functioned in detecting movements in the water nearby, either by physical disturbance of the water, or by weak electrical fields.

The curved notches on either side at the back of the head shield (Fig. 2.6a) are occupied by the pectoral fins (Fig. 2.5a), and pointed cornua run back on either side. The underside of the cephalaspid head shield (Fig. 2.6b) shows a large mouth at the front with a broad area of small ventral

23

scales behind. Around the edges of this scale field are nine or ten gill openings on either side.

The most remarkable features of the cephalaspid head shield are to be seen inside. The bony parts enclosed much of the brain and sensory organs, as well as parts of the blood circulatory system and digestive system. The brain and its associated cranial nerves, the major nerves that serve the various parts of the head region, have been reconstructed by the Swedish palaeontologist Erik Stensiö (1963) with a fair degree of confidence because of the extensive bony envelope (Fig. 2.6c). The large orbits and inner ear regions are quite clear. Even the semicircular canals of the inner ear, the organs of balance, can be seen. The brain stem itself is located in the midline, and it was made from the three main portions seen in primitive living fishes, the **medulla** at the back which leads into the spinal cord, the **pons** in the middle, and the **telencephalon** (forebrain) in front with an elongate **hypophysial sac** running forwards from it. The cranial nerves III (eye movement), V_2 (mouth and lip region), VII (facial), IX (tongue and pharynx), and X (gill slits and anterior body) have been identified by comparison with living vertebrates. The five broad canals running from the lateral sensory fields to the **vestibule** of the inner ear show clearly (Fig. 2.6c).

An analysis of the locomotion of cephalaspids (Belles-Isles 1987) has shown that they were capable of sustained swimming, short bursts of fast locomotion, and fairly delicate manoeuvring, rather like sharks that live on or just above the sea-bed. The shape of the head in side view is an aerofoil, so that forwards movement would have tended to produce lift. When the tail beat from side to side, it produced a resultant force that drove the fish forwards and slightly downwards (Fig. 2.6d). The downwards component was produced by the presence of the long upper lobe on the tail, but it was counteracted by lift at the head end, and possibly also by the pectoral fins.

Cephalaspid fossils have almost all been found in freshwater or near-shore sediments from streams, lakes, and deltas. They may have foraged for detrital matter on the bottoms of lakes, moving by pulling their bodies along with the muscular pectoral fins. However, they could apparently also swim for long distances in search of new feeding grounds, or rapidly to escape predators.

LIVING AGNATHANS

Two groups of jawless fishes are still in existence, the lampreys and the hagfishes, but they look very different from the Ordovician, Silurian and Devonian forms just described. Both groups consist of fishes with elongated bodies, no bony armour, no jaws, and no paired fins.

The 30 or so species of lampreys all spend some of their life in

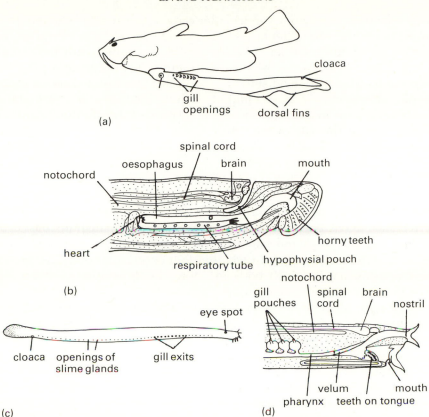

(a)

(b)

(c)

(d)

Figure 2.7 Living jawless fishes: (a) lamprey, feeding by attachment to a bony fish, and (b) longitudinal section of anterior end of body; (c) Pacific hagfish, external lateral view of body, and (d) longitudinal section of anterior end of body. (Figs (a) & (c) after Dean 1895; (b) & (d) modified from Jensen 1966.)

freshwaters where they breed. They feed by attaching themselves to other fishes (Fig. 2.7a) with their sucker-like mouths, and rasping at the flesh. The mouth and oesophagus are within a deep funnel which is lined with small pointed teeth that permit firm attachment to the prey. There is a fleshy protusible 'tongue' which also bears teeth and which is used in rasping at the flesh. Lampreys (Fig. 2.7b) have a single nasal opening on top of the head which runs into a pouch beneath the brain, large eyes, and two **semicircular canals** on each side. There is an internal skeleton consisting of a notochord and an attached cartilaginous skull and gill arches.

The marine hagfishes look superficially very like lampreys (Fig. 2.7c), but they live in burrows in soft sediments, feeding on worms and decaying carcasses on the sea-bed. Hagfishes have a single nasal opening at the very front that connects directly to the pharynx (Fig. 2.7d), quite unlike the lamprey nostril. The eyes are reduced and covered with thick skin, and there is only one semicircular canal on each side. The mouth is ringed with six strong tentacles, and inside it are two horny plates bearing numerous small teeth which can be protruded with the mouth lining. This apparatus can be turned in and out, producing a pinching action with which the hagfishes can grasp the flesh of a dead or dying animal. They remove a large lump of flesh by holding it in a firm grasp, and tying a knot in the tail, passing it forwards towards the head, and bracing against the side of their prey.

AGNATHAN RELATIONSHIPS

The relationships of the agnathan groups to each other, and to other vertebrates, have been hard to establish. One key dispute has been whether the lampreys and hagfishes are entirely distinct, and related to separate vertebrate lineages, or whether they form a monophyletic group, the Cyclostomata, in which both are more closely related to each other than to any other vertebrate group.

In the past, monophyly of the living agnathans was generally assumed (Fig. 2.8a), although often on the basis of 'negative' characters (absence of paired fins, jaws, ribs, ossification, fin muscles, and other internal features found in the jawed fishes and tetrapods). Various authors (e.g. Janvier 1981, 1984, 1986, Maisey 1986) argued that the lampreys are much more closely related to the gnathostomes than are the hagfishes, on the basis of 20 or 30 synapomorphies (Fig. 2.8b).

Schaeffer & Thomson (1980) have questioned the validity of these characters, and reaffirmed the monophyly of the Cyclostomata (Fig. 2.8a),

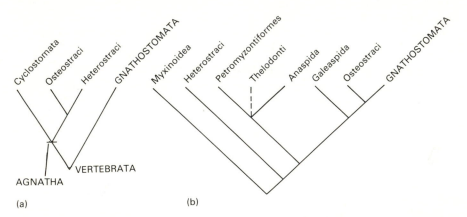

Figure 2.8 Cladograms showing two views of the relationships of the jawless fishes and the Gnathostomata (jawed fishes): (a) 'classical' view of Romer (1966); and (b) cladistic view of Janvier (1981, 1984, 1986); position of the Thelodonti uncertain; Janvier (1986) places the Astraspida as sister group of the Myxinoidea, and the Arandaspida as sister group of the Heterostraci.

26

a view supported by Yalden's (1985) reassessment of the feeding mechanisms of hagfishes and lampreys. He finds direct homologies in the mouth and tongue cartilages, the longitudinal tooth rows, and the mouth muscles, but most of these resemblances may be primitive characters of all craniates.

THE FIRST SHARKS

Sharks and rays, members of the class Chondrichthyes, or 'cartilaginous fishes', are generally reckoned to be the most primitive living gnathostomes, jawed vertebrates. A typical early shark, *Cladoselache* from the Late Devonian of Ohio (Fig. 2.9; Zangerl 1981), reaches a length of 2 m. The skin does not seem to have borne scales, although small multicusped tooth-like scales have been found on the edges of the fins, in the mouth cavity, and around the eye.

Externally the tail fin is nearly symmetrical, but internally the notochord bends upwards into the dorsal lobe only (the **heterocercal** tail condition). There are two dorsal fins, one behind the head, and the other half way down the body, and each has a spine in front. There are two sets of paired fins, the pectoral and **pelvic** fins, each set approximately beneath one of the dorsal fins, and each associated with girdle elements of the skeleton. *Cladoselache* was probably a fast swimmer, using sideways sweeps of its broad tail as the source of power, and its pectoral fins for steering and stabilization. As in modern sharks, the skeleton of *Cladoselache* is cartilaginous, with no true bone.

Cladoselache is usually assigned to the order Cladoselachida, a group restricted to the Late Devonian. Other shark remains are known from rocks of this age, but none as well as *Cladoselache*. Most major lineages arose in the subsequent Carboniferous period, so the chondrichthyans are more fully considered later, in Chapter 6.

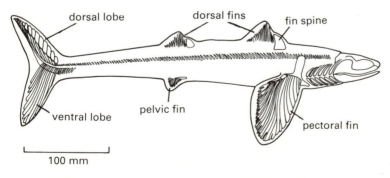

Figure 2.9 *Cladoselache*, one of the first sharks. (After Zangerl 1981.)

ORIGIN OF JAWS AND GNATHOSTOME RELATIONSHIPS

The Gnathostomata, all fishes other than agnathans, and the tetrapods, are marked by the possession of jaws, a feature that opened an enormous number of adaptive pathways in terms of diets and food-handling techniques that were closed to agnathans. Jaws allowed the development of a truly predatory mode of life for the first time: only jaws can grip a prey item firmly, allow it to be manipulated, cut cleanly, and ground up. How did jaws evolve?

The most-quoted theory for the origin of jaws is that they formed from modified anterior gill arches (Fig. 2.10a). In agnathans, the gill slits are

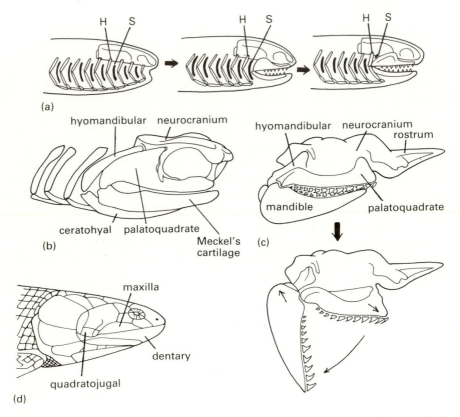

Figure 2.10 The evolution of jaws and jaw suspension: (a) the 'classical' theory for the evolution of jaws from the anterior two or three gill arches of a jawless form (left) to the fully equipped gnathostome (right); gill openings in black; H, hyomandibular; S, spiracular gill opening; (b) braincase, jaws, and gill supports of the Carboniferous shark *Cobelodus*, to show the amphistylic system of jaw attachment to the neurocranium; (c) braincase and jaws of the modern shark *Carcharhinus*, with the jaws closed (top) and open (bottom), to show the hyostylic system of jaw support and the highly mobile palatoquadrate; (d) head of the Devonian lobefin *Osteolepis*, to show the autostylic, or fused, system of jaw attachment. (Fig. (a) based on Romer 1933; (b) after Zangerl 1975; (c) after Moss 1972, courtesy of the Zoological Society of London; (d) after Moy-Thomas & Miles 1971.)

separated by bony or cartilaginous arches which are frequently in two or three sections. A hypothetical ancestral agnathan with eight gill slits and nine gill arches evolves into an early gnathostome by the loss of four gill slits, and the fusion and modification of the anterior three gill arches. The most anterior may form parts of the floor of the braincase. The second gill arch might have been modified to form the **palatoquadrate**, the main part of the upper jaw, and **Meckel's cartilage**, the core of the lower jaw (**mandible**). The third gill arch was then modified in part to provide a skull bone and a mandible bone that formed part of the jaw joint, the **hyomandibular** in the skull and the **ceratohyal** in the lower jaw.

The gill-arch theory for the origin of jaws is based on the anatomy of living vertebrates, and particularly on aspects of the early embryological development of mouth and pharyngeal structures. There are no relevant fossils showing the transition. Some anatomical evidence, however, suggests that the theory may not be so simple in reality (Schaeffer 1975). The agnathan gill arches are externally placed, and not homologous with the internal gill arches of gnathostomes, so that jaws arose after the transition from external to internal gill arches. In addition, the palatoquadrate and Meckel's cartilage may never have been gill arches at all, but were always associated with the mouth from their first appearance.

The palatoquadrate in gnathostomes may be attached to the **neurocranium**, the main portion of the skull which enclosed the brain and sensory organs, in various ways. In early sharks, such as *Cobelodus* from the Late Carboniferous of North America (Fig. 2.10b), there is a double attachment with links fore and aft, the **amphistylic** condition. This pattern has been modified in two main ways. In most modern fishes, the palatoquadrate contacts the neurocranium at the front only, and the jaw joint is entirely braced by the hyomandibular. On opening the jaw, the palatoquadrate can slide forwards, which increases the gape. This is the **hyostylic** jaw suspension condition (Fig. 2.10c). The second modification has been to exclude the hyomandibular from support of the jaw, and to fuse the palatoquadrate firmly to the neurocranium, the **autostylic** condition. This is typical of certain fish groups (the chimaeras and lobefins; Fig. 2.10d), as well as the land vertebrates.

Living gnathostomes are grouped in the classes Chondrichthyes and Osteichthyes (bony fishes and tetrapods), and two extinct groups are the class or subclass Acanthodii of the Silurian to Permian, and the class Placodermi of the Devonian. Acanthodians and osteichthyans share various features of the braincase (Maisey 1986) and they are tentatively regarded as sister-groups in the cladogram (Fig. 2.11).

The placement of placoderms has been more difficult. They are currently regarded by different authors as the sister group of chondrichthyans, of osteichthyans, or of both groups together, so the three groups are shown in an unresolved branching pattern (Fig. 2.11).

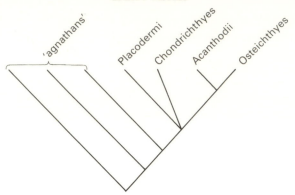

Figure 2.11 Cladogram showing the relationships of the main groups of jawed fishes.

THE ARMOUR-PLATED PLACODERMS

The osteostracans and heterostracans (see above) were not the only heavily armed fishes in Devonian seas. The placoderms, found largely in the Devonian, bore similar bony carapaces over the regions of their heads and shoulders, but in all cases these bony shells were mobile. There was a special neck joint that allowed the anterior portion of the head shield to be lifted.

The class Placodermi falls into nine orders (Denison 1978, Gardiner 1984a, Goujet 1984) of which the following are reasonably well known; the Arthrodira, Acanthothoraci, Petalichthyida, Rhenanida, Ptyctodontida, Phyllolepida, and Antiarchi. The relationships of these groups are much debated at present. The arthrodires form the largest group, and will be described in most detail.

Order Arthrodira

The arthrodires, nearly 200 genera, make up more than half of all known placoderms. The Middle Devonian form *Coccosteus* (Miles & Westoll 1968) has a trunk shield that covers only part of the dorsal surface, and it extends back as far as the shoulder region below (Fig. 2.12a). There is a pectoral fin and a pelvic fin, both supported by limb girdles, but much smaller than in sharks (Fig. 2.9). The tail is heterocercal, and there is a long dorsal fin. The posterior part of the body is covered with small scales, but these are rarely preserved. It is likely that *Coccosteus* was a powerful swimmer, achieving speed by lateral sweeps of its tail and posterior trunk. However, its flattened shape suggests that it probably lived near the bottom of seas or lakes.

The head and trunk shields (Fig. 2.12b & c) consist of several plates, and there is a gap (the **nuchal gap**) between the head and trunk shields at

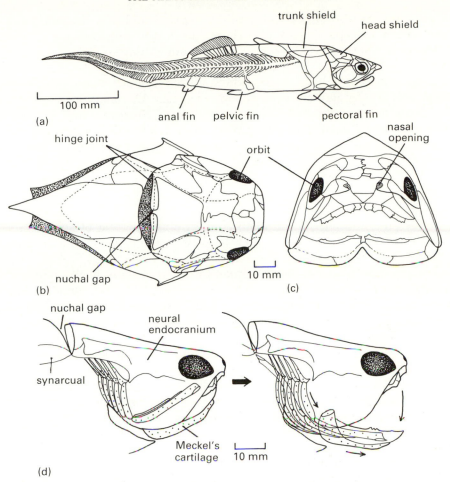

Figure 2.12 The arthrodire placoderm *Coccosteus* from the Middle Devonian of Scotland: (a) whole body in lateral view, (b) head shield in dorsal and (c) anterior views, and (d) jaw opening movements showing the position of the head and visceral and branchial skeletons, with the jaws closed (left) and open (right): (Fig. (a) after Moy-Thomas & Miles 1971; (b) & (c) after Miles & Westoll 1968; (d) after Miles 1969.)

the line of hinging. The lower jaw plates join weakly in the middle (Fig. 2.12c), and their dorsal margin is worn to a sharp edge against a series of eight small plates in the upper jaw. These are not teeth, but they wear into equally effective sharp beak-like plates that would have been capable of an effective cutting, puncturing, and crushing action.

The jaws open (Fig. 2.12d) by an upwards swing of the skull and dropping of the lower jaw (Miles 1969). The skull hinges about the ball and socket joints on the lateral margins of the dorsal part of the head shield, and the size of the gape is limited in the width of the nuchal gap. It has been suggested that placoderms employed a head-lifting form of jaw opening in feeding on the bottom of the sea and lakes. It would have been easier to capture prey by driving the lower jaw forwards in the bottom mud and lifting the head, than by attempting to drop the lower jaw.

Later arthrodires have even more reduced armour than *Coccosteus*, often only a very limited trunk shield. Two Late Devonian families, the Dinichthyidae and the Titanichthyidae of North American and northern Africa, achieved giant size: at lengths of up to ten metres, these were the largest predators in Devonian seas, and the largest vertebrates yet to evolve.

Diverse placoderms

The acanthothoracans have a head shield rather like that of some early arthrodires. One odd feature is that the plates were separate in juveniles, but appear to have fused in the adults.

The petalichthyids are another small group of bottom-dwelling forms. *Lunaspis* from the Early Devonian of Europe (Fig. 2.13a) is flattened, with a short trunk shield and long cornual plates. The anterior part of the head shield, around the eyes and nostrils, is covered by numerous tiny scales, as is the long trunk.

The rhenanids have a body covering of small tesserae instead of the more typical large plates. *Gemuendina* from the Early Devonian of Germany (Fig. 2.13b) looks superficially like a ray with its very flattened body, broad pectoral fins, and narrow whip-like tail, and it may have swum by wave-like undulations of the pectoral fins. There are large bone plates in the midline, around the eyes, nostrils, and mouth, and on the sides of the head, which are divided by a mosaic of small plates that extends on to the trunk and pectoral fins.

Ptyctodonts show very reduced armour plating. They are generally small, usually less than 200 mm in length, with long whip-like tails, a long posterior dorsal fin, and a high anterior dorsal fin supported by a spine on the trunk shield. *Ctenurella* from the Late Devonian of Australia and elsewhere (Fig. 2.13c) has much reduced armour. Some ptyctodonts have **claspers**, elongate elements associated with the pelvic fins that are assumed to have been involved with the process of internal fertilization. Claspers are seen in male chondrichthyans, but the structure of the ptyctodont clasper is different from that of a shark.

The phyllolepids have large bony heads and trunk shields made from plates with a very clear ornament of concentric ridges. *Phyllolepis* (Fig. 2.13d) has a flattened body with a rounded snout and a long narrow tail portion, although this is not well known from fossils.

The antiarchs were a diverse group from the Middle and Late Devonian which retained a heavy armour covering, and specialized in a bottom-dwelling mode of life, feeding by swallowing mud and extracting organic

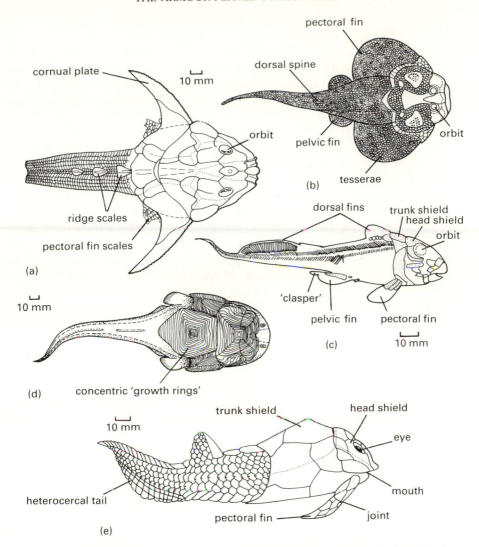

Figure 2.13 Diverse placoderms: (a) the petalichthyid *Lunaspis* in dorsal view; (b) the rhenanid *Gemuendina* in dorsal view; (c) the ptyctodont *Ctenurella* in lateral view; (d) the phyllolepid *Phyllolepis* in dorsal view; (e) the antiarch *Pterichthyodes* (Figs (a) & (b) after Moy-Thomas & Miles 1971; (c) & (d) modified from Stensiö 1969; (e) after Hemmings 1978.)

matter. *Pterichthyodes*, an early form (Fig. 2.13e), has a high domed trunk shield made from a small number of large plates. The pectoral fin is entirely enclosed in bone, and it was movable against the trunk shield by a complex joint. There is also a second joint about half way along the fin. This fin was probably of little use in swimming, and it may have served to shovel sand over the back of the animal so that it could bury itself (P. Janvier, personal communication).

THE ACANTHODIANS

The acanthodians were generally small fishes, mostly less than 200 mm long, that include the oldest known gnathostomes. The first acanthodians date from the Early Silurian, but they became abundant only in the Devonian. Several lines survived through the Carboniferous and one into the Early Permian.

Acanthodians have slender bodies with one or two dorsal fins, an anal fin, and an heterocercal tail fin (Fig. 2.14a & b). The pectoral and pelvic fins have been modified to long spines, and there may be as many as six pairs of spines along the belly of early forms. The other fins just noted are all supported by a spine on the leading edge. The name 'acanthodian' refers to these liberal arrays of spines. The internal skeleton is rarely seen.

The body is covered with small closely-fitting scales which were made from bone and dentine (Fig. 2.14c). These show concentric lines that record the growth of the scale. It seems that young acanthodians had a fixed number of scales over most of the body, and each scale grew by addition of bone and dentine at the margins as the animal grew larger.

Most acanthodians have teeth. Toothless forms probably fed on small food particles which they may have filtered from the water. Only some of the later forms may have taken larger prey. They had a wide gape and **gill rakers**, sharpened spikes in the throat region which are attached to the hyomandibular and ceratohyal elements. One specimen has been found with a bony fish in its body cavity, presumably swallowed whole. The large eyes of acanthodians suggest that they lived in open deep water, and they may have fed at middle depths. The fin spines and other spines may have had a primarily defensive function in making acanthodians unpleasant for larger fishes to swallow. Later forms, like *Acanthodes* itself, seem to have been able to erect their pectoral spines,

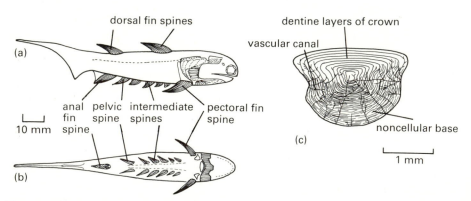

Figure 2.14 Acanthodian diversity and anatomy: (a) *Climatius* in lateral view; (b) *Euthacanthus* in ventral view, showing the fin spines; (c) single scale of *Acanthodes*. (After Moy-Thomas & Miles 1971.)

which would have effectively caused them to stick in the gullet of a would-be predator!

OLD RED SANDSTONE FISHES IN SCOTLAND

Some of the most prolific collections of Devonian agnathans, placoderms, acanthodians, and lobefinned fishes, have come from the Old Red Sandstone of the north of Scotland, deposits laid down in a large subtropical lake that covered much of Caithness, the Moray Firth, Orkney, and Shetland (Fig. 2.15a). Lake levels rose and fell as a result and wet and dry climatic conditions, some following annual cycles, others longer-term Milankovitch cycles of 20 000 and 90 000 years. These fluctuations affected the oxygen content and salinity of the water. The sediments frequently occur in repeated cycles that occupy thicknesses of about 10 m of the rock column, and repeat through a total thickness of 2–4 km of rock (Trewin 1986). In places, annual varves, generally less than 1 mm thick, may be detected.

Fossil fishes occur in the Scottish Old Red Sandstone both as scattered fragments and in great concentrations within 'fish beds'. Mortality horizons, single layers containing high concentrations of fish carcasses,

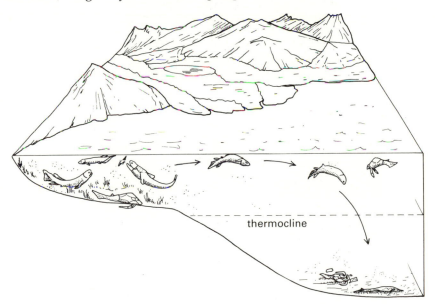

thermocline

Figure 2.15 The Old Red Sandstone lakes of the north of Scotland: topographic sketch showing sediment source from alluvial fans and plains derived from erosion of the uplands, and the cycle of life, death, and fossilization of the fish fauna; from left to right: fishes living in shallow areas of the lake, carcasses float out to the middle of the lake, and sink into the cold anoxic conditions beneath the thermocline where they are preserved in laminated muds on the deep lake floor. (After Trewin 1985, courtesy of Blackwell Scientific Publications Ltd.)

seem to have formed during deoxygenation events which may have occurred every ten years or so when the lake was deepest. Repeated mortality events of this kind occurred over thousands of years, and built up major fish beds in several places. These could have either followed an algal bloom, when decaying algae removed oxygen from the water, or a severe storm which stirred up deep anoxic waters to the surface. Other likely causes of fish kills in the Old Red lakes include rapid changes in salinity and cold shock. The carcasses floated for some time near the surface, buoyed up by gases of decay. After a few days the gas escaped, possibly by rupturing the body walls, and the carcasses fell to the anoxic lake-floor where they were buried by fine sediments. This process yields

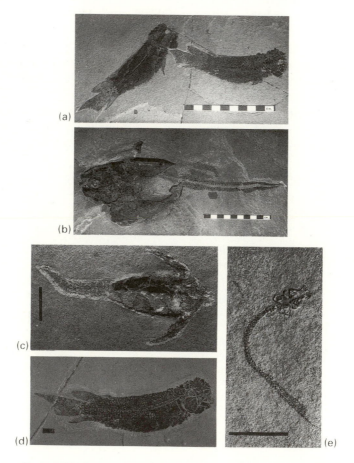

Figure 2.16 Typical Old Red Sandstone fishes from Achanarras Quarry, Caithness: (a) pair of *Dipterus*; (b) *Coccosteus*, showing preservation of the dermal head shield, and the deeper portions of the body and tail skeleton; (c) juvenile *Pterichthyodes*; (d) *Dipterus*, showing slight separation of head elements on fossilization; (e) *Palaeospondylus*, an enigmatic small fish. Scales: 100 mm rulers in (a) & (b); scale bars are 10 mm in (c) & (e). (Photographs courtesy of Dr N. H. Trewin.)

extensive beds of fish remains representing several species, and the carcasses are often in good condition (Fig. 2.16) because they have not been scavenged, and because of the low-energy bottom conditions.

The Old Red Sandstone food chains reconstructed by Trewin (1986) are based on lakeside plants (mosses, reedy horsetails, and scale trees) and phytoplankton which were eaten by lobefins such as *Dipterus* and *Osteolepis* (see below). There is also evidence for small arthropods around the lake margins, and these may have been a source of food for these fishes as well. The smaller fishes were preyed on by carnivorous forms such as *Coccosteus* (Fig. 2.12b–d) and the bony fish *Cheirolepis* (Fig. 2.18) which have been found with remains of acanthodians and of *Dipterus* in their stomachs. The heavier placoderms such as *Pterichthyodes* (Fig. 2.13e) scavenged for organic matter – decaying plant and animal remains – on the shallower oxygenated parts of the lake-bed. The top carnivore seems to have been the lobefin *Glyptolepis* which reached lengths of over 1 m. It may have been a lurking predator like the modern pike, hiding among water plants and launching itself rapidly at passing prey.

EARLY BONY FISHES

The bony fishes, class Osteichthyes, arose early in the Devonian, or even in the Silurian, but the first fossils are only isolated scales. From the start, the bony fishes fall into two quite distinctive lineages, the subclass Actinopterygii and the subclass Sarcopterygii. These are distinguished readily by their fins (Fig. 2.17) – actinopterygians have 'ray fins' which are supported by a series of narrow cartilaginous or bony rods called radials, while sarcopterygians have fleshy 'lobe fins' supported by a single basal bone and with muscles that can modify the posture of the fin.

Good actinopterygian remains are known from the Early Devonian, and by Late Devonian times the group had begun to diversify. A typical early form is *Cheirolepis* from the Middle Devonian of Scotland (Fig. 2.18), typically 250 mm in length (Pearson & Westoll 1979). The body is slender and elongate, and the tail is strongly heterocercal, although the tail fin beneath makes it nearly symmetrical. There are large triangular dorsal and anal fins and paired pectoral and pelvic fins.

(a) (b) (c)

Figure 2.17 The fins of (a) an actinopterygian, *Amia*, to show the simple basal skeleton, (b) the lobefin *Eusthenopteron*, an osteolepiform, and (c) the lobefin *Epiceratodus*, a lungfish, to show the more complex skeleton which supports a muscular lobe in the middle of the fin. (Modified from Romer & Parsons 1970.)

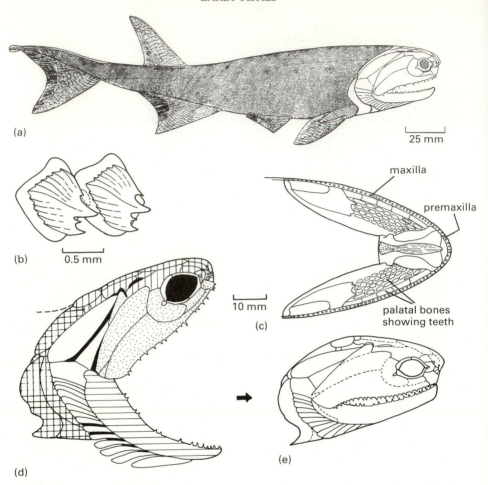

Figure 2.18 The Middle Devonian bony fish *Cheirolepis*: (a) reconstruction of the body in lateral view; (b) two trunk scales; (c) ventral view of the palate showing the teeth; (d) & (e) opening and closing of the jaws, showing the five major mobile units, as described in the text. (After Pearson & Westoll 1979.)

The body is covered with small overlapping lozenge-shaped scales (Fig. 2.18b) which articulate with each other by means of a peg and socket arrangement in the tail region. The scales are arranged in sweeping diagonal rows that run backwards and downwards. There are larger ridge scales on the dorsal edge of the tail which act as a cutwater. The fin rays (**ceratotrichia**) are covered with jointed dermal bones, the **lepidotrichia**. These provide a covering for the fin and they also stiffen it in comparison with sharks, for example, which have only ceratotrichia. The scales are composed of layers of bone, dentine, and an enamel-like substance on the outside.

The skull is relatively heavy, with a bony braincase and palatal elements inside, and an outer bony box made from numerous thin dermal bone plates. There is a large eye and two nostrils on each side, and a broad mouth lined with irregularly spaced sharp teeth (Fig. 2.18c–e). The teeth are borne on three bones around the edges of the mouth, the **maxilla** and **premaxilla** in the skull, and the **dentary** in the lower jaw, and these are the main tooth-bearing elements in subsequent vertebrates. The palatoquadrate is inside the maxilla, and it is covered by palatal bones which also bear rows of teeth. At the back of the skull are the outer dermal elements of the shoulder girdle, attached to the gill region.

The head skeleton of *Cheirolepis* is highly **kinetic**, that is, composed of several mobile units which can move against each other (Fig. 2.18d & e). When the jaws open, a very wide gape is possible because the five units move apart. The skull roof moves back, the gill region expands and moves back and down, and the shoulder girdle moves downwards.

Cheirolepis was a fast-swimming predator which presumably used its large eyes in hunting, and possibly even in transfixing its prey before capture (Pearson & Westoll 1979). Its great gape would have enabled *Cheirolepis* to engulf prey up to two-thirds of its own length; such prey would include the abundant acanthodians, and small lobefins and placoderms found in association. The sharp teeth of *Cheirolepis* might not seem suitable for cracking open placoderms, but there were shorter teeth on the palatal bones which might have been capable of moderate crushing activity.

Cheirolepis was capable of powerful and prolonged swimming using sideways beats of its tail region to produce thrust. It used its pectoral fins for steering, but these were not highly mobile, and *Cheirolepis* was probably rather clumsy when trying to turn rapidly. The paired fins also functioned to prevent rolling.

Devonian actinopterygians like *Cheirolepis* are known from all parts of the world, but only ten genera have been found so far. The actinopterygians radiated dramatically in the Carboniferous and later, and they are the dominant fishes in the seas today (see Chapter 6).

THE LOBEFINS

The Sarcopterygii were a much more significant group in the Devonian than the Actinopterygii, although they have since become much rarer. There are four principal sarcopterygian groups, the osteolepiforms, porolepiforms, coelacanths (Actinistia), and lungfish (Dipnoi), all of which arose in the Devonian. All sarcopterygians share muscular lobed paired fins with bony skeletons (Fig. 2.17), as well as several skull features not seen in other vertebrates.

The osteolepiforms had their heyday in the Devonian, although certain forms survived through the Carboniferous and into the Early Permian.

Osteolepis from the Middle Devonian of Scotland and elsewhere (Andrews & Westoll 1970b) has a long slender body with large midline fins (two dorsals, one anal), and lobed paired fins (pectoral and pelvic). The tail is heterocercal, with fins above and below (Fig. 2.19a & b).

The porolepiforms, represented by *Holoptychius* (Fig. 2.19c), generally have larger rounded scales, and longer pointed pectoral fins with more extensive lobed portions. The body shape is different, with porolepiforms being deeper and shorter than osteolepiforms, and having a shorter skull and a smaller eye.

The head of sarcopterygians is generally like that of early actinopterygians. The osteolepiform *Eusthenopteron* (Fig. 2.20a & b) has a complex of thin dermal bone plates covering the outer portions of the head, gill region, and attached shoulder girdle. Small teeth are borne on the maxilla, premaxilla, and dentary, as well as on several bones of the palate

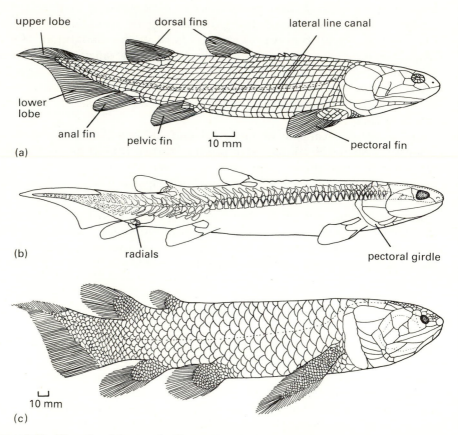

Figure 2.19 Diversity of the osteolepiforms (a) & (b) and porolepiforms (c): (a) & (b) lateral views of *Osteolepis*, with and without scales, (c) lateral view of *Holoptychius*, with scales. (Fig. (a) after Moy-Thomas & Miles 1971; (b) after Andrews & Westoll 1970a; (c) after Andrews 1973.)

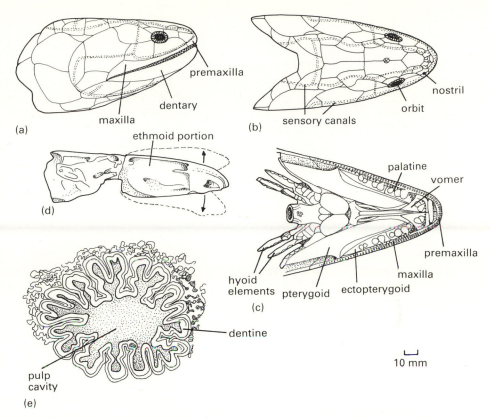

Figure 2.20 The skull of the osteolepiform *Eusthenopteron* in (a) lateral, (b) dorsal, and (c) ventral views; (d) lateral view of the braincase, showing the postulated range of movement about the middle joint; (e) cross-section of a tooth to show the labyrinthine infolding of the enamel (tooth diameter, 5 mm). (After Moy-Thomas & Miles 1971.)

(Fig. 2.20c). Some of the palatal teeth are heavy, and they have complex, or labyrinthine, internal patterns of infolding (Fig. 2.20e), the so-called labyrinthodont type of tooth, found also in early tetrapods. As in early actinopterygians, the skull is highly kinetic, being jointed in order to allow the mouth to open wide. Even the braincase (Fig. 2.20d), deep within the skull, is jointed in order to permit greater flexibility.

The coelacanths arose in the Middle Devonian, and are represented by fossils up to the Late Cretaceous, when it was thought they had died out. Of course the discovery of *Latimeria*, a living coelacanth, in 1938 is now well known. Typical coelacanths (Fig. 2.21) have short bodies with large dorsal, anal, and paired fins, all of which are lobed except for the anterior dorsal. The tail is characteristically divided into three parts – a dorsal and ventral portion separated by a small middle lobe at the end of the notochord. The skull is short overall, although the snout portion is longer than in the osteolepiforms. *Latimeria* (Forey 1988) lives in deep oceans off

41

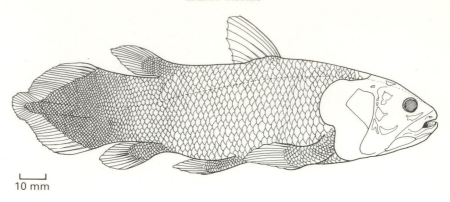

10 mm

Figure 2.21 The coelacanth *Latimeria* from the modern seas of the Indian Ocean. (After Andrews 1973.)

the coast of Comoro Islands where it feeds on fishes. It swims slowly by beating its paired fins in a pattern like the locomotion of a tetrapod, and it can achieve fast thrust by beating its tail.

The lungfishes (Thomson 1969, Bemis *et al.* 1986) were particularly diverse in the Devonian, but they have dwindled in importance ever since, leaving only three genera still living. *Dipterus* from the Middle Devonian of Scotland (Fig. 2.22) has a long body, as in the osteolepiforms, but the fins and skull bones are very different. The fins are pointed, with long central lobes supported, in the paired fins, by a rather symmetrical array of bones (Fig. 2.17). The tail is heterocercal and bears a narrow fin beneath. The skull bears a complex array of small bones around the large eyes and mouth. There are no teeth on the margins of the jaws as in other bony fishes, only a pair of large dentine-covered grinding plates in the middle of the palate (Fig. 2.22b), and a scattering of smaller tooth-like structures in front. These paired plates are typical of later lungfishes and indicate a crushing function for feeding on tough and hard food.

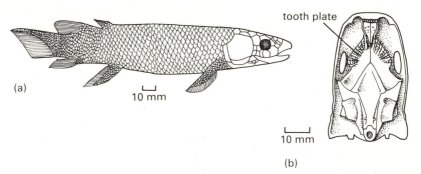

(a)

10 mm

tooth plate

(b)

10 mm

Figure 2.22 The Devonian lungfish *Dipterus*, (a) in lateral view, and (b) ventral view of the palate, showing the tooth plates. (Fig. (a) after Moy-Thomas & Miles 1971; (b) modified from Westoll 1949.)

42

Several lineages of lungfishes radiated in the Carboniferous, and two continued into the Mesozoic and Cenozoic. Many changes took place over this time: fusion of the palatoquadrate to the skull (the autostylic condition), elaboration of the crushing tooth plates, and the development of a special hypermineralized dentine, all of which increased the crushing power of the jaws. The three genera of living lungfishes have reduced the bony parts of their skeletons. The braincase and parts of the backbone remain cartilaginous, and the outer skull bones are reduced in number and weight. The Australian lungfish *Neoceratodus* is deep-bodied and has broad pectoral and pelvic fins, while the South American *Lepidosiren* and the African *Protopterus* have long eel-like bodies and very much reduced paired fins which are very slender.

The relationships of the sarcopterygian groups to each other have been controversial, not least because it has long been assumed that the tetrapods, the land vertebrates, arose from within the sarcopterygian clade. The problems concern the relationships of the osteolepiforms, porolepiforms, coelacanths, and lungfishes to each other, and the identification of one of these as the closest outgroup of the Tetrapoda.

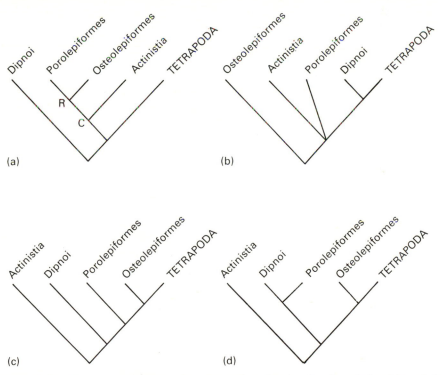

Figure 2.23 Cladograms showing four competing theories for the relationships of the sarcopterygian fishes and tetrapods, according to (a) Romer (1966) and other 'classical' sources; (b) Rosen *et al.* (1981); (c) Panchen & Smithson (1987); and (d) Janvier (1986). Abbreviations: C, Crossopterygii; R, Rhipidistia.

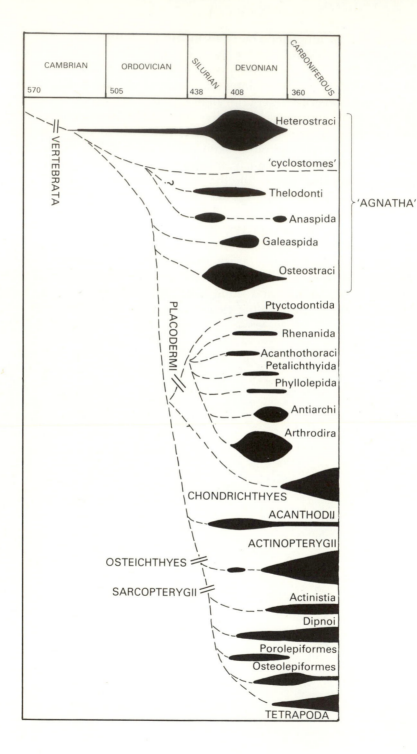

Figure 2.24 The evolution of early and mid Palaeozoic fishes. The pattern of relationships (indicated with dashed lines) is based on cladograms described above, and the 'balloon' shapes indicate the span in time of known fossils (vertical axis) and their relative abundance (horizontal axis).

The 'classical' view (e.g. Romer 1966) has been that the osteolepiforms and porolepiforms form the Rhipidistia which, together with the Actinistia, forms the Crossopterygii. This is treated as the sister group of the Tetrapoda, with the Dipnoi as outgroup (Fig. 2.23a). Rosen *et al.* (1981) presented a detailed cladistic argument that the lungfishes are the sister group of the tetrapods (Fig. 2.23b). The 'Crossopterygii' and 'Rhipidistia' are split up into their constituent parts, Osteolepiformes, Porolepiformes, and Actinistia, as outgroups. Panchen & Smithson (1987) presented a third view (Fig. 2.23c) in which the 'Crossopterygii' and 'Rhipidistia' are also split up, but the Osteolepiformes are regarded as the sister group of the Tetrapoda. This corresponds in part to the view of Janvier (1986), who pairs the Dipnoi and Porolepiformes (Fig. 2.23d). The biological implications of these schemes for the origin of the tetrapods will be discussed further in Chapter 3.

EARLY FISH EVOLUTION AND MASS EXTINCTION

The evolution of fishes was slow during the Ordovician: for the first 100 Myr or so after the first fish appeared, only sporadic remains of heterostracans are known. Then, in the Silurian, an explosion of evolutionary activity seems to have taken place (Fig. 2.24) – four groups of agnathans and the acanthodians came on the scene. Further, in the Devonian, the seven major placoderm orders arose, as well as the first sharks, and six important groups of bony fishes, including their derivatives, the tetrapods.

During the Late Devonian, a time span of 14 Myr (374–360 Myr), many of these groups disappeared – most of the agnathans and placoderms, as well as several families of acanthodians and bony fishes. Other groups which survived into the Carboniferous seem to have been heavily depleted. Of the 46 families of fishes currently recognized at the start of the Late Devonian (Carroll 1987), 35 died out during the next 14 Myr, a total extinction rate of 76%, which is high by any standards. However, there is no evidence at present that the extinctions took place instantaneously, or even within 1–2 Myr, and new fossils continue to extend the ranges of some families well beyond the date of major depletion. Nevertheless, multi-phase Late Devonian extinction events have been recognized among marine invertebrates such as corals, brachiopods, and ammonoids, as well as phytoplankton, and it is not unexpected then to find evidence for a possible series of staged extinctions of the fishes, and their replacement in the Carboniferous by new groups (see Chapter 6).

CHAPTER THREE

The Amphibians

The first vertebrates to move on to land, the amphibians, arose during the Devonian period. The name amphibian refers to the fact that the modern forms – frogs, newts, and salamanders – live both in the water and on land, and it is assumed that many of the fossil forms had similar double lifestyles. The amphibians radiated extensively during the Carboniferous and Early Permian, some as small semi-aquatic forms, but many as larger fish- and meat-eaters that could, in some cases, live fully terrestrial lives. A selection of papers on these topics may be found in Panchen (1980). The modern groups arose as early as the Triassic, and radiated thereafter.

In this chapter, the major anatomical and physiological changes that were necessary when a lobefin fish became a **tetrapod** (literally 'four feet') are reviewed, and the evolution and biology of the extinct and living amphibian groups is described.

PROBLEMS OF LIFE ON LAND

The major problems with life on land relate to weight and structural support as much as to the physiology of breathing air. A fish is buoyed up by the water and its body weight may be effectively zero. On land, however, the body has to be held up by some form of limbs, and the skeleton and all of the internal organs have to become structurally modified in order to cope with the new downwards pull of gravity. The backbone of a fish is adapted for the stresses of lateral stretching and bending during swimming, but the main forces to which a tetrapod is subject are caused by gravity. The vertebrae and the muscles around the backbone have to become modified to prevent the body from sagging between the limbs.

The mode of locomotion of a tetrapod on land is generally different from that of a fish in water. Instead of a smooth gliding motion, the limbs have to operate in a jerky fashion producing steps to propel the body forwards. The paired fins of sarcopterygian fishes already had internal bones and muscles that produced a form of 'walking'. However,

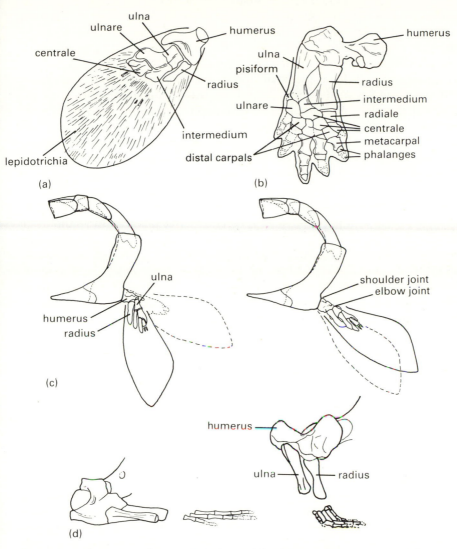

Figure 3.1 The origin of tetrapod limbs and land locomotion: (a) pectoral fin of the osteolepiform fish *Eusthenopteron* showing interpreted identities of the bones; (b) equivalent forelimb of the primitive amphibian *Eryops*; (c) possible movements of the forelimb of *Eusthenopteron*; (d) step cycle of the forelimb of the primitive amphibian *Proterogyrinus*. (Figs (a) & (b) after Gregory 1935; (c) after Andrews & Westoll 1970a; (d) after Holmes 1984.)

profound modifications had to occur in the lobed fin before it was a moderately effective land limb.

The pectoral fin of *Eusthenopteron* (Fig. 3.1a) contains the major bones of a tetrapod limb (Fig. 3.1b): the single upper arm bone, the **humerus**, the two forearm bones, the **radius** and **ulna**, the main wrist bones, the **ulnare** and **intermedium**, and two other wrist bones, the **centralia** (singular, centrale). The amphibian has additional elements in the wrist, the **radiale**, distal **carpals** 1–5, sometimes an additional bone at the side, the **pisiform**, and the four or five fingers which are composed of **metacarpals** and **phalanges**. The limb bones of *Eusthenopteron* are to be found in early tetrapods, and indeed most of them are still present in our arms, although it is speculative to attempt to draw homologies for all of the bones of the wrist and ankle. The osteolepiform pelvic fin also contains the basic tetrapod bones of the hindlimb, the thigh bone (**femur**), the lower leg bones (**tibia**, **fibula**), and the major ankle bones (**fibulare**, **intermedium**). Although close anatomical similarities exist, there were major functional differences: *Eusthenopteron* could not have walked properly on land on its fins.

How can we compare the locomotor abilities of an osteolepiform and an early amphibian? It is possible to reconstruct the possible movements of vertebrate limbs by manipulating the fossil bones, or casts of them. The maximum amount of rotation and hinging at each joint can be assessed since this depends on the shapes of the ends of the limb bones. The likely distribution of limb muscles can also be reconstructed by studies of the shapes of the limb bones. Certain muscles attach to particular knobs or ridges (**processes**) on the limb bones and the limb girdles, while others may leave characteristic markings on the bone surface (**muscle scars**). In *Eusthenopteron*, the limbs point backwards and only a very little sideways, and the limb skeleton could swing back and forwards through only 20–25° (Andrews & Westoll 1970a). The main motion was at the shoulder joint, with a very slight elbow bend (humerus–ulna/radius hinge). The lepidotrichia of the remainder of the fin were flexible, and they might have increased the size of the swing, but only slightly (Fig. 3.1c).

In evolving the ability to walk, the tetrapod limb had to alter considerably both in structure and in orientation, when compared to the osteolepiform fin (Holmes 1977). New bones appeared, and the joints (elbow, wrist) became more clearly defined. The humerus lengthened and the shoulder joint swung round so that the humerus pointed partly sideways as well as backwards. The elbow joint became more of a right-angle and the lower part of the limb was directed downwards. The wrist acted as a hinge, and the new bones in the hand allowed it to spread out widely and fulfil its role as a weight-supporting surface. In walking (Fig. 3.1d), the humerus swung back and forwards in a horizontal plane. During a stride, it also twisted so that the radius and ulna were swung down from a near-horizontal orientation.

The limb girdles became heavily modified with the change in limb function. The pectoral girdle of osteolepiforms and other fishes is effectively part of the skull (Fig. 3.2a) since the outer elements are attached to the gill and throat bones. When the osteolepiforms began to use their pectoral fins in walking, additional forces were applied. At every step, the pectoral girdle takes up the impact of the weight of the

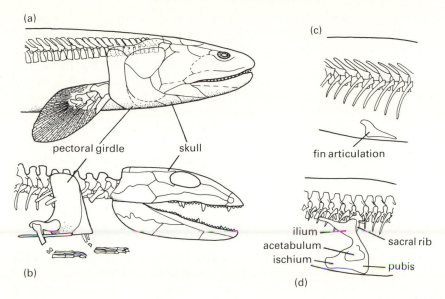

(a)

(c)

pectoral girdle   skull

fin articulation

ilium

acetabulum   sacral rib

ischium

pubis

(b)

(d)

Figure 3.2   The transition from osteolepiform fish (a) & (c) to primitive amphibian (b) & (d): (a) & (b) the separation of the skull from the shoulder girdle; (c) & (d) the enlargement of the pelvic girdle and its firm attachment to the vertebral column via the ilium and sacral rib. (Figs (a), (c) & (d), after Stahl 1974 and Gregory, W. K., 1951; (b) after Godfrey, S. J., *in* Carroll 1987.)

front part of the body as each hand hits the ground. In an osteolepiform, these impacts would be transmitted from the pectoral girdle directly to the skull, and the whole head would reverberate in time to the walking steps. Hence, the pectoral girdle became separated from the skull in the earliest tetrapods (Fig. 3.2b). The pelvic girdle was also much modified. Whereas in fishes it is a small unit that embedded within the body wall (Fig. 3.2c), it is firmly attached to the vertebral column in tetrapods (Fig. 3.2d). This is because of the additional forces imposed by the role of the hindlimb in walking and support. The weight of the body would simply force the pelvic girdle of a fish upwards into its body cavity if it moved on to land, and the girdle would twist about during walking. A firm set of bracing bones evolved in the first amphibians to prevent this happening.

In addition to the problems of locomotion on land, the earliest tetrapods had to modify the structures associated with feeding and respiration. The skulls of osteolepiforms were highly kinetic (see p. 41), but this mobility was largely lost in the early amphibians. The jaw movements of amphibians are also much simpler than those of most fishes. The lower jaw hinges at one point at the back of the skull, on a roller joint between the **articular** bone in the lower jaw and the **quadrate** in the skull. The first tetrapods presumably fed on small fishes and the increasing numbers of terrestrial invertebrates; millepedes, spiders, cockroaches, dragonflies, and the like.

Air-breathing needs lungs, or some equivalent supported vascular surface, instead of gills. Lungs contain internal folds and pouches lined with heavily vascularized skin and bathed in fluid. Air is drawn in, passed into the fine pouches, and oxygen passes through the moistened walls into the bloodstream. Living lungfishes have functional lungs of course, and the same is assumed for osteolepiforms and indeed most other early bony fishes. The first tetrapods may have been only marginally better than their fish ancestors at air-breathing.

A further physiological problem with life on land is the maintenance of water balance. In the air, water can evaporate through the moist skin of the body, the lining of the mouth and nostrils, and the early tetrapods risked desiccation. The earliest amphibians probably remained close to fresh water which they could drink in order to avoid this problem. However, certain forms evolved semipermeable skin coverings, sometimes with dermal scales, which would have cut down water loss.

Sensory systems had to change too. The lateral line system could only be used in the water (it is retained in many aquatic amphibians). Eyesight was as important on land as in shallow ponds, and the sense of smell may have improved, but we have no evidence of that in the fossils. Early tetrapods had a poor sense of hearing in air, as did their ancestors. The main bone associated with hearing in modern amphibians and reptiles, the **stapes**, is present in early amphibians, but it is too massive to be effective in hearing high-frequency sound. The stapes is a modified version of the hyomandibular element which forms part of the jaw-hinging apparatus in most fishes (see p. 29).

Amphibians betray their ancestry in their mode of reproduction. Even highly terrestrial forms have to lay their eggs in water where the young hatch out as aquatic larvae, tadpoles. After some time living in the water, breathing through gills, the tadpoles metamorphose into the adult form. Fossil tadpoles are rare, probably because they are so small and their bones are poorly developed, but sufficient specimens have been found in Carboniferous and Permian rocks to confirm that the first amphibians passed through larval stages similar to those of modern forms (Fig. 3.3).

Why did some sarcopterygian fishes venture on to land in the first place if they faced so many structural and physiological problems? The classical theory (e.g. Romer 1966), was that fishes moved on to land in order to escape from drying pools. The Devonian was supposedly a time of seasonal droughts, and the freshwater fishes probably found themselves in stagnant and dwindling pools. Terrestrial locomotion evolved as a means of staying in the water! This theory has been criticized on the grounds of lack of evidence for droughts, and because the theory would explain only moderate terrestrial adaptations, not the much-modified tetrapod limb for example. The simplest hypothesis is that vertebrates moved on to land because there was a rich and untapped supply of food there. Waterside plants and terrestrial invertebrates diversified in the

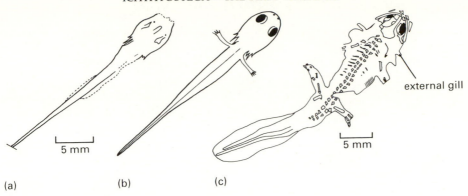

(a)   (b)   (c)

external gill

5 mm

5 mm

Figure 3.3 Fossil 'tadpoles' of Carboniferous and Permian amphibians; drawings of fossils from (a) France and (c) North America; (b) reconstruction of an intermediate stage. (Fig. (a) after Heyler 1975; (b) after Boy 1974; (c) after Milner 1982.)

Devonian, and it was inevitable that some group of organisms would exploit them sooner or later.

ICHTHYOSTEGA – THE FIRST TETRAPOD

Ichthyostega, from the Late Devonian of Greenland, is generally identified as the oldest-known tetrapod. Recently, a number of even older amphibian remains have been reported, but the fossils are not so good. They include footprints possibly from the Early Devonian of Australia and the Middle to Late Devonian of Brazil, and skeletons from the Late Devonian of Greenland, the USSR, and Australia. However, *Ichthyostega* is the only form described from nearly complete skeletal remains (Jarvik 1955, 1980).

Ichthyostega is about 1 m long. It has an osteolepiform body outline with a 'streamlined' head, deep vertebrae, and a tail fin (Fig. 3.4a). There is no question that *Ichthyostega* has powerful limbs and limb girdles of the tetrapod type. The ribs are unusually massive, and they have broad plate-like processes along their posterior margins which overlap slightly and form a near-solid side wall. These may have served as a partial support for the internal organs, since *Ichthyostega* probably still had a weak osteolepiform-type vertebral column.

The skull of *Ichthyostega* looks very like that of an osteolepiform in side view (cf. Figs 3.4b and 2.20a), and it retains the buried lateral line canals. However, *Ichthyostega* has lost certain elements at the back that covered the gill region, and the pectoral girdle is now separate. In dorsal view, it can be seen that *Ichthyostega* (Fig. 3.4c) has a broader and shorter skull than an osteolepiform (Fig. 2.20b), with the eyes placed further back. Ventrally (Fig. 3.4d), the arrangement of bones and teeth is still osteolepiform (cf. Fig. 2.20c).

51

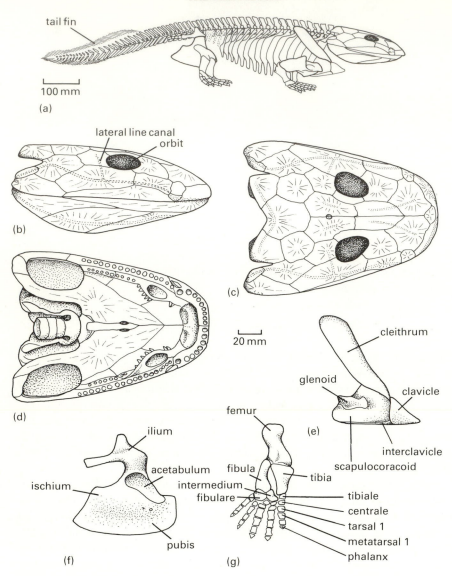

Figure 3.4 The anatomy of *Ichthyostega*, an early amphibian: (a) body; (b)–(d) skull in lateral, dorsal, and ventral views; (e) shoulder girdle in lateral view; (f) pelvic girdle in lateral view; (g) leg and foot in anterior view. (After Jarvik 1955.)

The pectoral girdle of *Ichthyostega* (Fig. 3.4e) is separate from the skull, and it is simplified in some respects when compared with that of an osteolepiform. In *Ichthyostega*, there are four main elements; a **cleithrum** above and a **scapulocoracoid** below, the latter bearing the joint surface or **glenoid** for the humerus, and a **clavicle** and **interclavicle** in front of and below the scapulocoracoid respectively. The pelvic girdle (Fig. 3.4f) shows the typical paired elements seen in all tetrapods on each side; an **ilium** above, and a **pubis** and **ischium** below, the pubis lying to the front. The joint surface for the head of the femur, the **acetabulum**, is borne in part on all three of these bones. The pelvis is attached to the vertebral column by elongate ribs of the **sacral vertebrae** which meet the inner surfaces of the ilia on each side. The pubes and ischia also meet their opposite numbers in the midline ventrally, thus making the pelvic girdle a firm all-round basket that holds the acetabula in immovable positions, and supports the posterior part of the trunk and the tail. The glenoid and acetabulum face sideways and backwards, the characteristic of tetrapods, instead of simply backwards as in osteolepiforms. The hindlimb (Fig. 3.4g) shows the standard tetrapod elements: femur, tibia, and fibula in the leg, fibulare, intermedium, **tibiale**, three centralia, and five distal **tarsals** (1–5) in the ankle, and five toes, each of which has a **metatarsal** and a number of phalanges. Counting outwards from toe 1 (the 'big toe' on the inside), the phalanges number 3,3,3,3,2.

Ichthyostega still ventured into the water, as is shown by the retention of a tail fin and a lateral line system. The vertebral column was flexible, as in an osteolepiform, and *Ichthyostega* could have swum by powerful sweeps of its tail. The strong limbs show, however, that *Ichthyostega* could waddle about on land, but the weight of its large skull and heavy ribcage probably meant that it had to rest its belly and head on the ground from time to time. Although showing terrestrial adaptations, it would probably be wrong to envisage *Ichthyostega* as being particularly fleet of foot! The sharp fangs show that *Ichthyostega* ate flesh, probably mainly fish, with the occasional large worm or millepede; nothing too fast!

THE CARBONIFEROUS WORLD

The main phases of early amphibian evolution took place in the Carboniferous period (360–286 Myr). By that time, most of the continents were coalescing into a supercontinent, and land was continuous from Europe to North America, South America, and Africa, with no inter-vening Atlantic Ocean (Fig. 3.5). Much of Europe and North America lay around the Carboniferous equator, and tropical conditions prevailed in Carboniferous amphibian localities. Damp forests of vast trees and lush undergrowth became widespread. The plants included giant club mosses, 40 m tall lycopods such as *Lepidodendron*, horsetails up to 15 m tall such as *Calamites*, ferns, and seed ferns. As these trees and bushes died, they built up thick layers of decaying trunks, leaves, and roots which were buried and eventually turned into coal. The trees provided new habitats for flying insects, including some giant forms like dragonflies with the wingspans of pigeons. The decaying plant matter and undergrowth provided even richer habitats for ground-dwelling insects, spiders, and millepedes (some up to 1.8 m long).

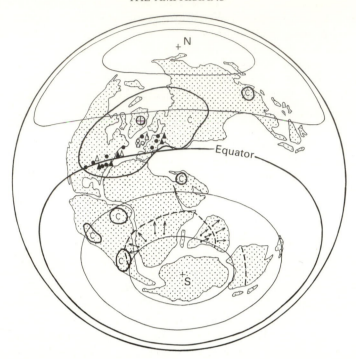

Figure 3.5 Map of the world in Carboniferous times, showing the north (N) and south (S) poles, and the postulated continental positions. Coal forests are marked C, and the main amphibian localities are shown with symbols as follows: Late Devonian (⊕), Early Carboniferous (●), Late Carboniferous (○) for temnospondyls and anthracosaurs, and Early Carboniferous (▲) and Late Carboniferous (△) for lepospondyls (mainly microsaurs). The dashed line over South America, southern Africa, and India, shows the known edge of Carboniferous glacial deposits, and the arrows show directions of glacier movement. (Modified from Pough *et al.* 1989.)

These new habitats opened up great possibilities for the early amphibians, and they diversified extensively (see box). Some forms continued to exploit freshwater fishes by becoming secondarily aquatic, while others became adapted to feed on the insects and millepedes.

DIVERSITY OF CARBONIFEROUS AMPHIBIANS

The amphibians radiated into about 20 families in the Carboniferous. Classically (e.g. Romer 1966), the amphibians have been assigned to three major lineages, the Labyrinthodontia, characterized by the labyrinthodont tooth structure (shared with osteolepiforms, Fig. 2.20e), large body size, and compound vertebrae; the Lepospondyli, characterized by small size, simple tooth structure, and fused spool-like vertebrae; and the Lissamphibia, the modern groups such as frogs and salamanders. It has become clear, however, that the 'Labyrinthodontia' and 'Lepospondyli' are not mono-phyletic groups (Panchen & Smithson 1988). The major Carboniferous amphibian groups will be introduced in approximate order of their appearance in the fossil record, and their relationships will be considered later.

Family Crassigyrinidae

Crassigyrinus from the Early Carboniferous of Scotland (Panchen 1985) has a large skull (Fig. 3.6a–c) with heavily sculptured bones. The deep embayments in the side of the skull at the back just behind the eyes are generally called otic notches, and it has been assumed that these accommodated a **tympanum**, or ear drum, which was linked to the inner ear by the stapes. However, in primitive forms such as *Crassigyrinus*, this space

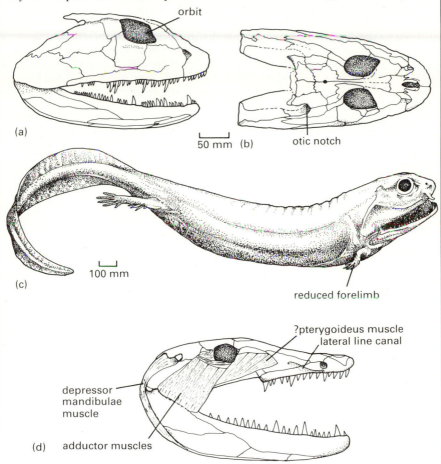

Figure 3.6 Early amphibians: (a)–(c) *Crassigyrinus*; (a) & (b) skull in lateral and dorsal views; (c) whole-body restoration, based on new Scottish material, but the tail is imaginary; (d) skull of *Megalocephalus* in lateral view, with a tentative restoration of the main jaw muscles (Figs (a) & (b) after Panchen 1985; (c) after Milner *et al.* 1986; (d) after Beaumont 1977.)

might equally have accommodated a **spiracle**, a remnant of an anterior gill slit still seen today in sharks. With its deep skull and sharp fangs, *Crassigyrinus* was clearly a meat-eater with powerful jaws which could have seized large fishes and resisted their struggles in the mouth. *Crassigyrinus* has minute forelimbs, a long narrow body, and probably a flattened tail bearing a broad fin.

Family Loxommatidae

The loxommatids are known from various Early Carboniferous localities (Beaumont 1977). *Megalocephalus* (Fig. 3.6d) has a small rounded orbit that runs into an unusual large pointed opening in front, which might have housed a gland, or have been a site for muscle attachments. There are traces of lateral line canals. The jaws are lined with short pointed teeth, and there are about six larger 'fangs' set into the bones of the palate. The skull is very low; in fact it is only about as deep as the lower jaw, so that accommodation for the brain was clearly not a priority! Large jaw muscles probably ran from the side of the skull to the upper surface of the lower jaw, and these **adductor muscles** acted to close the jaw. The jaw opened by means of a smaller jaw **depressor muscle** which ran behind the jaw joint. Muscles can only pull, and the solution of placing a jaw opener *behind* the pivot joint of the jaw is adopted in most tetrapods. *Megalocephalus* was no doubt a fish-eater, like *Crassigyrinus*, but its mode of life is hard to reconstruct in detail because almost nothing is known of its **postcranial skeleton**, the skeleton behind the head region.

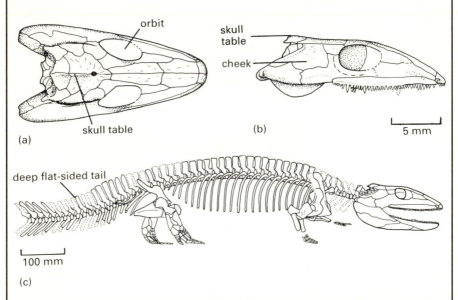

Figure 3.7 The early anthracosauroid *Proterogyrinus*: (a) & (b) skull in dorsal and lateral views; (c) restoration of the skeleton. (After Holmes 1984.)

Suborder Anthracosauroideae

The anthracosauroids, a group that arose in the Early Carboniferous, and survived into the Late Permian, include some 15 genera of moderate-sized fish-eaters. Some were apparently terrestrial, while others became secondarily adapted to life in the water.

Proterogyrinus, from the Early Carboniferous of West Virginia, USA (Holmes 1984) and Scotland, is about 1 m long and has an elongate skull (Fig. 3.7a & b). The skull table, the square area at the back of the skull (Fig. 3.7a), is set off from the cheek area, and there is a line of weakness between the two units which presumably allowed the skull to flex during jaw opening, as in osteolepiforms. There is a moderate notch at the back which may have accommodated an ear drum. *Proterogyrinus* (Fig. 3.7c) has large vertebrae, a short neck, and a flat-sided tail. The limbs are well developed for moving rapidly on land, but the flattened tail shows that *Proterogyrinus* could swim well.

Family Colosteidae

The colosteids, such as *Greererpeton* from the Early Carboniferous of West Virginia, USA (Smithson 1982), have an elongate body with 40 vertebrae in the trunk and neck, a broad tail, and short limbs (Fig. 3.8a). The skull (Fig. 3.8b & c) is very different from that of anthracosaurs: the eyes are placed further forward, the skull and lower jaw are lower and flatter, and there is no otic notch. The lateral line canals are also well developed, suggesting an aquatic lifestyle.

Order Temnospondyli

The temnospondyls are the main Carboniferous amphibians, a group that survived in abundance through the Triassic, and with much reduced

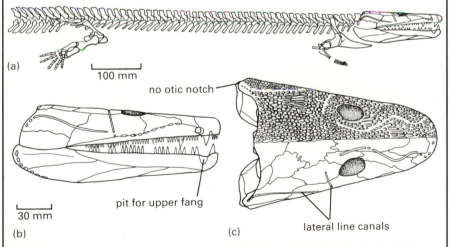

(a)

100 mm

no otic notch

pit for upper fang

30 mm

(b)

(c)

lateral line canals

Figure 3.8 The colosteid *Greererpeton*: (a) whole-body restoration; (b) & (c) skull in lateral and dorsal views, showing the sculpturing of the skull bones on the left side of the dorsal view (c) only. (Fig. (a) after Godfrey *in* Carroll 1987; (b) & (c) after Smithson 1982.)

Box – continued

diversity into the Early Cretaceous, a total span of over 150 Myr. During this time, 170 genera in over 30 families have been recorded. *Dendrerpeton* from the Late Carboniferous of Nova Scotia, Canada (Godfrey *et al.* 1987) has a broader skull (Fig. 3.9) than most anthracosauroids, and it has a rounded front margin, which is typical of temnospondyls. The palate (Fig. 3.9b) shows several characteristic temnospondyl features; a broad open space in the middle, the **interpterygoid vacuity**, which is very small in anthracosauroids, a long narrow process from the braincase which runs forward across the interpterygoid vacuity, and a pair of broad flat vomers at the front.

Order Aïstopoda

The aïstopods, a group of six genera from the Carboniferous and Early Permian of North America and Europe (Baird 1964), were snake-like animals, ranging in length from 50 mm to nearly 1 m, with up to 230 vertebrae, and no limbs or limb girdles (Fig. 3.10a). *Phlegethontia* has a very light skull (Fig. 3.10b) with large orbits, and the bones which normally form the back of the skull have been reduced or lost. Because the skull is small, the braincase seems relatively large, and it is exposed in all views. Each vertebra (Fig. 3.10c) is formed from a single element, unlike those of most other early amphibians, a condition termed **holospondylous**. The upper portion of the vertebra, the neural arch, which encloses the spinal cord and provides sites for muscle attachment, is fused to the main body of the vertebra, the **centrum**.

The aïstopods are assumed to have lost their limbs secondarily, rather than to have evolved directly from a limbless fish ancestor. Their habits are hard to imagine, but may have been equatic.

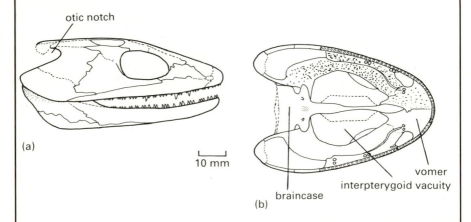

(a)

10 mm

(b)

otic notch

braincase

vomer

interpterygoid vacuity

Figure 3.9 The early temnospondyl *Dendrerpeton*: skull in (a) lateral and (c) ventral views. (After Godfrey *et al.* 1987.)

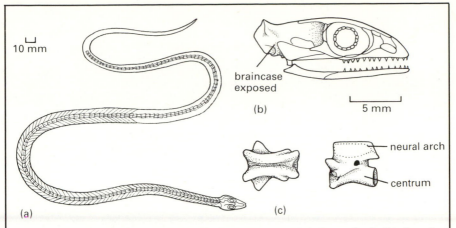

Figure 3.10 The aïstopod *Phlegethontia*: (a) reconstructed skeleton; (b) skull in lateral view; (c) trunk vertebra in dorsal (left) and lateral (right) views. (After Gregory 1948, courtesy of the American Journal of Science.)

Order Nectridea

The nectrideans (A. C. Milner 1980) are an aquatic group, known from the Late Carboniferous and Permian. Many are newt-like in appearance, with very long flattened tails which were presumably used in swimming. *Diplocaulus* and *Diploceraspis* from the Early Permian of Oklahoma and Texas (Fig. 3.11), have remarkably expanded skulls (Olson 1951, Beerbower 1963). The skull (Fig. 3.11b & c) has enormous 'horns' growing out at the sides which gives the head a boomerang-like appearance. The horns are formed from massive outgrowths of the squamosal and tabular bones, which normally form relatively small parts of the back corners of the amphibian skull. Juveniles have almost no horns at all, but a study of hundreds of specimens of *Diplocaulus* at all stages of growth (Olson 1951) shows how they grew out more and more as the animals became older (Fig. 3.11d).

The function of the horns is more of a problem. Recent biomechanical studies (Cruickshank & Skews 1980) on models of the head of *Diplocaulus* have shown that its hydrofoil shape provided lift when it was held roughly horizontal or just tipped up in even very weak currents. Cruickshank & Skews (1980) suggest that *Diplocaulus* and *Diploceraspis* fed on fishes which they caught from a lurking position on the bottom. They flicked their tails sharply, rushed up from beneath, grabbed a fish, and rapidly sank to the bottom again to enjoy their feast.

Order Microsauria

The microsaurs, the largest group of 'lepospondyls', consisting of 11 families of Late Carboniferous and Early Permian animals (Carroll & Gaskill 1978), were mainly terrestrial in habits. *Tuditanus*, an early form from the Late Carboniferous of Ohio, USA (Fig. 3.12a & b), was a highly terrestrial animal, having the proportions of a lizard, with powerful limbs and a strong skull with short teeth adapted for crushing and piercing the tough skins of invertebrates such as insects, spiders, and millepedes.

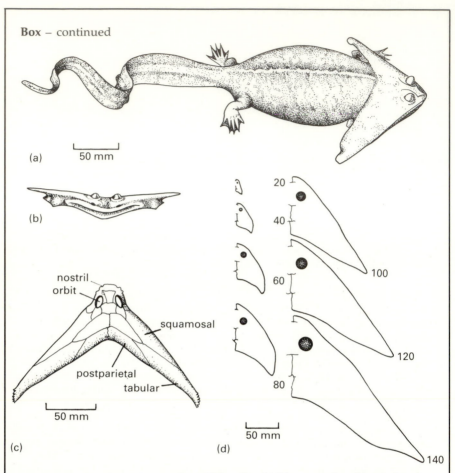

(a)

50 mm

(b)

nostril
orbit

squamosal

postparietal
tabular

50 mm

(c)   (d)

Figure 3.11 The 'horned' nectridean *Diplocaulus*: (a) life restoration; (b) anterior view of head; (c) dorsal view of skull; (d) sequence of growth stages, from juvenile (top left) to adult (bottom right), showing the growth of the projecting 'horns'. The figures 20, 40, 60, etc. are measurements, in mm, of total body lengths. (Figs (a) & (b) after Cruickshank & Skews 1980; (c) altered from Romer 1952; (d) after Olson 1951.)

Other microsaurs, such as *Microbrachis* from the Late Carboniferous of Czechoslovakia (Fig. 3.12c), seem to have been secondarily aquatic, with long slender bodies, and reduced limbs and limb girdles. Some microsaur lineages show reduced skull bones, massive occiputs (the posterior part of the skull roof), long bodies, and short legs, which suggest that they were burrowers or leaf-litter foragers.

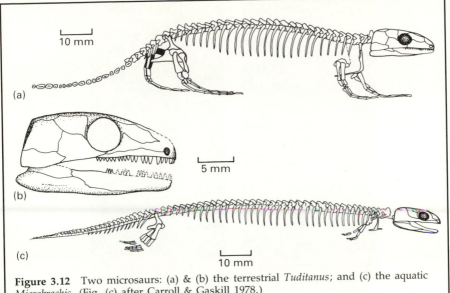

Figure 3.12 Two microsaurs: (a) & (b) the terrestrial *Tuditanus*; and (c) the aquatic *Microbrachis*. (Fig. (c) after Carroll & Gaskill 1978.)

THE NÝŘANY AMPHIBIAN COMMUNITIES

Carboniferous amphibians are known from a number of rich deposits in both Europe and North America. One of the most diverse faunas has come from Nýřany, a small mining town in western Czechoslovakia. Fossil amphibians were first reported from coal mines in this area in the 1870s (A. R. Milner 1980), and since then many hundreds of specimens have been collected and studied.

The fossil amphibians nearly all came from a 300 m thick sequence of coalified shales and mudstones near the base of the Nýřany Gaskohle Series (Westphalian D, Late Carboniferous in age, c. 300 Myr). These sediments were laid down in an enclosed lake under gentle conditions, and they contain remains of plants such as *Calamites*, a giant horsetail that grew in up to 1 m of water. There are also rare fossils of small sharks, acanthodians, and actinopterygians, as well as water-living arthropods and terrestrial millepedes. At the time of the deposition of these beds, the lake was small and poorly aerated, and the sediments represent a fairly rapid accumulation.

The fossil amphibians are generally very well preserved, and they occasionally show traces of soft parts here, and in similar localities elsewhere (Fig. 3.13). The cadavers seem to have sunk to the bottom rapidly, with relatively little decomposition and no scavenging. It may be that the animals swam a little too deep in the lake, and encountered anoxic bottom waters which suffocated them.

Figure 3.13 Skeleton of the trimerorhachoid temnospondyl *Saurerpeton* from the Late Carboniferous of the United States, showing excellent preservation of the delicate bones, and of the body outline: (a) dorsal slab; (b) ventral slab of the same specimen. The original specimen is 48 mm long. (Photographs courtesy of Dr A. R. Milner.)

A census of most of the 700 or so Nýřany tetrapod specimens currently housed in museums around the world (A. R. Milner 1980) shows that there were 20 species of amphibians, with representatives of most major groups, and four species of reptiles. These fall into three main ecological associations (Fig. 3.14):

(a) Open-water/lacustrine assocation
Three very rare forms from Nýřany, an eogyrinid anthracosauroid and two loxommatids presumably fished in the open water.

(b) Terrestrial/marginal association
Representatives of 13 species lived on or close to the shores of the lake. These include primitive temnospondyls, anthracosaurs, an aïstopod, four microsaurs, and three primitive reptiles.

(c) Shallow-water/swamp-lake association
The remaining seven amphibians from Nýřany all appear to have been partially aquatic, and to have swum rapidly about in shallow parts of the lake where plants grew in the water and where the bottom was covered

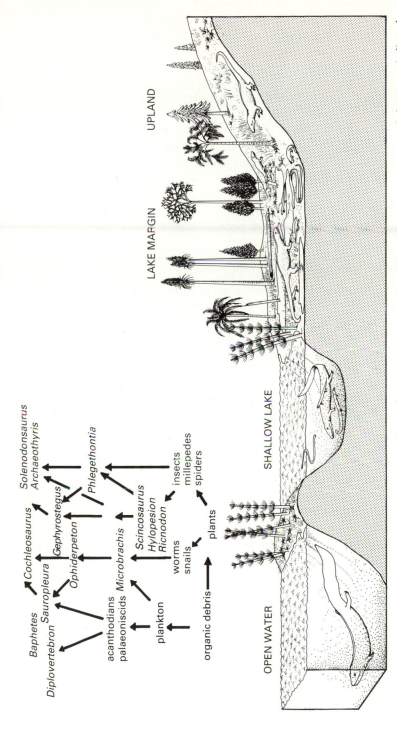

Figure 3.14 A Late Carboniferous amphibian community, based on the Nýřany locality, Czechoslovakia. Four main habitats are indicated, with representative vegetation and amphibians, from left to right: open water (eogyrinid, *Baphetes*); shallow lake (*Ophiderpeton*, *Sauropleura*, *Microbrachis*, *Scincosaurus*); lake margin (*Gephyrostegus*, *Amphibamus*, *Phlegethontia*, *Ricnodon*, etc.); possibly upland (*Scincosaurus*). The food-web on the left shows what eats what (the arrows run from the base of the food chains – the plants – through various invertebrates and fishes to the predatory amphibians, and terminating at the top of the diagram with the 'top' carnivores which feed on other amphibians). (Based on A. R. Milner 1980, and other sources.)

with plant debris. They include two temnospondyls, a branchiosaur, two nectrideans, a microsaur, and an aïstopod. Most of these presumably fed on small fishes or small tetrapods.

PRIMITIVE AMPHIBIANS AFTER THE CARBONIFEROUS

Several of the major Carboniferous lineages of amphibians survived into the Permian (286–245 Myr) and beyond. These include groups such as the Anthracosauroideae, Aïstopoda, Nectridea, and Microsauria, most of which had died out by the end of the Early Permian, as well as the Temnospondyli which lasted much longer, and two groups that were typically Permian, the Seymouriamorpha and the Diadectomorpha. These last three groups will be described now.

Order Temnospondyli (Permian–Cretaceous)

A dozen or so temnospondyl families were in existence during the Early Permian, and these include a number of terrestrially-adapted forms. *Eryops*, from the Early Permian of North America (Fig. 3.15a), shows heavier limbs and a more massive skeleton than its earlier relatives. This 2 m long animal was one of the top carnivores in its faunas, feeding on the smaller amphibians and reptiles which have been found with it, as well as on fishes. The dissorophid temnospondyls were probably fully terrestrial in habit. They have short skulls (Fig. 3.15b) with huge orbits and a large ear drum. Other Early Permian temnospondyls were gharial-like fish-eaters.

The branchiosaurs represent an interesting side-branch in temno-spondyl evolution in the Late Carboniferous and Early Permian of central Europe in particular. They are small animals, 50–100 mm long, which show many larval characters (Fig. 3.15c & d): obvious external gills, and many of the elements of the wrist and ankle are not **ossified** (i.e. they were still cartilaginous and had not turned into bone). At one time, the branchiosaurs were identified as the tadpole larvae of temnospondyls such as *Eryops*, but Boy (1972) has concluded that, while some may be larvae (cf. Fig. 3.3), most are in fact paedomorphic adults, sexually mature animals with juvenile bodies (see p. 11). The anatomy of the *Branchiosaurus* skull in particular (Fig. 3.15d) shows so many synapo-morphies with the dissorophids (cf. Fig. 3.15b) that A. R. Milner (1982) has interpreted the branchiosaurs as a paedomorphic sister group.

The Triassic to Cretaceous temnospondyls include 15 families that had mostly died out by the end of the Triassic. Two lineages survived into the Jurassic and one even into the Early Cretaceous. Most were aquatic animals which had broad flat skulls and reduced limbs. It is likely that the terrestrial amphibians of the Early Permian gave way to the burgeoning

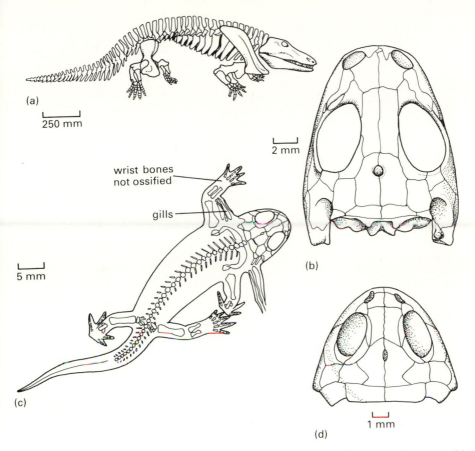

Figure 3.15 Permian temnospondyls: (a) *Eryops*; (b) dorsal view of skull of *Doleserpeton*; (c) & (d) the larval amphibian *Branchiosaurus*, showing tadpole-like characters, such as gills and poorly ossified bones: (c) reconstructed skeleton; (d) skull in dorsal view. (Fig. (a) after Gregory, W. K. 1929; (b) after Bolt 1977; (c) based on Boy 1972; (d) after Bulman & Whittard 1926.)

new groups of terrestrial reptiles in the Late Permian and Triassic (see Chapters 4 and 5), and only the aquatic and piscivorous temnospondyls survived.

Suborder Seymouriamorpha

The seymouriamorphs are a small group of terrestrial and aquatic forms. *Seymouria*, from the Early Permian (Fig. 3.16a), was a 600 mm long active terrestrial animal that lived in fair abundance in the southern midwestern United States (Heaton 1980). It has powerful limbs, and the body was held higher off the ground than in most amphibians so far considered.

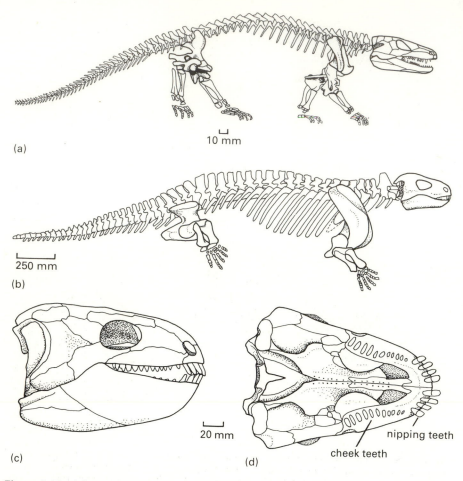

(a)

10 mm

(b)

250 mm

(c)

(d)

nipping teeth

cheek teeth

20 mm

Figure 3.16 Advanced reptiliomorph amphibians: (a) *Seymouria* skeleton; (b)–(d) *Diadectes*: (b) skeleton; (c) skull in lateral and (d) ventral views, showing the herbivorous adaptations of the dentition. (Fig. (a) after White 1939; (b) altered from Romer 1944; (c) & (d) after Carroll, 1969b.)

Order Diadectomorpha

The diadectomorphs, Late Carboniferous and Early Permian terrestrial forms, show characteristics that have always suggested they are on the evolutionary borderline between amphibians and reptiles. *Diadectes* from the western United States (Fig. 3.16b–d) is rather heavily built, with massive limb girdles, short limbs, and very heavy vertebrae and ribs (Heaton 1980). Its key features are, however, seen in the skull. *Diadectes* was one of the first terrestrial vertebrates to adopt an herbivorous diet: there are eight short peg-like teeth at the front of the jaw (Fig. 3.16c & d) which were used for nipping off mouthfuls of vegetation, and rows of broad blunt cheek teeth which were used to grind it up.

66

The amphibians of the Carboniferous and Early Permian were diverse and varied in their appearance. The classic assignment of most of them to major subclasses, the Labyrinthodontia for the larger ones, and the Lepospondyli for the smaller ones, has been widely criticized as a result of new cladistic analyses.

The Amphibia as currently defined is a paraphyletic group, since it includes the ancestors of reptiles. It includes the basal members of the wider clade Tetrapoda (Fig. 3.17), which share many derived characters in comparison to osteolepiforms, such as the modified limbs (parallel lower limb elements, full complement of wrist and ankle bones, digits, hinge-like wrist and knee joints, rotary ankle joint), the pelvic girdle attached to the sacral vertebrae, a stapes, and other features (Gaffney 1979, Panchen & Smithson 1988).

The amphibians are divided into two broad groupings, a batrachomorph clade, or the 'true' amphibians, and a reptiliomorph clade, or those amphibians on the line to the reptiles (Fig. 3.17). *Ichthyostega* may be the sister-group of all other tetrapods, or it may be a batrachomorph.

The batrachomorph clade includes the nectrideans, colosteids, micro-saurs, temnospondyls, and lissamphibians. The batrachomorphs (and possibly *Ichthyostega*) have a shallow skull and a fused skull roof with no kinesis with the cheek. Batrachomorphs have only four fingers in the

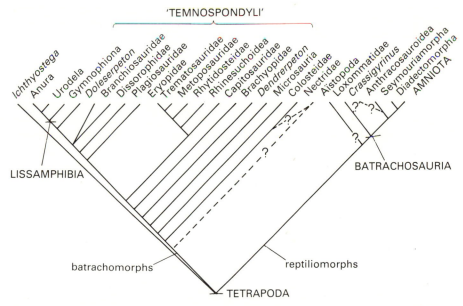

Figure 3.17 Cladogram showing the relationships of the major groups of amphibians. (Based on Panchen & Smithson 1988, A. R. Milner 1988, and other sources.)

hand, representing the permanent loss of one finger. The sequence of groups shown in the batrachomorph clade in Figure 3.17 is tentative.

The reptiliomorph clade is characterized by a mobile basal articulation (where the braincase rotates against the palatal bones) and a specialized retractor pit for the eye muscles on the front of the braincase (Panchen & Smithson 1988). The loxommatids may be primitive reptiliomorphs, followed by *Crassigyrinus*. The anthracosauroids, seymouriamorphs, and diadectomorphs form successively closer outgroups to the Amniota (= reptiles + birds + mammals), but it is uncertain whether the anthracosauroids are the sister group of *Crassigyrinus* or of the seymouriamorphs (Panchen and Smithson 1988).

EVOLUTION OF MODERN AMPHIBIANS

Modern amphibians, the Lissamphibia, are diverse, being represented by about 4000 species which fall into three distinctive groups, the order Anura (frogs and toads), the order Urodela (newts and salamanders), and the order Gymnophiona (limbless caecilians). The history of each of these will be outlined briefly before a consideration of their origins and relationships.

Order Anura

The frogs and toads are so distinctive in their anatomy that there is no problem in their identification. The skeleton (Fig. 3.18a) is highly modified for their jumping mode of locomotion: the hindlimb is extremely long, with the addition of a flexible pelvis and elongate ankle bones giving it a 'five-crank' hindlimb; the ilia run far forwards and the posterior vertebrae are fused into a rod called a **urostyle**, making a strong pelvic basket; the forelimbs and pectoral girdle are impact absorbers for when the frog lands; and there are no ribs and a short stiffened vertebral column with only four to nine vertebrae in the trunk. The head is short and flat, and the upper jaw is lined with small gripping teeth for processing insects or other prey.

The specialized characters of the frog skeleton can be detected even in one of the earliest forms, *Vieraella* from the Early Jurassic of South America (Fig. 3.18a), which has elongate hindlimbs, reduced numbers of vertebrae, and a flattened skull. It is primitive in having more vertebrae than in most modern frogs (nine), small traces of ribs, and slightly heavier limb bones, but it offers few guides to ancestry. Some of the 23 modern families may be traced back as far as the Jurassic or Cretaceous, but most have very short fossil records, or none at all (Estes & Reig 1973). *Triadobatrachus* from the Early Triassic of Madagascar (Fig. 3.18b) is the oldest-known frog: it has a reduced number of vertebrae, reduced ribs, elongate ilia, and a frog-like skull.

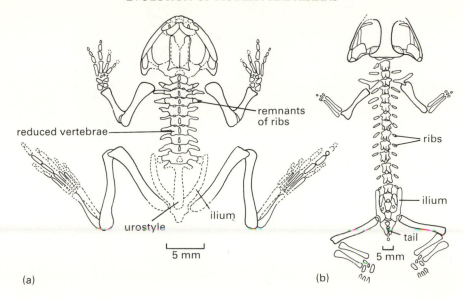

Figure 3.18 Early frogs: (a) the Jurassic *Vieraella*, showing most modern frog adaptations; (b) the first frog *Triadobatrachus*. (After Estes & Reig 1973, by permission of the editor, copyright © 1973 by the Curators of the University of Missouri.)

Order Urodela

The newts and salamanders show far fewer obvious specializations than the frogs. The body is elongate, and there are generally four short walking limbs and a flattened swimming tail. The fossil record of salamanders, like that of frogs, is patchy (Estes 1981). The oldest well-preserved salamander, *Karaurus* from the Late Jurassic of the USSR (Fig. 3.19a), has a broad flattened skull with large orbits and rows of small teeth around the jaws. The skull roof is covered with heavily ornamented bone. The skull of a modern salamander (Fig. 3.19b) shows many changes; the bones are generally lighter, and the braincase has become fused with the parietal bones and is partially exposed on the skull roof.

Order Gymnophiona

The Gymnophiona, or caecilians, are strange little amphibians that look like earthworms (Fig. 3.19c). They have lost their legs, hence an alternative name, apodans (literally 'no feet'), and they live by burrowing in leaf litter or soil, or swimming in ponds, in tropical parts of the world. The skull is solidly built, and can be used for burrowing by battering the soil with the snout (Fig. 3.19d). There may be as many as 200 vertebrae in the trunk region, but the tail is generally short. Until recently no fossils were known. Single vertebrae have now been reported from the Late Cretaceous and Palaeocene of Brazil, and the oldest caecilian from the

69

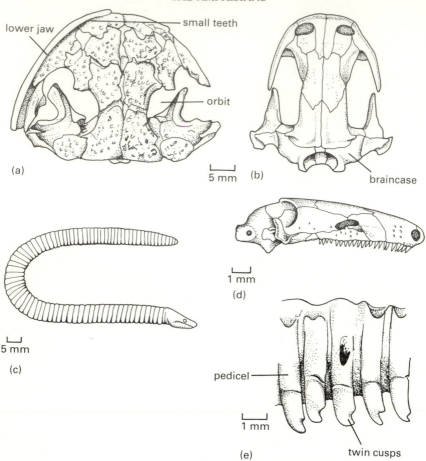

Figure 3.19 Salamanders and caecilians: (a) dorsal view of the skull of the Jurassic salamander, *Karaurus*; (b) similar view of a modern salamander skull; (c) a typical modern caecilian; (d) skull of the modern caecilian *Grandisonia*; (e) pedicellate teeth of the salamander *Amphiuma*. (Fig. (a) after Ivakhnenko 1978; (b) altered from Romer & Parsons 1970; (c) modified from Pough *et al.* 1989; (d) after Carroll & Currie 1975; (e) altered from Parsons & Williams 1963.)

Early Jurassic of North America, an animal with reduced legs, has just been found.

Origin of the modern orders

Many biologists regard the three modern groups as members of a clade Lissamphibia (e.g. Parsons & Williams 1963, Bolt 1977, Rage & Janvier 1982, A. R. Milner 1988) on the basis of their possession of bicuspid (two-cusped) **pedicellate teeth** (Fig. 3.19e), in which the base and crown are separated by a zone of fibrous tissue, and other characteristics (A. R.

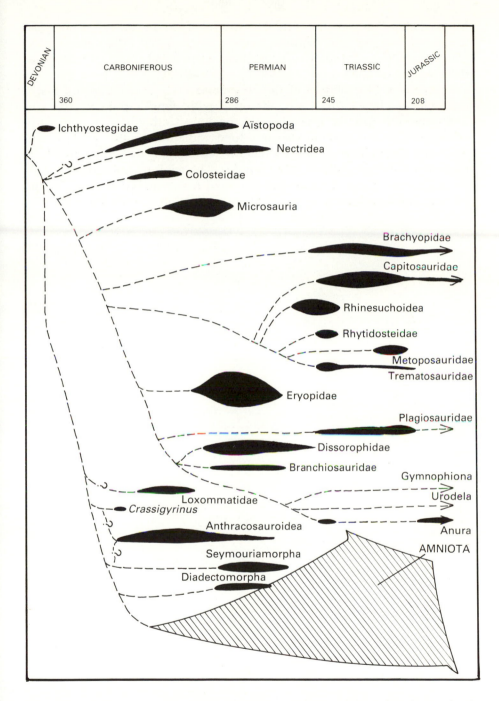

Figure 3.20 Evolutonary tree showing the major amphibian groups, their known fossil records (vertical scales), postulated relationships (dashed lines; based on cladogram in Fig. 3.17), and relative diversity (width of 'balloon').

Milner 1988). The ancestry of the Lissamphibia, according to this analysis, is placed among the temnospondyls (Fig. 3.17). The dissorophid *Doleserpeton* (Fig. 3.15b) has pedicellate teeth (Bolt 1977), while other temnospondyls show other lissamphibian features.

The pattern of evolution of the amphibians (Fig. 3.20) shows a major radiation in the Early Carboniferous, with new reptiliomorph and temnospondyl groups appearing in the Late Carboniferous and Early Permian. The temnospondyls continued radiating in the Triassic and dwindled through the Jurassic, while the reptiliomorphs had by then long made the transition fully to being reptiles. The modern amphibian groups probably arose in the Triassic, but scattered fossils are known only in the Jurassic and Cretaceous, before a major expansion in the Cenozoic.

CHAPTER FOUR

The evolution of early reptiles

During the Late Carboniferous, the temnospondyl and anthracosaur amphibians dominated most terrestrial landscapes, especially the damp forests. However, small lizard-sized tetrapods were also in existence, creeping in and out of the vegetation in drier areas, in search of insects and worms. They laid eggs that did not have to hatch in water. These were the first reptiles, and they included the ancestors of all of the major groups of amniotes (i.e. reptiles, birds, and mammals) that were to dominate the Earth from Permian times onwards.

In this chapter, the early reptiles will be described, and key biological problems of living a life completely divorced from the water will be considered. The early radiations in the Late Carboniferous and Permian form an essential prelude to understanding the better-known ages of the dinosaurs and of the mammals.

HYLONOMUS AND *PALEOTHYRIS* – BIOLOGY OF THE FIRST REPTILES

The oldest reptile currently known has been reported from the Early Carboniferous (350 Myr) of Scotland, but it is not as completely known as the later *Hylonomus* and *Paleothyris* from the mid Carboniferous (310 Myr, 300 Myr, respectively) of Nova Scotia (Carroll 1964, 1969a). The body (Fig. 4.1a) is slender, and is about 200 mm long, including the tail. Unlike many amphibians, the head is relatively small, being about one-fifth of the trunk length rather than one-third to one-quarter. The skull of *Hylonomus* is incompletely known, with uncertainty about the posterior view and the palate, but *Paleothyris* is represented by better skull remains (Fig. 4.1b–f; see box).

The light construction of the skull, and the small sharp teeth, suggest that *Hylonomus* and *Paleothyris* fed on invertebrates such as insects and millepedes that had tough skins. The teeth could readily pierce the tough skin to reach the flesh inside.

One of the key features of the skull of *Paleothyris* which relates to

ANATOMY OF AN EARLY REPTILE

The skull consists of a thin outer covering of dermal roofing bones with a modest braincase, loosely attached, inside. The outer covering is broken by two large orbits and two nostrils. The array of bones in the skull of *Paleothyris* is similar to that of advanced reptiliomorph amphibians (cf. Fig. 3.16), but it has no otic notch, and the bones at the back of the skull table (supratemporal, tabular, postparietal) are very much reduced and seen mainly in the posterior view of the skull on the occiput (Fig. 4.1e). The skull and jaw bones may be divided into five main sets which relate to the standard views:

(a) *Cheek* (Fig. 4.1b): from the front, the side of the skull shows the following bones: premaxilla and maxilla, both bearing teeth, **lacrimal** and **prefrontal** in front of the **orbit**, and **postfrontal**, **postorbital** and **jugal** behind, and **squamosal**, **quadratojugal**, and **quadrate** making up the posterior angles of the skull.

(b) *Skull table* (Fig. 4.1c): paired **nasals**, **frontals**, and **parietals** form the dorsal surface of the skull, with the nasals lying between the nostrils, and the frontals between the orbits.

(c) *Palate* (Fig. 4.1d): paired **vomers** lie behind the palatal extensions of the premaxillae, and behind them the **pterygoids** which run back and sideways to meet the quadrates. These elements are attached to the maxillae and jugals at the side by the **palatines** and **ectopterygoids**. The main ventral element of the braincase, the **parasphenoid**, lies behind and between the pterygoids, and it sends a long process forwards in the midline in the interpterygoid vacuity. Several of the palatal bones (palatine, pterygoid, parasphenoid) bear teeth, and these tend to be lost in the course of amniote evolution.

(d) *Occiput* (Fig. 4.1e): the back of the skull shows how the braincase fits inside the cranium: the **postparietals**, **tabulars**, and **supratemporals** of the skull table form the dorsal margin and are fused firmly to the **supraoccipital**, the dorsal braincase element. The other elements of the braincase, the **opisthotics** and **exoccipitals**, which support the semcircular canals of the inner ear, lie to either side of the **foramen magnum**, the broad passage through which the spinal cord passes back from the brain. The opisthotic also runs sideways towards the squamosal, quadratojugal, and quadrate of the cheek region, and a robust stapes makes a link to the quadrate. The lower margin of the braincase is formed by the **basioccipital**, which also provides a ball-like **occipital condyle** which articulates with the first vertebra in the neck.

(e) *Lower jaw (mandible)* (Fig. 4.1b & f): the main lower jaw element in side view (Fig. 4.1b) is the dentary at the front which bears the teeth. Behind it are the **surangular** above and the **angular** below. In medial (inside) view (Fig. 4.1f), it can be seen that the angular wraps round under the jaw, and the main bones are the **splenial** in front and the **prearticular** behind, with a small **coronoid** between and forming a peak in the jaw margin. The jaw joint lies on the articular bone, a small complex element at the back.

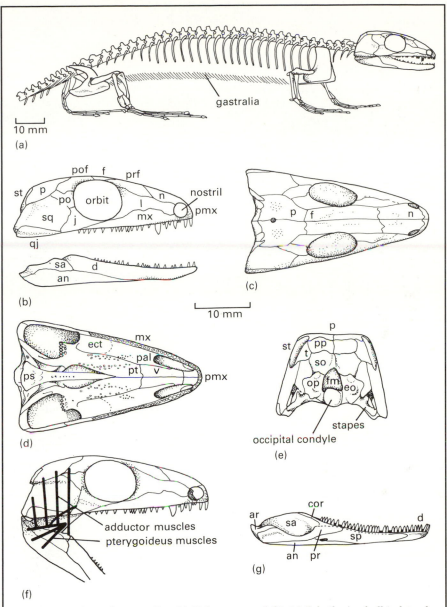

Figure 4.1 The earliest reptiles: (a) *Hylonomus*; and (b)–(g) *Paleothyris*; skull in lateral (b), dorsal (c), ventral (d), and occipital (e) views; (f) restoration of the main jaw closing muscles; (g) lower jaw. Abbreviations: an, angular; ar, articular; cor, coronoid; d, dentary; ect, ectopterygoid; eo, exoccipital; f, frontal; fm, foramen magnum; j, jugal; l, lacrimal; max, maxilla; n, nasal; op, opisthotic; p, parietal; pal, palatine; pmx, premaxilla; po, postorbital; pof, postfrontal; pp, postparietal; prf, prefrontal; ps, parasphenoid; pt, pterygoid; qj, quadratojugal; sa, surangular; so, supraoccipital; sp, splenial; sq, squamosal; st, supra-temporal; t, tabular; v, vomer. (Fig. (a) after Carroll & Baird 1972; (b)–(g) after Carroll 1969a.)

Box – continued

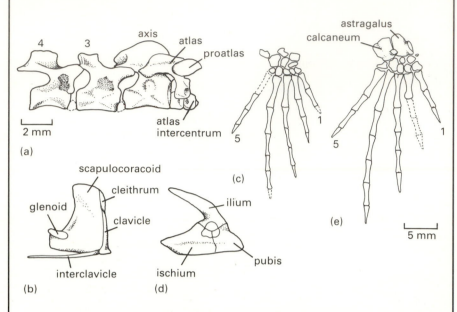

(a)

2 mm

4   3   axis   atlas   proatlas

atlas
intercentrum
5

astragalus
calcaneum

1

5

(c)

scapulocoracoid

cleithrum

glenoid

clavicle

ilium

(e)

5 mm

interclavicle   ischium   pubis

(b)   (d)

   The skeleton of *Hylonomus* (Fig. 4.1a) and *Paleothyris* is lightly-built. The vertebrae consist of spool-like pleurocentra with small crescent-shaped intercentra between. The first two **cervical** vertebrae (Fig. 4.2a), the vertebrae of the neck, are highly modified to make the junction with the occipital condyle of the skull. Vertebra 1, the **atlas**, consists of six separate elements, the intercentrum which fits beneath the occipital condyle, the pleurocentrum behind it, and a **proatlas** and atlas arch on each side above the occipital condyle. Vertebra 2, the **axis**, is a large element with the pleurocentrum and neural arch fused to each other, and a small intercentrum behind. The remaining three or four cervical vertebrae follow a similar pattern, but they have rather smaller neural arches than the axis.

   The dorsal vertebrae, those lying in the trunk region, number about 21 in *Hylonomus* and 27 in *Paleothyris*, making totals of about 26 and 32 **presacral** vertebrae (cervicals + dorsals) respectively. The cervicals bear short ribs, while the dorsals bear longer ones. Behind the presacral vertebrae are two sacrals in the hip region which are attached to the ilia by specialized ribs, and then an unknown, but large, number of **caudal** vertebrae in the tail.

   The limbs and limb girdles are basically the same as in the Carboniferous amphibians (cf. Figs 3.1b, 3.4e–g). The pectoral girdle (Fig. 4.2b) consists of a large fused scapulocoracoid which bears a screw-shaped glenoid for the head of the humerus. The cleithrum and clavicle are reduced to thin strips of bone in front of the scapulocoracoid, and the interclavicle is a long T-shaped element beneath. The arm is short (Fig. 4.1a), and the hand (Fig. 4.2c) long and slender. It shows all the wrist bones seen in *Eryops* (Fig. 3.1b), and the phalangeal formula of the hand is 2,3,4,5,3 – a typical value for anthracosaurs and reptiles.

The pelvis (Fig. 4.2d) consists of a narrow ilium, and a heavy pubis and ischium beneath, which meet each other in the midline as in amphibians (cf. Fig. 3.4f). The hindlimb and foot are longer than the forelimb. The ankle bones have changed in one respect from those of *Ichthyostega* (Fig. 3.4g), apart from becoming more slender. The tibiale, intermedium, and a centrale of primitive amphibians have fused into a larger element termed the **astragalus**. The fibulare is also larger, and is termed the **calcaneum**. The phalangeal formula of the foot is 2,3,4,5,3.

There are no scales in the skin of *Hylonomus* or *Paleothyris*, but these animals have chevron-like **gastralia**, or abdominal 'ribs', closely spaced in the belly region (Fig. 4.1a).

feeding is an increase in the strength of the jaws when compared to amphibians, sufficient to nip through the toughest arthropod skin. A new muscle group, the pterygoideus, supplements the adductors in pulling the jaw up and forwards (Fig. 4.1g). The palatal teeth in *Paleothyris* are smaller than those on the premaxilla and maxilla, and they presumably played a less important role, probably in holding the food and in further crushing it after it had been cut up. The tongue was probably toughened on its upper surface, and worked against the palatal teeth.

The stapes in *Paleothyris* is heavy, as in the amphibians, and it probably had a limited function in hearing. Low-frequency sounds could be transmitted as vibrations from the throat region through the stapes to the braincase. It is unlikely that *Paleothyris* had a tympanum since there is no otic notch.

Restorations of the life appearance of *Hylonomus* and *Paleothyris* (Fig. 4.3) show that they probably looked very like modern terrestrial insectivorous lizards. Both are lightly-built, so it is remarkable how well their remains have been fossilized. This is explicable because of the unique conditions of preservation: both *Hylonomus* and *Paleothyris* have been found in fossilized tree trunks.

Hylonomus comes from mudstones, sandstones, and coals, deposited in shallow freshwater lakes and rivers of the Cumberland Group of Joggins, Nova Scotia. In the 1840s geologists discovered abundant upright tree stumps of the lycopod (club moss) tree *Sigillaria*. The first fossil vertebrates were collected there in 1852, and since then over 30 productive tree stumps were discovered, and the contained bones removed for study. The total haul included skeletons of hundreds of amphibians (six species of microsaurs, one temnospondyl, and one embolomere) as well as three reptile species, *Hylonomus* and two less well-preserved forms (Carroll 1970).

It seems that in mid Carboniferous times the Joggins area was covered with lush forests of *Sigillaria*, up to 30 m tall. Occasionally, the lakes flooded and the forests were inundated with sediment. The trees died and fell, leaving only their roots and buried lower trunks in place. As new forests became established above, the centres of the lycopod tree trunks

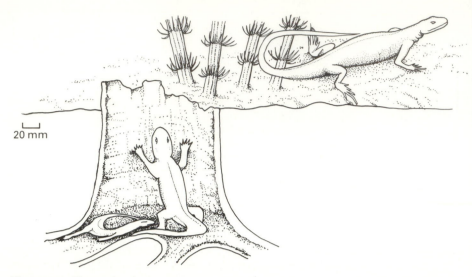

Figure 4.3 The mode of preservation of the early reptiles *Hylonomus* and *Paleothyris* which were trapped in hollow tree stumps in the mid Carboniferous of Nova Scotia. (After Carroll 1970 and other sources.)

rotted, and millepedes, snails, and small tetrapods fell in (Fig. 4.3). These animals lived for some time in their traps as the amphibians and reptiles fed on the invertebrates and left faecal remains, but eventually they died. The tree trunks then provide a concentration of the small terrestrial animals of the time.

THE CLEIDOIC EGG – A PRIVATE POND

Modern amniotes are set apart from the amphibians primarily by the fact that they lay eggs which have semipermeable shells and which contain sufficient fluid and food for the embryo to develop fully into a terrestrial hatchling. The eggs are not laid in water, and the aquatic larval stage, the tadpole, is omitted. Amniotes generally lay far fewer eggs than do amphibians or fishes because more reproductive energy has to be invested in each egg, and because the young are protected from predation to a much later stage in development. Reproduction also takes place on dry land, so that internal fertilization is essential.

The egg of amniotes, called the **amniotic** or **cleidoic** (literally 'closed') egg (Fig. 4.4a) has two key features:

(a) a semipermeable shell, usually calcareous, but leathery in snakes, some lizards, and some turtles, which allows gases to pass in (oxygen) or out (waste carbon dioxide), but keeps the fluids inside; and

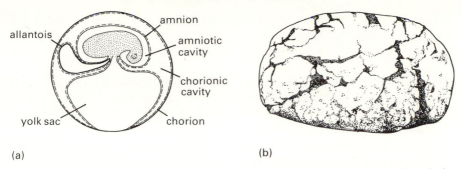

Figure 4.4 The cleidoic egg: (a) diagram showing the semi-permeable shell and the extraembryonic membranes; (b) the oldest supposed fossil cleidoic egg from the Early Permian of Texas; about 40 mm long. (Fig. (a) altered from Pough *et al*. 1989; (b) after Kirsch 1979.)

(b) extraembryonic membranes, specialized membranes that lie 'outside' the embryo, the chorion, amnion, and allantois. The chorion surrounds the embryo and yolk sac, while the amnion surrounds the embryo more closely. Both function in protection and gas transfer. The allantois forms a sac which is involved in respiration and stores waste materials. As the embryo develops, the yolk sac, full of highly proteinaceous food, dwindles and the allantois fills up.

Fossil eggs are rare. The oldest supposed egg (Fig. 4.4b), from the Early Permian (c. 270 Myr) of Texas, may in fact be an inorganic nodule or the remains of a soft-shelled reptilian egg (Kirsch 1979), rather than a calcareous egg shell as at first assumed. In the absence of Carboniferous eggs, how can we identify *Hylonomus* and *Paleothyris* as the oldest amniotes? The argument is phylogenetic. The intricate features of the amniotic egg of all living amniotes develop in the same way. *Hylonomus* and *Paleothyris* already lie on one of the major amniote lineages (see p. 99, so the amniotic egg must have arisen once only, and at a point in the cladogram *below* those two early reptiles.

RADIATION OF THE REPTILES IN THE CARBONIFEROUS

Two major amniote lineages became established in the Late Carboniferous, distinguished on the basis of their **temporal fenestrae** – openings behind the orbit which probably function in reducing the weight of the skull and in conserving calcium. The argument is that bone is costly to produce and maintain, as well as being heavy, and it can be advantageous to dispense with it where it is not required. Much of a skull is under stress from the movements of the jaws and neck muscles, but some spots, in the cheek region and palate, are under no stress, and cavities may develop without reducing the effectiveness of the skull. Fenestrae also provide additional attachment edges for specific jaw muscles.

The groupings of amniotes are as follows (Fig. 4.5):

(a) 'Anapsida' (a paraphyletic group as presently defined): reptiles with no temporal fenestrae. Includes early forms like *Hylonomus* and *Paleothyris*, as well as several lineages in the Permian and Triassic, and the turtles.
(b) *Synapsida*: reptiles with one lower temporal fenestra, surrounded by the postorbital, jugal, and squamosal. Includes the extinct mammal-like reptiles and the mammals.
(c) *Diapsida*: reptiles with two temporal fenestrae, a lower one as in synapsids, and an upper one surrounded by the postorbital, squamosal, and parietal. Includes the lizards, snakes, crocodilians, and birds, as well as numerous extinct groups such as the dinosaurs and pterosaurs.

A fourth pattern is added here for convenience, although it arises in reptiles only much later than the Carboniferous:

(d) 'Euryapsida' (a polyphyletic group, consisting of motley probable descendants of diapsids): reptiles with one upper temporal fenestra, surrounded by the postorbital, squamosal, and parietal, possibly having evolved from diapsid reptiles by the loss of their lower temporal fenestra. Includes the marine nothosaurs, plesiosaurs, and ichthyosaurs of the Mesozoic.

The basal anapsids, synapsids, and diapsids of the Late Carboniferous include 20 or so genera of small and medium-sized insectivores (see box), whose relationships will be considered later.

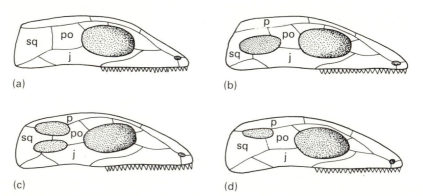

Figure 4.5 The four main patterns of temporal fenestrae in amniote skulls: (a) anapsid; (b) synapsid; (c) diapsid; (d) euryapsid.

THE CARBONIFEROUS REPTILES

Family Protorothyrididae

The basal anapsids include animals such as *Hylonomus* and *Paleothyris* (Figs 4.1–4.3) and six other genera from the Late Carboniferous and Early Permian of North America and Czechoslovakia which are assigned to the family Protorothyrididae (Carroll & Baird 1972, Clark & Carroll 1973). The protorothyridids show little variation during the 30 Myr or so of their evolution. They seem to have been agile insectivores, rather like many modern lizards in their ecology.

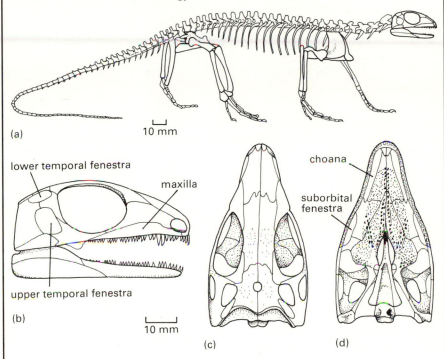

Figure 4.6 The first diapsid reptile, *Petrolacosaurus*: (a) skeleton; (b)–(d) skull in lateral, dorsal, and ventral views. (After Reisz 1981.)

Family Petrolacosauridae

Petrolacosaurus, a slender 400 mm long reptile from the Late Carboniferous (295 Myr) of Kansas, USA, is the oldest-known diapsid reptile (Reisz 1981). The body (Fig. 4.6a) has similar proportions to *Hylonomus*, but with a relatively smaller head, less than one-fifth of the body length, a longer neck, and longer limbs. The skull (Fig. 4.6b–d) is also similar, but with larger orbits, two temporal fenestrae, and more small teeth on the palatal bones.

Box – continued

The teeth are small and sharp and clearly indicate a diet of insects and other small animals. There is an extra opening in the palate, the **suborbital fenestra** (Fig. 4.6c), in addition to the **choana**, or internal nostril, of all tetrapods, through which the air passages from the nasal cavity pass into the mouth.

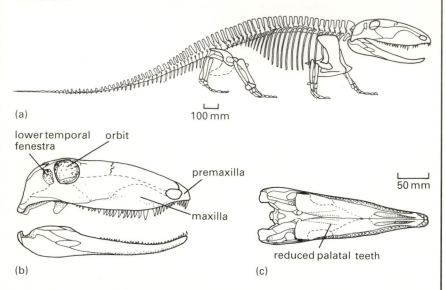

(a)

⊢⊣
100 mm

lower temporal fenestra   orbit

premaxilla

⊢⊣
50 mm

maxilla

reduced palatal teeth

(b)

(c)

Figure 4.7 The ophiacodont *Ophiacodon*: (a) skeleton; (b) & (c) skull in lateral and ventral views. (After Romer & Price 1940.)

Family Ophiacodontidae

The third lineage of Late Carboniferous reptiles includes forms with synapsid skulls, and the majority of them fall in the family Ophiacodontidae, a group of six or seven genera that survived into the Early Permian. The first ophiacodont, *Archaeothyris* from the Morien Group of Nova Scotia, which also yielded *Paleothyris*, is incompletely known. Its relative *Ophiacodon*, from the Late Carboniferous and Early Permian of New Mexico (Romer & Price 1940, Reisz 1986), is larger than the reptiles so far described, being 1.5–3 m in length (Fig. 4.7). The skull is relatively very large. It has a long high narrow snout region which makes up three-fifths of the total length, and the orbit and temporal fenestra are small and placed high. The limb bones are massive. *Ophiacodon* was a meat-eater, and it may have fed on fishes and tetrapods rather than mainly on insects.

During the Permian, the continents moved into even closer contact than in the Carboniferous (Fig. 3.5), and the supercontinent Pangaea (literally 'whole world') came fully into being. A southern ice cap developed in the Late Carboniferous, and disappeared in the Early Permian as Antarctica drifted north. Most finds of Late Carboniferous and Early Permian amphibians and reptiles are from the northern hemisphere, probably because of the cold and temperate climates of southern continents. By Late Permian times, however, rich deposits of fossil tetrapods are known, from southern Africa in particular.

In the northern hemisphere, Early Permian climates in fact became hot and arid in many parts, with the development of extensive evaporite deposits in North American and western and central Europe. Major floral changes took place as a result of these climatic changes. The lush damp tropical Carboniferous forests disappeared as the previously dominant club mosses and horsetails died out. They were replaced by seed-bearing plants of rather more modern type, conifers in the northern hemisphere, and glossopterids in the south. The replacement of amphibians by reptiles as the dominant terrestrial tetrapods during the Permian must be related, in part at least, to these major climatic and floral changes.

PELYCOSAURS – THE SAIL-BACKED SYNAPSIDS

The most diverse tetrapods of the Early Permian, representing up to 70% of all genera, were the pelycosaurs (Romer & Price 1940, Reisz 1986), six families of basal synapsids (see box). The first family to appear, the ophiacodonts (see p. 82), were most important in the Late Carboniferous, although they lived on into the Permian. At least two of the other five families arose in the Late Carboniferous, but achieved their greatest diversity in the Permian. Two recent books on pelycosaurs and other mammal-like reptiles are Kemp (1982) and Hopson & Barghusen (1986).

PELYCOSAUR DIVERSITY

The eothyrid *Eothyris*, a small animal from the Early Permian of Texas (Fig. 4.8), has a low skull with a much shorter and broader snout than that of *Ophiacodon*. The two caniniform teeth are very large, and *Eothyris* was clearly a powerful predator.

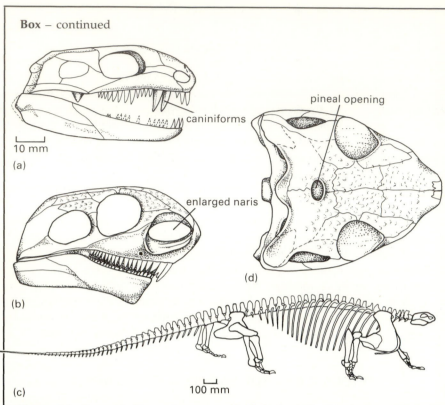

(a)

caniniforms

10 mm

pineal opening

enlarged naris

(d)

(b)

(c)

100 mm

Figure 4.8 Two primitive pelycosaurs: (a) *Eothyris* skull; (b)–(d) *Cotylorhynchus* skeleton and skull in lateral and dorsal views. (After Romer & Price 1940.)

The caseids, herbivorous pelycosaurs from the mid Permian of North America and Europe, include small and large forms. *Cotylorhynchus* from Texas and Oklahoma (Fig. 4.8b–d), is the largest pelycosaur, at a length of 3 m, but its skull looks as if it comes from an animal one-quarter of the size! The key caseid characters are seen in the skull (Fig. 4.8b & d): greatly enlarged nostrils, a pointed snout that extends well in front of the tooth rows, reduced numbers of teeth with no caniniforms, and a very large pineal opening (the circular opening between the parietals). There are several indications that *Cotylorhynchus* was a herbivore: the teeth are spatulate in shape rather than pointed, and they have crinkled edges, the jaw joint is placed *below* the level of the tooth rows, an adaptation that gives greater strength to the jaw adductors, and the barrel-shaped ribcage presumably contained massive guts which were necessary for digesting large quantities of rough plant food.

The varanopseids, a small group of Early Permian pelycosaurs from North America, include four or five genera of small carnivores. The limbs are long and the skeleton lightly built, so that they are interpreted as active and agile in their habits.

The edaphosaurids, such as *Edaphosaurus* from the Early Permian of New Mexico and Texas (Fig. 4.9) were herbivores. They have enormously

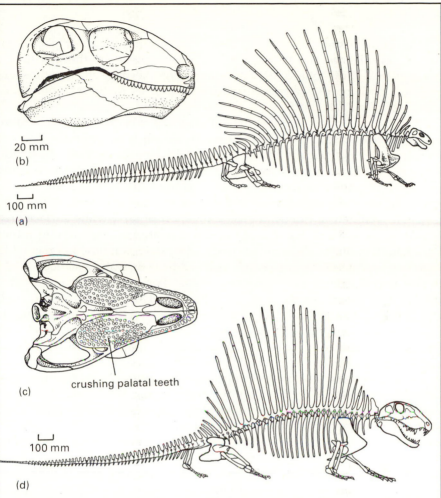

Figure 4.9 The advanced pelycosaurs *Edaphosaurus* (a)–(c) and *Dimetrodon* (d): (a) & (d) skeleton; (b) & (c) skull in lateral and ventral views. (After Romer & Price 1940.)

elongated neural spines of the cervical and dorsal vertebrae, which were probably covered by skin, hence the popular term 'sail backs'. The skull of *Edaphosaurus* is relatively small in comparison to the body size, and it shows several adaptations to herbivory: peg-like teeth, a deep lower jaw, and extensive palatal teeth (Fig. 4.9c) which are large and form a broad crushing surface.

The largest pelycosaur family, the sphenacodontids, includes eight genera (Reisz 1986) of carnivores from the Late Carboniferous and Early Permian of North America and Europe. *Dimetrodon* from the Early Permian of Texas, New Mexico, and Oklahoma, has a large sail, and it reaches a length of about 3 m. It has a large skull, with a small orbit and a high temporal fenestra (Fig. 4.9d).

Pelycosaur physiology

The function of the pelycosaur sail is a fascinating topic for palaeobiological speculation, and its importance is underlined by the fact that it arose at least three times independently, in the herbivorous edaphosaurs and the carnivorous sphenacodontids, as well as in the contemporary temnospondyl *Platyhystrix*.

During growth, the neural spines of *Dimetrodon* became relatively more and more elongated, and the increase in length is in proportion to the increase in body weight rather than body length. A basic biological principle is the volume to surface area relationship; as animals become larger, their volume or weight increases as the cube of their body length, while the surface area increases as the square. This, of course, reflects the fact that volume is a three-dimensional aspect of size, while surface area is two-dimensional. If the size of the sail is related to body weight rather than length or surface area of the body, a physiological function related to volumetric problems is suggested, such as heat or water balance.

The neural spines have grooves at the base which were probably occupied by blood vessels. Further, when fossil skeletons are excavated, the neural spines generally lie in a neat fence-like array, which suggests that they were held together by a tough covering of skin in life. The 'sail' then was composed of heavily vascularized skin, and its function seems to have been thermoregulatory (Haack 1986).

The pelycosaurs, like modern reptiles, were probably **ectotherms**: they did not generate much heat within their bodies, and had to rely on external sources. In the early morning, when air temperatures were rising, *Dimetrodon* still had a low body temperature after the cold night. Without a sail, Haack (1986) calculated that it would take a large 250 kg *Dimetrodon* 12 hours of basking in the sun to increase its body temperature from 25°C to 30°C. With its sail, the same animal took three hours, since the sail gave an additional surface area for heat uptake during basking. As body temperatures increase, reptiles generally become more active, and the sail may have been of high selective advantage to forms that had it in speeding up the move from nighttime torpor to daytime activity. As a top predator, these few extra hours may have permitted *Dimetrodon* to grab its still-torpid prey which lacked sails. The sail was also useful to some extent in dissipating heat if the animal became overheated, but seemingly less so (Haack 1986) than had been thought.

The weakness of these arguments is the fact that most pelycosaurs, and other contemporaries, lacked sails, and yet seemed to survive quite well!

THE THERAPSIDS OF THE LATE PERMIAN

The pelycosaurs gave way in the mid Permian to a major synapsid clade, the Therapsida. These include a diverse assemblage of carnivores and herbivores of the Late Permian, some of which continued into the Triassic. Therapsid derived characters include an enlarged temporal fenestra, loss of the supratemporal bone, a sheet-like reflected lamina on the angular bone (Fig. 4.10a), the jaw joint shifted forwards, reduction of the palatal teeth, as well as modifications of the shoulder and pelvic

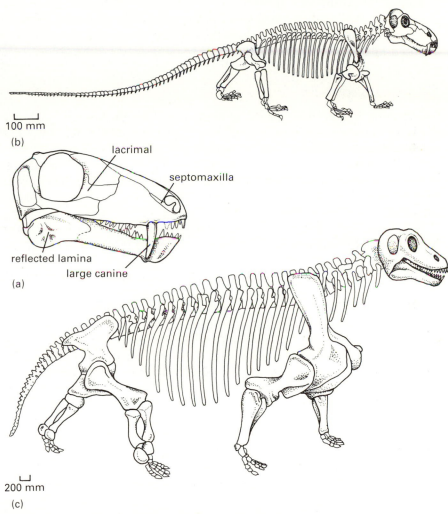

100 mm

(b)

lacrimal

septomaxilla

reflected lamina

large canine

(a)

200 mm

(c)

Figure 4.10 Early therapsids: (a) skull of *Biarmosuchus* in lateral view; (b) herbivorous dinocephalian *Titanophoneus*; (c) herbivorous dinocephalian *Moschops*. (Fig. (a) after Sigogneau & Chudinov 1972; (b) after Orlov 1958; (c) after Gregory 1926.)

girdles and of the hindlimb (Kemp 1982, 1988a, Hopson & Barghusen 1986). The major therapsid groups of the Late Permian are reviewed here.

Basal therapsids – Biarmosuchia

Several rather poorly-known genera of synapsids from the mid Permian of Russia are regarded as the most primitive therapsids (Hopson & Barghusen 1986). *Biarmosuchus* (Fig. 4.10a), for example, was a small carnivore that resembled the sphenacodontids in most respects. However, the occiput slopes back rather than forwards, and the supratemporal bone is absent. The numbers of teeth are reduced and there is a prominent single canine, as well as a few small palatal teeth. An additional element, the septomaxilla, present within the nostril of pelycosaurs, is now exposed on the side of the skull.

Suborder Dinocephalia

The dinocephalians include 40 genera of synapsids known only from the Late Permian of Russia and South Africa, which fall into both carnivorous and herbivorous lineages (Kemp 1982, King 1988). A carnivorous form, *Titanophoneus* from the mid Permian of Russia (Fig. 4.10b), is a large animal with short limbs and a heavy skull. The incisors and canines are well developed, and they were presumably used for grasping and piercing prey.

   The Tapinocephalidae includes a range of herbivorous forms, some quite bizarre in appearance. *Moschops* from South Africa (Fig. 4.10c) is a large animal about 5 m long with a massive ribcage and heavy limbs, but tiny hands and feet. The hindlimbs are held close under the body in the erect posture, while the forelimbs still stick out sideways in the primitive sprawling posture. The head is also relatively small compared to the body, reminiscent of the herbivorous pelycosaurs (cf. Figs 4.8b, 4.9a). The skull of *Moschops* has a rounded snout, but the posterior part is elevated in a broad, square heavily-built structure. What was its function? The roofing bones of the cranium are exraordinarily thick (up to 100 mm), and it has been suggested (Barghusen 1975) that this was an adaptation for head butting, as is observed today among sheep and goats (Fig. 4.11). The main force of the butt hit the thickened dorsal shield of the skull, and was transmitted round the sides to the occipital condyle. The occiput was also thickened and placed well beneath the skull, and the occipital condyle lay in a direct line with the butting point. The impact was then transmitted down the thick vertebral column of the neck to the massive shoulder region.

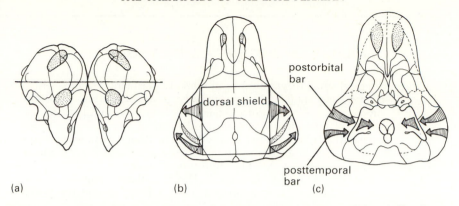

(a) (b) (c)

postorbital bar

dorsal shield

posttemporal bar

Figure 4.11 Head-butting behaviour in *Moschops*: (a) lateral view of the skulls of two butting individuals showing the line of transmission of the impact through the occipital condyle; (b) & (c) dorsal and ventral views of the skull showing the broad thickened dorsal shield, and transmission of forces from it through the postorbital and post-temporal bars to the occipital condyle. (After Barghusen 1975.)

Suborder Dicynodontia

The dicynodonts, a group of over 70 genera, were dominant herbivores in the Late Permian (Kemp 1982, King 1988), and nearly all died out at the end of the Permian. Several lines radiated in the Triassic, and some were large, being 3 m or so long, and they must have had an ecological role similar to large modern grazing mammals. *Kannemeyeria* (Fig. 4.12) has a narrow pointed snout and the parietal forms a very high crest. The ribcage is vast and the limbs and girdles powerfully built. Dicynodont biology has been studied in some detail (see box).

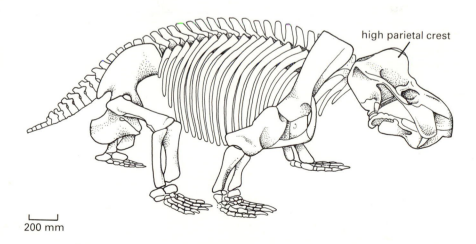

high parietal crest

200 mm

Figure 4.12 Skeleton of the large Triassic dicynodont *Kannemeyeria*. (After Pearson 1924.)

89

BIOLOGY OF *EMYDOPS* – A DICYNODONT

Emydops, a typical small dicynodont from the Late Permian of South Africa, has a 40–60 mm long skull with a particularly short snout (Fig. 4.13a–d). *Emydops*, unlike many dicynodonts, retains a few teeth in addition to the canines, about six postcanines in the maxilla and in the dentary. These small sets of teeth worked against each other, and they are worn to form effectively a single grinding surface. The rest of the jaw margins are made of sharp bone, presumably covered by a horny beak in life.

Emydops had a very mobile jaw joint. The articulating surface of the articular is nearly twice as long as that of the quadrate, so that the lower jaw could slide some distance back and forwards during a jaw opening cycle. Crompton & Hotton (1967) reconstructed the jaw actions of *Emydops* using a complete and undistorted skull. By manipulating the jaws and studying patterns of tooth wear, they were able to work out with some confidence how it operated. They drew a series of four diagrams of the main stages which took place when *Emydops* opened its mouth and bit off some food (Fig. 4.14a–d). Firstly, the jaw opened fully, then moved forwards by sliding at the joint. The food was taken in between the tips of the jaws as the lower jaw closed completely, and was then pulled back firmly with the jaw joint sliding back. This last **retraction** phase was the most powerful and

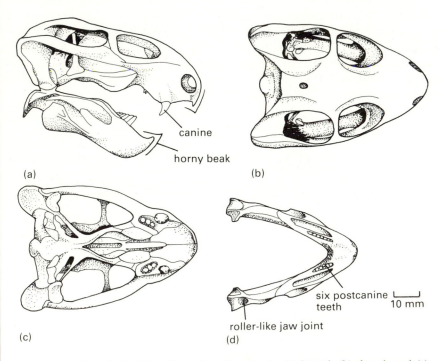

Figure 4.13 The skull of the dicynodont *Emydops* in (a) lateral, (b) dorsal, and (c) ventral views, and (d) the mandible in dorsal view. (After Crompton & Hotton 1967.)

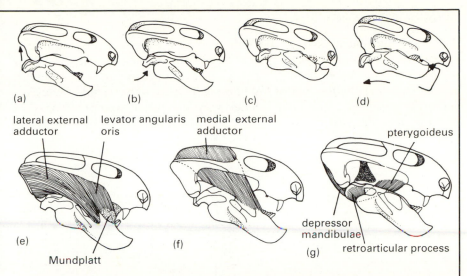

(a)   (b)   (c)   (d)

lateral external adductor   levator angularis oris   medial external adductor   pterygoideus

(e)   (f)   depressor mandibulae

Mundplatt   (g)   retroarticular process

Figure 4.14 (a)–(d) A single chewing cycle of *Emydops*, (a) as it lowers its jaw and moves it back, (b) moves it forward, (c) up for the bite, (d) and backwards to tear the food; (e)–(g) restoration of the jaw muscles of *Emydops*, drawn as if at progressively deeper levels, from (e) to (g). (After Crompton & Hotton 1967.)

had the effect of tearing the food at the front of the mouth and grinding any food that was between the cheek teeth.

The jaw muscles of *Emydops* were also reconstructed (Fig. 4.14e–g) by an analysis of the shape of the jaw bones and patterns of the surface. Most of the jaw adductors ran nearly horizontally, and their contraction would have powered the retraction phase of the jaw action. These key muscles include a major lateral external adductor which ran from the outside of the squamosal and quadratojugal to a long ridge on the side of the dentary (Fig. 4.14e), a medial external adductor which ran inside the zygomatic arch from the parietal and postorbital to the top of the dentary (Fig. 4.14f). Other features include a flexible sheet of tissue in the cheek region, the Mundplatt, which delimited the jaw and was kept taut by the levator angularis oris muscle (Fig. 4.14e), a small pterygoideus muscle which pulled the jaw forward, and the jaw opening muscle, the depressor mandibulae (Fig. 4.14g), which ran from the back of the squamosal to the **retroarticular process**, the part of the lower jaw behind the jaw pivot.

Emydops fed on vegetation which it snipped off with its horny beak and passed back, probably with a muscular tongue, to the cheek region for grinding and crushing before it was swallowed. The tusks of certain dicynodonts show wear striations when they are examined under high magnifications, which suggests that they were used for scraping in the soil for plant material, and the diet may have consisted of roots, horsetail stems, club mosses, and ferns.

In the Late Permian, dicynodonts like *Emydops* made up 80–90% of their faunas (Benton 1983a). Often, five or six dicynodonts of different sizes were present in a fauna, and they were preyed on by carnivorous dinocephalians or gorgonopsians. The huge success of these Late Permian dicynodonts must relate to their specialized jaw apparatus.

Suborder Gorgonopsia

The dominant carnivores in the Late Permian were the gorgonopsians (Fig. 4.15a & b), a group of 35 genera from South Africa and Russia. Their anatomy is remarkably similar, most forms being about 1 m long and with a skull superficially like that of the early carnivorous therapsids. A typical form, *Arctognathus*, could have opened its jaws with a gape of 90° or so in order to clear its vast canines. The jaws then accelerated shut on to the

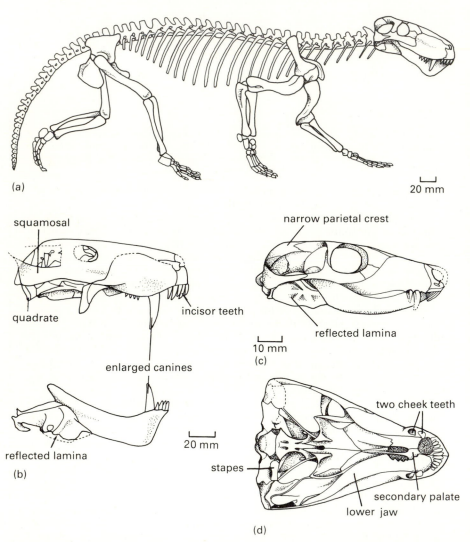

Figure 4.15 The gorgonopsians (a) *Lycaenops*, and (b) *Arctognathus*; (c) & (d) the therocephalian *Theriognathus*. (Fig. (a) after R. Broom 1932; (b) after Kemp 1969; (c) & (d) after Brink 1956, courtesy of the Bernard Price Institute.)

prey animal, and the large fangs passed each other but did not touch, thus effectively piercing the skin and flesh, and disabling its victim. The jaw then shifted forwards and the incisors interlocked, thus removing swallowable chunks of flesh (Kemp 1969). The gorgonopsians are reminiscent of sabre-toothed cats which arose much later on, but had similar enlarged canines and vast gapes (see p. 268). The gorgonopsians may have owed their success to the ability to prey on large thick-skinned dinocephalians and dicynodonts, and when these groups dwindled at the end of the Permian, so too did their predators.

Suborder Therocephalia

The therocephalians, another group of carnivorous therapsids, survived from the Late Permian into the Triassic. They ranged in size from small insectivores to large carnivores, and include some herbivores too in the Early Triassic. *Theriognathus*, a small carnivorous form from the Late Permian of South Africa, has a 75 mm long skull (Fig. 4.15c & d) with large orbits and temporal fenestrae. It shows several advanced characters in comparison to the gorgonopsians; the reflected lamina placed near the back of the jaw, the secondary palate made from vomer, premaxilla, maxilla, and palatine (Fig. 4.15d), and the narrow parietal crest which was extensively covered with the jaw adductor muscles.

Suborder Cynodontia

The cynodonts include the ancestors of the mammals. They arose at the end of the Permian and radiated mainly in the Triassic. The Permian forms are described here, and later cynodont evolution will be considered later (see Chapter 9) as a prelude to the origin of the mammals.

Procynosuchus, from the latest Permian of southern Africa (Kemp 1979), has a long-snouted skull with an expanded temporal region (Fig. 4.16). *Procynosuchus* shows a large number of advances over the therocephalians that are generally mammalian in character (Kemp 1982, 1988a, Hopson & Barghusen 1986): the wide lateral flaring of the zygomatic arches which allowed an increased mass of jaw adductor muscles; a specialized pit, the adductor fossa, on the upper margin of the dentary behind the tooth row; an enlarged dentary making up more than three-quarters of the length of the lower jaw; an enlarged nasal bone; the frontal excluded from the margin of the orbit; a double occipital condyle (Fig. 4.16d); and the beginnings of a secondary palate composed largely of the palatines (Fig. 4.16c), rather than the vomers and maxillae, as in therocephalians. The size of *Procynosuchus*, and the nature of its teeth, suggest that it ate insects.

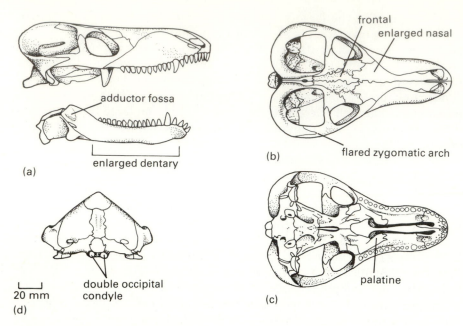

Figure 4.16 The early cynodont *Procynosuchus*, skull in (a) lateral, (b) dorsal, (c) ventral, and (d) occipital views. (After Kemp 1979.)

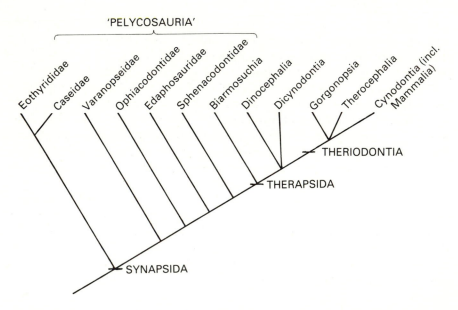

Figure 4.17 Cladogram showing the postulated relationships of the main groups of mammal-like reptiles, according to recent studies.

Relationships of the synapsid groups

The classification of the synapsids is a difficult matter. Most authors have accepted that the pelycosaurs are primitive with respect to the therapsids, and that the latter is a well-defined clade. Further, the cynodonts have been seen as the 'pinnacle' of mammal-like reptile evolution, leading directly to the mammals. Most other postulated relationships (Fig. 4.17) have been contentious (Reisz 1986, Kemp 1982, 1988a, Hopson & Barghusen 1986, King 1988), and the positions of several groups are left undecided in the cladogram.

RADIATION OF NON-SYNAPSID REPTILES
IN THE PERMIAN

The synapsids dominated most Permian terrestrial faunas, but other groups were important too (see box). The diapsids began to radiate, and a number of other lineages came and went. The range of types of reptiles was greater in the Permian than subsequently because of these early 'experimental' lineages that died out in the Permian or Triassic.

DIVERSITY OF PERMIAN NON-SYNAPSID REPTILES

Family Captorhinidae

The captorhinids are known from the Early Permian of North America primarily, with late survivors in the Late Permian of Africa and China. *Eocaptorhinus* (Fig. 4.18a) and *Captorhinus* (Fig. 4.18b & c) were small animals, about 400 mm long, but with relatively large heads (Holmes 1977; Heaton & Reisz 1986). The skull is heavy and bears surface sculpture. The proportions are similar to the earlier reptiles, but the skull is much broader at the back. The main pecularities of captorhinids are seen in the dentition. The peg-like teeth are present in multiple rows (Fig. 4.18b & c) that seem to slope diagonally across the width of the jaw, and five or six can be distinguished in one jaw bone. Ricqlès and Bolt (1983) argue that the teeth were budded off from a dental lamina, the gum tissue which produces teeth, that lay medially. As the animal grew in size, the maxilla added bony tissue plus teeth from the inside, and bone was removed and teeth worn on the outside. Thus, over time, the inner teeth in each row will come into wear at the jaw edge as older teeth are lost. This complex system of tooth replacement is interpreted as an adaptation for piercing or grinding tough plant material or hard-shelled invertebrates.

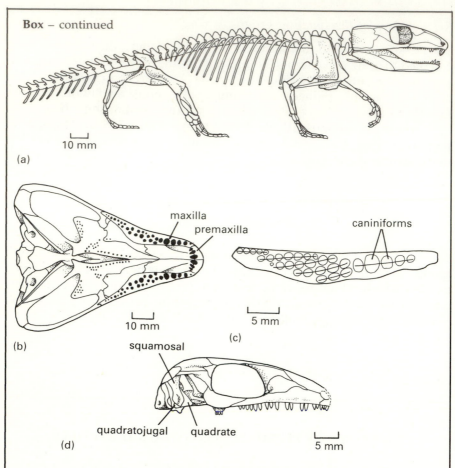

Figure 4.18 The captorhinomorphs (a) *Eocaptorhinus*, and (b) & (c) *Captorhinus*: (a) skeleton in walking posture; (b) ventral view of the palate; (c) ventral view of the maxilla, showing the multiple rows of teeth; (d) the millerettid *Millerosaurus*. (Fig. (a) after Heaton & Reisz 1986; (b) & (c) after Ricqlès & Bolt 1983; (d) after Carroll 1987.)

Family Millerettidae

The millerettids from the Late Permian of South Africa show some superficially lizard-like features in the skull (Fig. 4.18d). There is usually a temporal fenestra, but its lower bar is often broken, and the squamosal, quadrate, and quadratojugal may have been mobile. *Millerosaurus* was a small active insectivore with a 50 mm skull, and it probably lived rather like a modern lizard.

Family Procolophonidae

The procolophonids arose in the Late Permian and lived for about 50 Myr to the end of the Triassic. *Procolophon* from the Early Triassic of South Africa

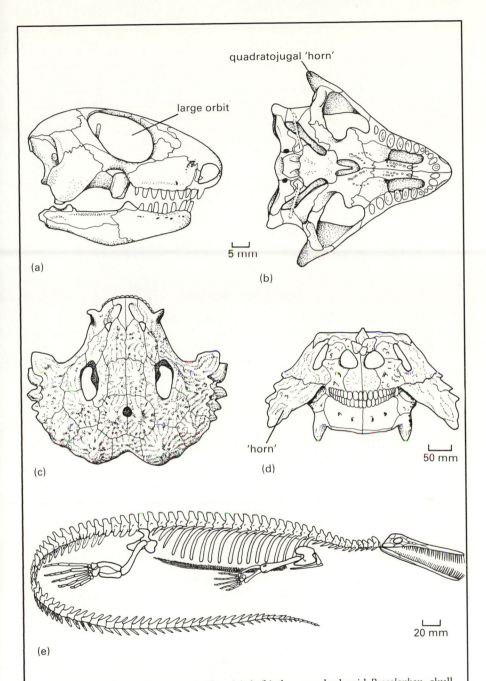

Figure 4.19 Diverse Permian reptiles: (a) & (b) the procolophonid *Procolophon*, skull in lateral and ventral views; (c) & (d) the pareiasaur *Scutosaurus*; (e) the mesosaur *Mesosaurus*. (Figs (a) & (b) after Carroll & Lindsay 1985; (c) & (d) after Kuhn 1969; (e) after Carroll 1987.)

and Antarctica (Carroll & Lindsay 1985), 300–400 mm long, has a relatively large broad skull (Fig. 4.19a & b). The very large orbits include a posterior portion that was associated with the jaw adductor muscles, and the quadratojugal is expanded into a 'horn'. The teeth are blunt and peg-like, present only in small numbers, and clearly associated with a herbivorous diet.

Family Pareiasauridae

The pareiasaurs were restricted to the Late Permian. Most were large, typically 2–3 m long and heavily built. The Russian *Scutosaurus* (Fig. 4.19c & d) has massive elephantine limbs with short feet, and a muscle 'hump' over the shoulders associated with massive neck muscles. The skull is broad and heavy and covered with thickened knobs and incised sculpture. The teeth suggest that pareiasaurs were plant-eaters that fed on softer vegetation than did the procolophonids.

Family Mesosauridae

The mesosaurs are the first-known marine amniotes, represented by abundant small skeletons, about 1 m long in all, from the Early Permian of South America and South Africa, areas which were in contact at the time. The body (Fig. 4.19e) is elongate, with a long neck and an especially long flat-sided tail which was used in swimming. The long thin jaws are lined with needle-like teeth that interlock as the jaws close. They provide a kind of straining device that allowed *Mesosaurus* to take a mouthful of small arthropods or fish and strain the water out before swallowing.

Subclass Diapsida

The diapsid reptiles, descendants of *Petrolacosaurus* (see p. 81), radiated in the Late Permian after what seems to be a gap in their record during the Early Permian. The most remarkable were the gliding weigeltisaurs of Europe and Madagascar (Evans & Haubold 1987). These small reptiles have enormously elongated ribs which stick out sideways forming horizontal 'wings' (Fig. 4.20a), but could be folded back when the animal was running about. The ribs were presumably covered with skin, and *Coelurosauravus* could have glided from tree to tree as the living lizard *Draco* does. The skull (Fig. 4.20b) is diapsid (the lower temporal bar is broken), and the squamosal and supratemporal have remarkable 'toothed' margins at the back.

The other Late Permian diapsids were less exotic. They include forms such as *Protorosaurus* and *Youngina* (Fig. 4.20c), only 350–400 mm long, which was probably an active lizard-like insectivore and carnivore (Gow 1975). The skull is similar to that of *Petrolacosaurus* (cf. Fig. 4.6), but with rather larger temporal fenestrae. The neck is short and the limbs are long. *Youngina* had a number of other terrestrial and aquatic relatives in the Late Permian, the Younginiformes, some of the latter with deep flattened tails and paddle-like feet.

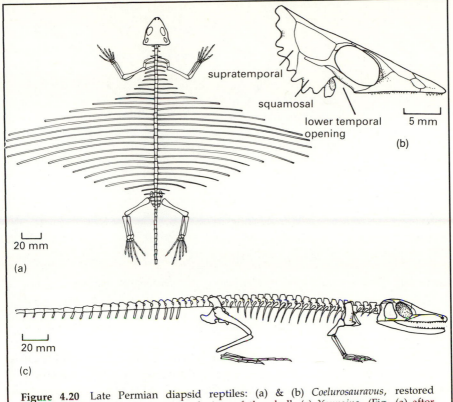

supratemporal

squamosal

lower temporal
opening

5 mm

(b)

20 mm

(a)

20 mm

(c)

Figure 4.20 Late Permian diapsid reptiles: (a) & (b) *Coelurosauravus*, restored skeleton in dorsal view, and lateral view of the skull; (c) *Youngina*. (Fig. (a) after Carroll 1978; (b) after Evans & Haubold 1987; (c) after Gow 1975; (a) & (c) courtesy of the Bernard Price Institute.)

The Permian reptiles fall into three main groups: the synapsids, the diapsids, and the anapsids. The anapsids are the problem group since it is paraphyletic (the anapsid skull is a primitive character of amniotes), and since the members are so diverse and seem to share so few derived characters. The only pairing that has generally been given (e.g. Romer 1966, Clark & Carroll 1973, Carroll 1987) is of the protorothyridids and captorhinids, as the Captorhinomorpha, but there is little evidence for this. Cladistic analyses have so far yielded a few fairly confident conclusions, but many of the groups are still hard to place in a phylogeny.

The protorothyridids are placed on the diapsid line, with the synapsids as the sister group of those two (Fig. 4.21). The turtles (Testudines) appear to be the sister group of the Diapsida, forming the 'Sauropsida', with the Synapsida as the outgroup (Gauthier *et al.* 1988b, Gaffney & Meylan 1988). In addition, these authors find that the closest sister group of Testudines is the Captorhinidae. The remaining anapsid groups have

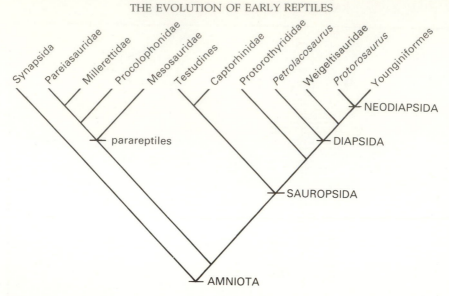

Figure 4.21 Cladogram showing the postulated relationships of the major groups of reptiles.

been placed tentatively in a clade termed the 'parareptiles' by Gauthier *et al.* (1988a).

The relationships of the diapsids seem a little clearer. The weigeltisaurs are primitive in many respects and form an outgroup to *Protorosaurus* and the Younginiformes, the Neodiapsida (Benton 1985a). These neodiapsids of the Late Permian include two lineages that rose to importance later, one leading to the dinosaurs, crocodilians, and birds, and the other to the lizards and snakes. These two divisions of the Neodiapsida are the Archosauromorpha and the Lepidosauromorpha respectively.

MASS EXTINCTION

The biggest mass extinction of all time took place at the end of the Permian (Maxwell 1989), and the tetrapods were involved. Of the 37 families that were present in the last 5 Myr of the Permian, the Tatarian Stage, 27 died out (a loss of 73%). These include six families of amphibians, captorhinids, millerettids, and pareiasaurs, as well as possibly the weigeltisaurs and younginiforms, and at least 15 families of synapsids (Fig. 4.22). Only about ten families of tetrapods survived (Benton 1985b, 1988).

This loss of tetrapod families must count as a mass extinction since so many groups died out, and across such a broad ecological spectrum, from small carnivores to massive herbivores. At the same time, 50% of marine invertebrate families died out, the highest rate of extinction in the history

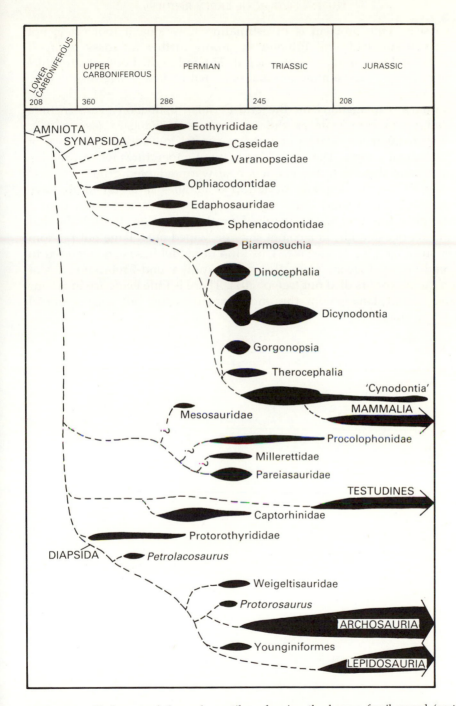

Figure 4.22 Phylogeny of the early reptiles, showing the known fossil record (vertical scale), relative abundance (horizontal dimension of 'balloons'), and postulated relationships (dashed lines).

of the seas. One problem is in estimating how long it took. It is not possible to subdivide the Tatarian stage any further for fossil tetrapod localities, and we cannot say whether all 27 families died out catastrophically at one time, or in a more gradual way, with a loss of five or six every million years.

Gradualist explanations for the end-Permian extinction event include the uniting of continents as the supercontinent Pangaea (see p. 83), which cut down the relative amount of coastal lowlands, and hence of overall habitat variety. The other idea is that temperatures fell in the Late Permian, and this led to the extinction of warm-adapted groups.

If the 73% extinction rate of tetrapod families is assumed to have happened within a short time, say one year, or even one hour, then a catastrophist interpretation is required for the extinction. It might be that the impact of asteroids, or a comet shower, wiped out a large but random selection of land life at a stroke. This kind of model has been extended to the end-Permian event from the better-known end-Cretaceous event when the dinosaurs died out (see p. 201). There is little evidence in favour of such an explanation for the end-Permian event, but such a model cannot be ruled out.

CHAPTER FIVE

Reptiles of the Triassic

Some of the key episodes in tetrapod evolution occurred during the Triassic period (245–208 Myr). It began with restricted faunas, of lower diversity than those of the Late Permian, depleted by the great mass extinction. On land, the mammal-like reptiles re-radiated during the Triassic, but they had already lost a number of their key adaptive zones to two new groups – the archosaurs and the rhynchosaurs. In the seas, several lines of fish-eating reptiles emerged, the nothosaurs, placodonts, and ichthyosaurs. The Late Triassic was a key episode in the evolution of tetrapods. Not only did the dinosaurs come on the scene, but a number of other major groups also arose: the crocodilians, the pterosaurs, the turtles, and the mammals (see Chapters 7 & 9). Current research on Triassic vertebrates is represented well in Padian (1986).

THE TRIASSIC SCENE

In many respects, the Triassic world was similar to that of Permian times. All continents remained united as the supercontinent Pangaea, and there is strong evidence that amphibians and reptiles could migrate widely (Fig. 5.1), and that faunas were similar worldwide. Triassic climates were warm, with much less variation from the poles to the equator than exists today. There is no evidence for polar ice caps, since the north and south poles both lay over oceans at the time. During the Triassic, there was apparently a broad climatic shift, at least in terms of the reptile-bearing rock formations, from warm and moist to hot and dry (Tucker & Benton 1982).

EVOLUTION OF THE ARCHOSAURS

The archosauromorph branch of the diapsids (see p. 100) includes several groups in the Triassic, the most important of which is the Archosauria. Most of the Triassic archosaurs are thecodontians, the paraphyletic group

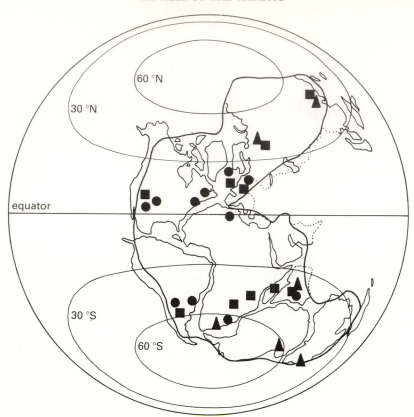

Figure 5.1 Map of the Triassic world, showing the arrangement of the present continents (light line), and the Triassic coastline (heavy line). Fossil reptile localities are indicated with symbols: ▲ Early Triassic, ■ Middle Triassic, ● Late Triassic. (Altered from Tucker & Benton 1982.)

that includes the ancestors of the crocodilians, pterosaurs, and dinosaurs.

The Early Triassic thecodontians took over the carnivorous niches formerly occupied by the gorgonopsids and titanosuchids that died out at the end of the Permian. *Proterosuchus* (Fig. 5.2) from South Africa (Cruickshank 1972) shows three archosaurian hallmarks: an opening in the side of the skull between the nostril and the eye socket, the **antorbital fenestra**; flattened (instead of rounded) teeth; and an additional knob-like muscle attachment on the femur, termed the fourth trochanter. *Proterosuchus* is a slender 1.5 m long animal that preyed on small and medium-sized mammal-like reptiles (therocephalians, dicynodonts) and procolophonids. It has short limbs that bend outwards at the knee, which shows that it had a sprawling posture, as in most Permian mammal-like reptiles and living lizards and salamanders.

An advanced little thecodontian from the end of the Early Triassic of South Africa heralds the beginning of the first major radiation of the

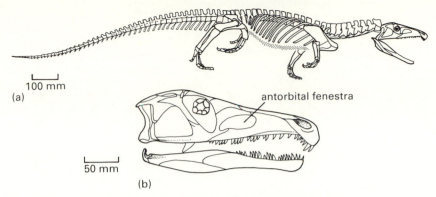

Figure 5.2 The proterosuchid *Proterosuchus*, skeleton in running posture (a), and skull (b). (Fig. (a) based on Greg Paul *in* Parrish 1986; (b) after Cruickshank 1972.)

archosaurs. *Euparkeria* (Ewer 1965), only 0.5 m or so in length (Fig. 5.3), may have been capable of walking both on all fours and bipedally. It has a short high-snouted skull with a large antorbital fenestra set in a pit, and large orbits and temporal fenestrae. The long teeth show that it was an efficient carnivore. It shows advances in the skull – the teeth are set in separate sockets as in all later archosaurs – and in the skeleton – long limbs and a reduced and flexible ankle (Fig. 5.3c).

The archosaurs underwent a major phase of evolutionary diversification at the end of the Early Triassic (Fig. 5.4). They branched into two major groups, termed the Crocodylotarsi and the Ornithosuchia (Gauthier 1986,

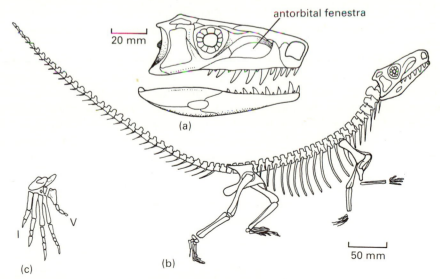

Figure 5.3 The agile thecodontian *Euparkeria* (a) skull in lateral view; (b) skeleton; (c) foot. (After Ewer 1965.)

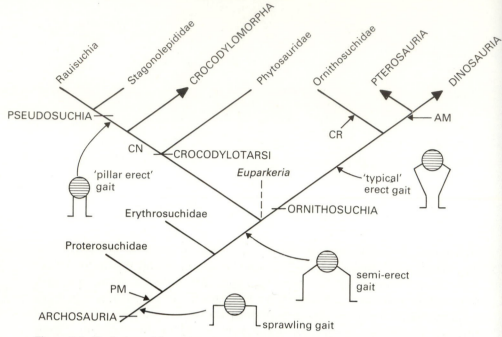

Figure 5.4 Evolution of the archosaurs in the Triassic, showing the postulated relationships of the major groups, and the evolution of locomotory adaptations – changing postures and ankle structures (AM, advanced mesotarsal; CN, 'crocodile normal'; CR, 'crocodile reversed'; PM, primitive mesotarsal ankle structures).

Benton & Clark 1988). The former assemblage includes the ancestors of the crocodilians, and the latter assemblage includes the dinosaurs and pterosaurs, and ultimately the birds.

The Crocodylotarsi show a number of major advances over *Proterosuchus*. In particular, their limbs are modified to allow them to run and walk more efficiently with the legs held slightly under the body in the semi-erect posture. All the joints had to be altered, and the ankle joint in particular is quite different from that of *Proterosuchus*, which has a simple hinge-like ankle joint, termed the 'primitive mesotarsal' (PM) condition (Fig. 5.5), whereas the crocodylotarsans have a more complex system in which the plane of bending runs diagonally, passing between the two main ankle bones, the astragalus and calcaneum. There is a peg on the astragalus and a socket on the calcaneum in this 'crocodile normal' (CN) ankle (Chatterjee 1982). Crocodylotarsans radiated in the Middle and Late Triassic (see box).

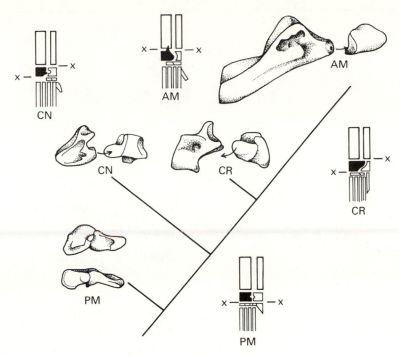

Figure 5.5 Evolution of the archosaurian ankle: the four forms (AM, CN, CR, PM, as in Fig. 5.4) and the lines of ankle flexure (marked ×–×) in four sketches of the lower leg and foot. The ankle bones illustrated from the front (astragalus on the left, calcaneum on the right) are, from bottom left to top right, *Proterosuchus* (PM), *Neoaetosauroides* (CN), *Riojasuchus* (CR), and an unidentified prosauropod dinosaur (AM). (Based on Cruickshank & Benton 1985.)

CROCODYLOTARSAN DIVERSITY

The most primitive crocodylotarsans, the phytosaurs, are so far known only from the Late Triassic of Germany and North America in particular. *Parasuchus* from India (Chatterjee 1978) is 2.5 m long and shows its crocodilian-like adaptations to fish-eating (Fig. 5.6a). The long narrow jaws are lined with sharp conical teeth that interlock in such a way that *Parasuchus* could seize a rapidly darting fish and pierce it with the long teeth, and then hold it firm while it expelled water from the sides of its mouth before swallowing. The nostrils of *Parasuchus* are raised on a mound of bone just in front of the eyes (not at the tip of the snout as in crocodilians), so that it could have lain just below the surface of the water with only its nostril-mound showing. *Parasuchus*, like many modern crocodilians did not only hunt fishes in the water. Two specimens of *Parasuchus* have been found with stomach contents of small tetrapods – the bony remains of prolacertiforms and a small rhynchosaur – which were probably seized on the river bank and dragged into the water.

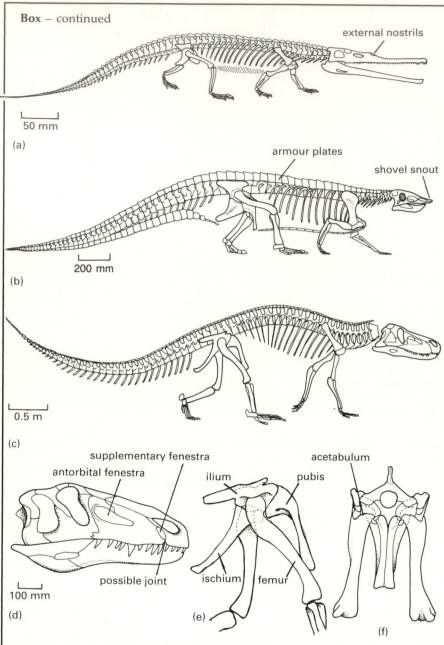

Figure 5.6 Crocodylotarsan thecodontians; (a) the phytosaur *Parasuchus*; (b) the aetosaur *Stagonolepis*: showing part of the armour, but most armour missing to show skeleton; (c)–(f) the rauisuchid *Saurosuchus*: (c) skeleton in walking pose; (d) skull in lateral view; (e) & (f) pelvic girdle and hind limbs in lateral and anterior views to show the 'pillar erect' gait. (Fig. (a) after Chatterjee 1978; (b) after Walker 1961; (c)–(f) after Bonaparte 1981.)

The aetosaurs (family Stagonolepididae) were the first herbivorous archosaurs, and they radiated nearly worldwide in the Late Triassic. *Stagonolepis* from Scotland (Walker 1961) is up to 2.7 m long with a tiny head, a powerful heavy tail and short stout legs (Fig. 5.6b). The skull has a blunt upturned snout which may have been used as a small shovel to dig around in the soil for edible tubers and roots. The body is encased in an extensive armour of heavy bony plates that are set into the skin, a necessary defence against the major carnivores of that time, the rauisuchians.

Saurosuchus, an advanced rauisuchid from the Late Triassic of Argentina (Bonaparte 1981), is one of the largest rauisuchids, reaching 6 or 7 m in length (Fig. 5.6c). The skull (Fig. 5.6d) shows a specialized slit-like opening just behind the nostril, and a possible joint just below. In addition, the hip bones are preserved in three dimensions, and Bonaparte (1981) was able to show how highly modified *Saurosuchus* was for a pillar-like erect gait (Fig. 5.6e & f). The ilium has a very low blade and in life it was oriented almost horizontally rather than vertically, which meant that the socket for the femur (the acetabulum) faced downwards rather than sideways. The skeletons of *Saurosuchus* were found in association with a rich fauna of aetosaurs, rhynchosaurs, small and large mammal-like reptiles (dicynodonts and cynodonts), and some rare temnospondyls and small dinosaurs (see p. 120). These probably all formed part of the diet of *Saurosuchus*, but the rhynchosaur *Scaphonyx* in particular, since it was extremely abundant in the Ischigualasto fauna, and was large enough (1–2 m long) to make a succulent meal.

OTHER ARCHOSAUROMORPHS

The archosaurs were the main group of archosauromorph diapsids to rise to prominence in the Triassic, but there were three other groups. *Trilophosaurus* from the Late Triassic of Texas (Gregory 1945) has an unusual heavily-built skull (Fig. 5.7a) with broad flattened teeth which are used for shearing through tough plant food.

Rhynchosaurs have been found in a number of Triassic faunas where they represent 40–60% of all skeletons found. The skull of *Hyperodapedon* (Benton 1983b) is triangular in plan view (Fig. 5.7b). The back of the skull is broader than the total length, and this vast width seems to have provided space for strong jaw-closing muscles. There are broad tooth plates on the maxillae in the palatal region (Fig. 5.7b) which bear several rows of teeth on either side of a midline groove. The lower jaw clamped firmly into the groove on the upper jaw, just like the blade of a penknife closing into its handle (Fig. 5.7c). This firm closing action of the jaw has been confirmed by studies of the jaw joint and of tooth wear: the jaw joint is a simple pivot with no scope for sliding back and forwards, and

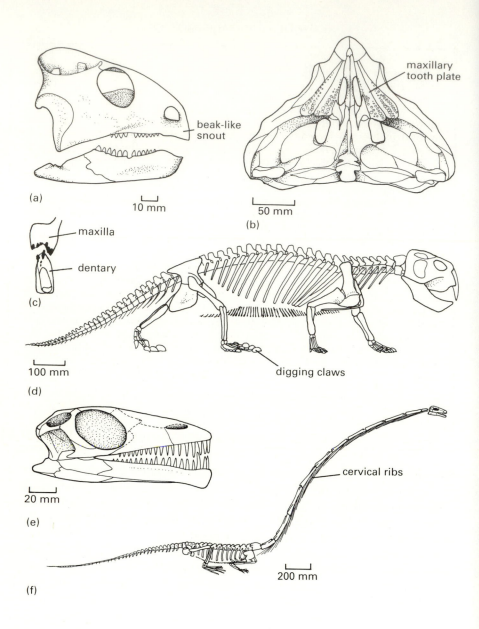

Figure 5.7 Archosauromorphs of the Triassic: (a) the trilophosaurid *Trilophosaurus*, skull in lateral view; (b)–(d) the rhynchosaur *Hyperodapedon*: (b) skull in ventral view, (c) vertical cross-section through the tooth-bearing bones of the skull (maxilla) and lower jaw (dentary) to show the precise fit, (d) skeleton; (e) & (f) the prolacertiform *Tanystropheus*: (e) skull, (f) skeleton of a large species, showing the enormously elongated neck. (Fig. (a) after Gregory 1945; (b)–(d) after Benton 1983b; (e) & (f) after Wild 1973.)

there are clear pits on the bone of the maxilla which show where the teeth of the lower jaw met the upper jaw. This kind of jaw action, with no sliding back and forwards, or from side to side, is the precision-shear system, just like a pair of scissors. Rhynchosaurs were herbivores that fed on tough plants, possibly seed-ferns. *Hyperodapedon* has massive high claws on its feet (Fig. 5.7d) which are very like those of living animals that dig up roots or ants' nests by scratching. *Hyperodapedon* then may have uncovered succulent tubers and roots by backwards scratching with its hindlimbs.

The prolacertiforms appeared first in the Late Permian and radiated in the Triassic. Some of the Triassic forms were broadly like large lizards, but by Middle Triassic times one of the most remarkable reptilian lineages had arisen within this clade. Prolacertiforms are characterized by long necks, but that of *Tanystropheus* from Central Europe (Wild 1973) was remarkable, being more than twice the length of the trunk (Fig. 5.7f). The neck was not greatly flexible since it is composed of only 9–12 cervical vertebrae. Each of these bears long thin cervical ribs which run back beneath the backbone and may have provided attachments for powerful neck muscles. Young *Tanystropheus* have relatively short necks, and as they grew larger the neck sprouted at a remarkable rate. Its function is a mystery. The sharp teeth (Fig. 5.7e) suggest that *Tanystropheus* fed on meat, while the limbs and other features may indicate a life in the water. Indeed, many of the specimens are found in marine sediments, and Wild (1973) reconstructs *Tanystropheus* as a coastal swimmer that fed on small fishes which it caught by darting its head about rapidly.

The trilophosaurids, rhynchosaurs, and prolacertiforms are regarded as archosauromorphs, on the line to the archosaurs themselves (Benton 1985a, Evans 1988). The prolacertiforms are closest to the archosaurs, sharing many characters, such as the long snout and narrow skull, the long nasal bones, the backwardly curved teeth, and the long thin cervical ribs. These phylogenetic placements (Fig. 5.8) represent a considerable change from those in most standard accounts (e.g. Romer 1966). Until recently, the rhynchosaurs and prolacertiforms were regarded as true lepidosaurs, thus on the other major diapsid branch. *Tanystropheus*, for example, apparently shares various features with the lizards, such as its incomplete lower temporal bar (Fig. 5.7f) and specialized teeth and vertebrae (Wild 1973). However, these are far outweighed by the evidence for archosauromorph affinity.

THE ELGIN REPTILES

Late Triassic reptile communities are important because they document a major transition from faunas dominated by mammal-like reptiles, thecodontians, and rhynchosaurs to the new dinosaur faunas. An example is from the Lossiemouth Sandstone Formation (Carnian,

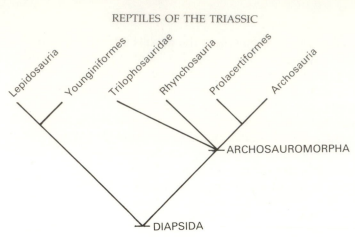

Figure 5.8 Cladogram showing the postulated relationships of the major diapsid groups.

c. 227 Myr) from Elgin in north-east Scotland, a fauna of eight reptile genera (Benton & Walker 1985). The fossil bones were found in a fine-grained whitish-buff sandstone which was evidently deposited in sand dunes by the wind, rather than in water, since it shows characteristic dune bedding, and the sand grains are well-rounded.

Initially, the Elgin reptiles were studied as they were found, without any preparation. Then a number of specimens were prepared by removing the rock with chisels, but this was not successful since the rock is hard and the bone is soft. A. D. Walker pioneered a casting technique in the 1950s and 1960s in which the soft and incomplete bone was removed by the use of acid, leaving near-perfect natural moulds in the rock. Casts were taken in flexible rubber and plastic compounds; these show exquisite detail of the bone surface (Fig. 5.9), and they are easier to work with than bone since they are not fragile! The details of the anatomy of *Stagonolepis*, *Ornithosuchus*, and *Hyperodapedon* described by Walker (1961, 1964) and Benton (1983b) have been obtained largely by these techniques.

The dominant animals in the Elgin fauna (Fig. 5.10) are the rhynchosaur *Hyperodapedon* (Fig. 5.7b–d) and the aetosaur *Stagonolepis* (Fig. 5.6b), both medium-sized herbivores that fed on tough vegetation which they may have dug up with their snouts and powerful feet. The main predator was the thecodontian *Ornithosuchus* (Fig. 5.12d), large specimens of which could have attacked either of the herbivores.

Three small reptiles, the procolophonid *Leptopleuron*, the sphenodontian (see pp. 190–1) *Brachyrhinodon* and the thecodontian *Scleromochlus*, were each about 150–200 mm long and represented 5–25% of the total fauna. *Leptopleuron* may be seen as a 'reptilian rodent' with its broad grinding back teeth and chisel-like front teeth. *Brachyrhinodon* has sharper teeth, probably for chopping small plants. *Scleromochlus* has long slender legs

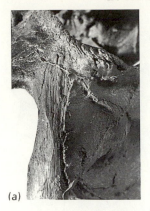

(a)

(b)

Figure 5.9 Close-up views of specimens from the Late Triassic of Elgin, north-east Scotland, to show the quality of preservation of surface detail: (a) premaxillary beak (left) and maxilla (right), showing striation on the premaxilla and blood vessel openings in the maxilla; photograph of PVC cast ($\times$ 1.3); (b) the lacrimal (tear) duct leading from the eye socket (bottom) into the nasal cavity (above); photograph of natural rock mould ($\times$3.0).

and it may have been able to leap around in the moving sands on the edges of the well-watered feeding grounds. These small animals were probably preyed on by young *Ornithosuchus* and by the very rare thecodontian *Erpetosuchus* and the possible dinosaur *Saltopus*.

IN TRIASSIC SEAS

There were three main groups of reptiles in Triassic seas, the nothosaurs, placodonts, and ichthyosaurs, all of which have the euryapsid skull pattern (see p. 80), with one (upper) temporal fenestra. Each has very different aquatic adaptations, and they represent a remarkable radiation of marine predators possibly from independent sources.

Order Nothosauria

The nothosaurs are elongate animals with small heads, long necks and tails, and paddle-like limbs (Fig. 5.11a). They are best known from the Middle Triassic of Central Europe where animals like *Pachypleurosaurus* have been found abundantly in marine sediments (Carroll & Gaskill 1985). These 0.2–4 m long animals were clearly mainly aquatic in adaptations, using wide sweeps of their deep tails to produce swimming thrust. The limbs may have been used to some extent in steering, but they were probably held along the sides of the body most of the time in order to reduce drag. The limb girdles are very much reduced (Fig. 5.11a), and they are only lightly attached to the sides of the body, so that they could not have supported the animal's weight on land.

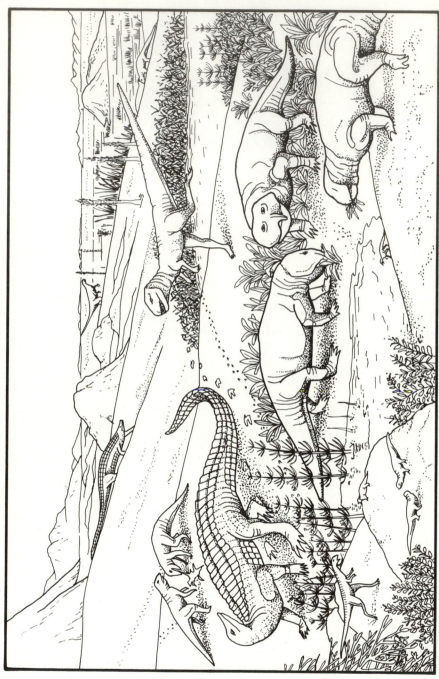

Figure 5.10 The Late Triassic Elgin fauna: three *Hyperodapedon* feed on low waterside plants in the lower right; behind them, an *Ornithosuchus* runs towards an armoured *Stagonolepis* (middle left); behind it, two *Erpetosuchus* feed on a carcase, and in the lower left, a tiny dinosaur, *Saltopus*, runs towards a tiny *Brachyrhinodon* and two *Leptopleuron*. The plants are based on similar localities elsewhere, since no plants have ever been found at Elgin. (Based on a drawing by Jenny Middleton, in Benton & Walker, 1985.)

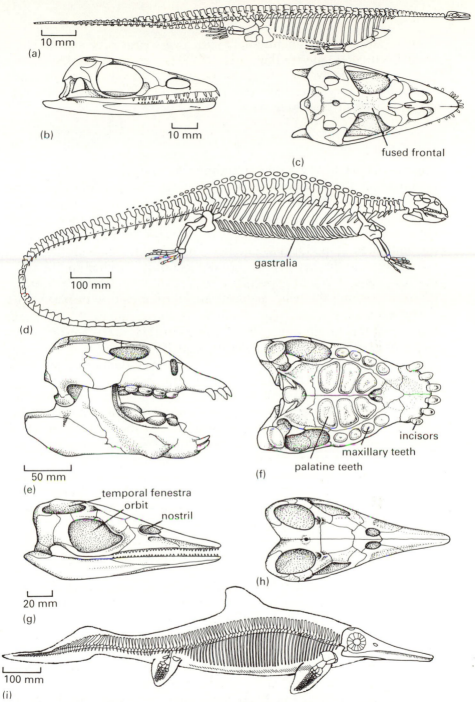

Figure 5.11 The marine reptiles of the Triassic: (a)–(c) the nothosaur *Pachypleurosaurus*: (a) skeleton; (b) & (c) skull in lateral and dorsal views; (d)–(f) the placodont *Placodus*: (d) skeleton in walking pose; (e) & (f) skull in lateral and ventral views; (g) & (h) the ichthyosaur *Grippia*, skull in lateral and dorsal views; (i) the ichthyosaur *Mixosaurus*. (Figs (a)–(c) after Carroll & Gaskill 1985; (d) after Peyer 1950; (e) & (f) after Peyer & Kuhn-Schnyder 1955; (g) & (h) after Mazin 1981; (i) after Kuhn-Schnyder 1963.

The skull is long and lightly-built with a very large orbit and nostril, but a small temporal fenestra (Fig. 5.11b & c). The pointed peg-like teeth are spaced fairly widely and project at the front of the jaws. They suggest a diet of fishes which the agile nothosaurs could have chased and snapped up with darts of their long necks.

Order Placodontia

The placodonts, a very different group of marine reptiles, were also most abundant in the Middle Triassic of Central Europe, and disappeared during the Late Triassic. *Placodus* (Fig. 5.11d) looks at first like a heavily-built land animal, but its remains are found in shallow marine beds. The tail is not deep, as might be expected if it were used in propulsion, and the limbs are not modified as paddles. However, the limb girdles, though heavier than in the nothosaurs, are not as firmly attached to the sides of the body as one would expect in a terrestrial form. The heavy array of gastralia covering the belly region is an unusual feature of placodonts.

The skull of *Placodus* (Fig. 5.11e & f) shows all of the remarkable features of this group (Sues 1987). The teeth (Fig. 5.11f) consist of three spatulate incisors on each premaxilla, four heavy teeth on each maxilla, three on each palatine, and three or four on each dentary. These palatal teeth are broad, flattened, and covered with heavy enamel. They were clearly used in crushing some hard-shelled prey, most probably molluscs, which were levered off the rocks in shallow coastal seas with the incisors, smashed between the massive palatal teeth and the flesh extracted. The broad triangular skull is such a shape that the maximum biting force occurs just in the region of the largest teeth on the palatine and dentary, and the extended squamosal probably bore powerful jaw adductors which ran forwards to the high process of the dentary. These muscles then ran nearly horizontally, and they would have provided a powerful backwards grinding pull to the lower jaw.

Order Ichthyosauria

The ichthyosaurs (literally 'fish lizards') were the most obviously aquatic reptiles of all with their dolphin-like bodies – no neck, streamlined form, paddles, fish-like tail. They arose in the Early Triassic and continued throughout the Mesozoic Era with essentially the same body form. The oldest ichthyosaurs include *Grippia* from Spitsbergen (Mazin 1981). Its skull (Fig. 5.11g & h) has a very large orbit, a nostril placed well back from the tip of the snout, and a single high temporal fenestra. The jaws are long and narrow and lined with uniform peg-like teeth. In later ichthyosaurs, the snout becomes longer, the orbit larger, and the bones at the back of the skull more 'crowded' backwards.

The ichthyosaurs radiated in the Middle and Late Triassic of Central

Europe, Spitsbergen, and the Far East. *Mixosaurus* (Fig. 5.11i) has advanced paddles with short limb bones and an excess number of phalanges. Some specimens even preserve blackened traces of the body outline which show the dorsal fin, the tail fin, and the mitten-like paddle coverings. Some Late Triassic ichthyosaurs reached lengths of 15 m. They had long bullet-shaped heads, teeth only at the front of the snout, a vast rib cage, and tremendously elongated limbs. The later ichthyosaurs (see pp. 198–9) were important in Jurassic and Cretaceous seas, but never reached this huge size.

Relationships

The nothosaurs, placodonts, and ichthyosaurs have frequently been associated within a subclass Euryapsida, even though their ancestry has been rather mysterious (e.g. Romer 1966). Several authors have argued that these marine groups are modified diapsids of one kind or another (e.g. Carroll 1987), and that seems to be the most fruitful area in which to seek ancestors. The placodonts are classified as neodiapsids by Sues (1987), tentatively somewhere on the lepidosauromorph line, while the ichthyosaurs are accorded a similar position by Mazin (1982) who argues that the Euryapsida are a true clade that evolved from a primitive diapsid stock. Other authors suggest separate diapsid origins for each of the three groups.

ARCHOSAUR EVOLUTION IN THE LATE TRIASSIC

The archosaurs underwent a second phase of radiation in the Late Triassic when a number of major groups arose which were to dominate for millions of years thereafter, the crocodilians, dinosaurs, and pterosaurs. The origins of these groups are introduced briefly here, and their subsequent evolution is described in more detail in Chapter 7.

Origin of the crocodilians

The crocodilians arose from thecodontians within the crocodylotarsan clade (see p. 106). The true crocodilians arose in the early Jurassic, but there were a number of 'crocodilomorphs' in the Late Triassic that seem most uncrocodilian at first sight. An example is the saltoposuchid *Terrestrisuchus* (Crush 1984) from South Wales, a lightly-built, delicate 0.5 m long animal (Fig. 5.12a). It has a long skull (Fig. 5.12b) with slender pointed teeth, and long hind limbs that suggest it was a biped. It probably fed on small reptiles, insects, and other invertebrates.

How can this fully terrestrial insectivorous biped be a close relative of the crocodilians? *Terrestrisuchus* has a number of diagnostic crocodilian

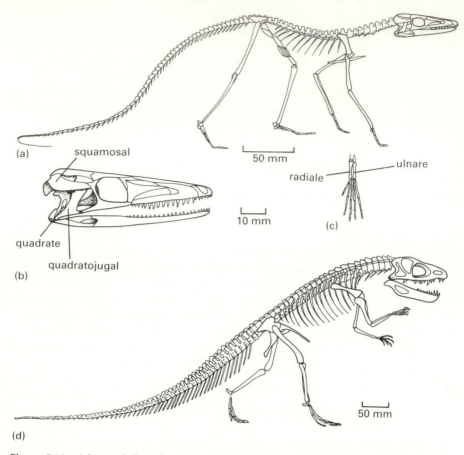

Figure 5.12 Advanced thecodontians: (a)–(c) the early probably bipedal crocodilomorph *Terrestrisuchus*: skeleton in quadrupedal walking pose, skull in lateral view, hand, showing typical crocodilian elongate wrist bones (radiale, ulnare); (d) the ornithosuchid *Ornithosuchus*. (Figs (a)–(c) after Crush 1984; (d) after Walker 1964.)

characters. The main bones of the wrist (radiale and ulnare) are elongated into rod-shaped elements, instead of being button-shaped (Fig. 5.12c), the lower element of the shoulder girdle (the coracoid) has a long backward-pointing spine, and the pelvis has an open acetabulum (hip socket). In addition there are a number of crocodilian-like specializations in the skull (Fig. 5.12b): the quadrate and quadratojugal are displaced inwards towards the braincase, and the cheek region is overhung by the squamosal.

The true crocodilians of the Jurassic, Cretaceous, and Cenozoic were generally larger, more heavily built, and typically aquatic.

Origin of the dinosaurs and pterosaurs

The dinosaurs arose from within the ornithosuchian clade of thecodontians (see p. 105). Their main relatives were the ornithosuchids from the Late Triassic of Scotland and South America, such as *Ornithosuchus* (Walker 1964) which ranged in length from 0.5–3.5 m (Fig. 5.12d). It has a slender build and long hindlimbs which were probably adapted for both quadrupedal and bipedal progression. In many respects, *Ornithosuchus* seems very dinosaur-like, but it has a specialized ankle type which is different from the dinosaur ankle. The ornithosuchid ankle (Fig. 5.5) is called 'crocodile reversed' (CR) since it has a peg on the calcaneum and a socket on the astragalus, the opposite of the CN type.

The closest sister group of the dinosaurs may be the pterosaurs (Fig. 5.4). This may seem an unusual pairing of groups, but Gauthier (1986) has noted a large number of striking similarities between the two groups, particularly in the hindlimb. For example, the four trochanter on the femur is an enlarged ridge located low down (Fig. 5.13c), the ankle joint is simplified to a hinge-like arrangement in which the astragalus and calcaneum are reduced and firmly attached to the lower leg bones (tibia and fibula), the 'advanced mesotarsal' (AM) ankle type (Fig. 5.5), and the middle three toes are elongated.

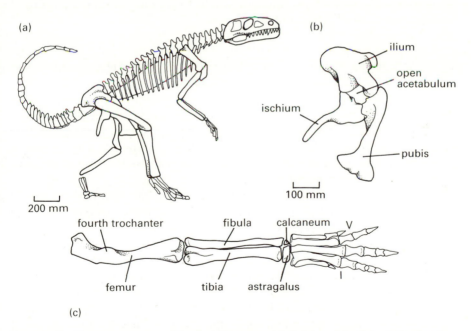

Figure 5.13 *Herrerasaurus*, one of the first dinosaurs: (a) skeleton; (b) pelvic girdle showing dinosaurian lay-out of the bones, and open acetabulum; (c) hindlimb, showing large fourth trochanter on femur, and long digitigrade foot. (Fig. (a) after Bonaparte 1978; (b) & (c) after Galton 1977.)

119

The oldest true dinosaurs are known from the early part of the Late Triassic (the Carnian Stage, 230–225 Myr) from various parts of the world. One is *Herrerasaurus* from Argentina (Fig. 5.13), a heavily-built 3 m long bipedal hunting animal. It shows a number of advances over *Ornithosuchus*, characters that are regarded as synapomorphies of the Dinosauria: it was an habitual biped (ratio of lengths of forelimb to hindlimb = 0.5), the acetabulum is fully open (Fig. 5.13b), and the head of the femur is bent inwards. The great changes in the hip and leg bones are associated with bipedality and with the fully erect gait of dinosaurs. This erect gait is like that of mammals, in which the head of the femur fits sideways into the acetabulum, rather than the pillar-like erect gait of the aetosaurs and rauisuchians, in which the femur fits straight up into a horizontal acetabulum.

Formerly, most palaeontologists regarded the dinosaurs as a diverse assemblage of archosaurs that arose from several ancestors – a polyphyletic group (e.g. Romer 1966, Chatterjee 1982). However, recent cladistic analyses (e.g. Gauthier 1986, Benton & Clark 1988) have strongly indicated that the Dinosauria is a monophyletic assemblage, because they all share such a complex of specialized characters. *Herrerasaurus* comes from a time in the early years of the dinosaurs when they were only rare elements in their faunas (1–2% of all skeletons), but before the end of the Triassic the dinosaurs had radiated widely to become the most abundant vertebrates on land. How did this happen?

RADIATION OF THE DINOSAURS – COMPETITION OR MASS EXTINCTION?

There are currently two ways of viewing the radiation of the dinosaurs in the Late Triassic. Either they 'took their chance' after a mass extinction event and radiated opportunistically, or they competed over a longer time-span with the mammal-like reptiles, rhynchosaurs, and thecodontians, and eventually prevailed.

Most authors (e.g. Charig 1984) have favoured the 'competitive' model for three reasons. Firstly, as mentioned above, many considered that the dinosaurs were a polyphyletic assemblage. Secondly, the origin of the dinosaurs was seen as a drawn-out affair, that started well-down in the Middle Triassic, and involved extensive and long-term competition (Fig. 5.14a). The dinosaur ancestors were regarded as superior animals, with advanced locomotory adaptations (erect gait) or physiological advances (warm-bloodedness, or cold-bloodedness: both cases have been argued!) which progressively competed with, and caused the extinction of, all of the mammal-like reptiles and thecodontians. Thirdly, the appearance of the dinosaurs has often been regarded as a great leap forward in evolutionary terms.

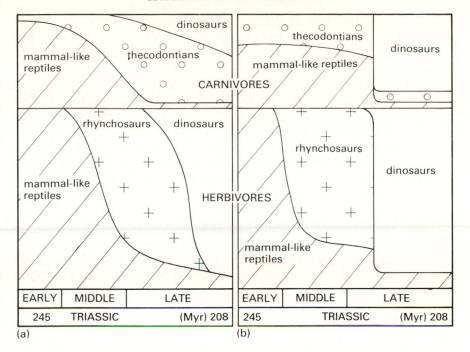

Figure 5.14 Two models for the replacement of mammal-like reptiles, thecodontians, and rhynchosaurs by dinosaurs: (a) a competitive replacement scenario; (b) an opportunistic mass extinction replacement model.

Several lines of evidence (Benton 1983a, 1986) have now suggested that this kind of competitive interpretation is incorrect:

(a) The pattern in the fossil record (Fig. 5.14b) does not support the competitive model. Dinosaurs appeared in the Late Triassic, and they radiated markedly after a major extinction event at the end of the Carnian, when various families of thecodontians, mammal-like reptiles, and the rhynchosaurs died out. Dinosaurs radiated further in the Early Jurassic after a second mass extinction at the very end of the Triassic when the remaining thecodontians, and other groups, died out.

(b) The 'superior adaptations' of dinosaurs were probably not so profound as was once thought. Many other archosaurs also evolved erect gait in the Late Triassic, and yet they died out (e.g. aetosaurs, rauisuchians, ornithosuchids, and some early crocodilomorphs). The physiological characters of dinosaurs – whether they were warm-blooded or not, for example – cannot be determined with confidence.

(c) The idea that simple competition can have major long-term effects in evolution is probably an over-simplification of a complex set of processes. The idea that families or orders of animals can compete with each other is very different from the ecological observation of competition within or between species. In palaeontological examples like this, competition has often been assumed to have been the mechanism, but the evidence has turned out to be weak in many cases (Benton 1987a).

(d) Cladistic analyses of the relationships of dinosaurs show that they came at the end of a long line of advanced thecodontians. Many characters which were regarded as uniquely dinosaurian (e.g. erect gait and all the anatomical changes that entails) were already present in other forms, such as the ornithosuchids and pterosaurs. The transition to Dinosauria is not marked by the sudden acquisition of a suite of remarkable new adaptations.

These arguments are presented in some detail as an example of the kinds of palaeobiological controversies that interest vertebrate palaeontologists. Unless my partisan bias has not already been detected by the reader, I favour the mass extinction and opportunism model for the radiation of the dinosaurs!

CHAPTER SIX

The evolution of fishes after the Devonian

Fish evolution after the Devonian followed two main paths. The Chondrichthyes (sharks and rays) radiated several times, but never achieved great diversity. The Osteichthyes (bony fishes) also radiated several times, and they have become a major element of marine life. Most recent fishes, the salmon, cod, herring, goldfish, sea horses, tuna, and so on, are bony fishes that form part of a vast radiation that began over 150 Myr ago. Several primitive fish lineages that were important in the Silurian and Devonian – the agnathans and lungfishes, as well as the coelacanths (Chapter 2) – have lived through the last 360 Myr since the beginning of the Carboniferous, but at low diversity. The purpose of this chapter is to explore the variety of sharks, rays, and bony fishes and to account for their relative success.

THE WEIRD SHARKS OF THE CARBONIFEROUS

About 20 families of sharks and their relatives lived during the Carboniferous, but many of these are known only from teeth and spines. Some recent finds have revealed some quite bizarre chondrichthyans in the Carboniferous, and the better-known groups are reviewed here.

Order Eugeneodontida

The eugeneodontids, or edestids, are known almost exclusively from their teeth which grew in spiral shapes (Fig. 6.1a), and are common fossils in the Carboniferous and Permian (Zangerl 1981). Each spiral consists of a series of teeth which are joined together in such a way that the largest teeth at the top are in use, and new teeth can rotate into place when the older ones are worn away. This system means that there is a constant supply of teeth available even when older ones break off. The tooth whorl fits between the two lower jaws (Fig. 6.1b) and operates

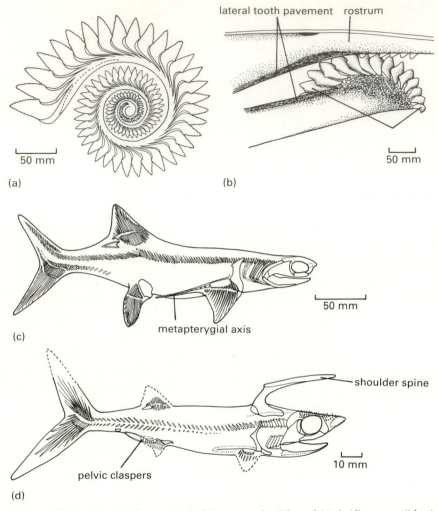

Figure 6.1 Early chondrichthyans: (a) & (b) eugeneodontids and (c) & (d) symmoriids: (a) tooth whorl of *Helicoprion*; (b) tooth whorl of *Sarcoprion* in place at the tip of the lower jaw and acting against a tooth pavement in the snout (rostrum); (c) *Denaea*; (d) male *Falcatus* with spine and claspers. (Figs (a) & (b) after Moy-Thomas & Miles 1971; (c) after Schaeffer & Williams 1977; (d) after Lund 1985.)

against similar sharp teeth in the upper jaw. The rest of the eugeneodontid skeleton is poorly known, but it is like *Cladoselache* (Fig. 2.9).

Order Symmoriida

One group of symoriids, represented by *Denaea* (Fig. 6.1c), has a body outline basically like *Cladoselache*. However, *Denaea* has no fin spines, unusual for a shark, and it has a whip-like extension to the pectoral fin called a **metapterygial axis**. The function of this is uncertain, whether it was used as a defensive structure or in some reproductive display.

The most remarkable symmoriids are the stethacanthids such as *Falcatus* (Lund 1985), a small shark up to 145 mm long (Fig. 6.1d). It looks like a dogfish except that a long shelf-like spine extends from roots deep in the muscles of the 'shoulder' region to run over the head, like a sun shade. The spine is present only in sexually mature males, identified by the presence of pelvic claspers, specialized skeletal elements that are used to grasp the female during fertilization. Male *Falcatus* sharks may have aggregated prior to the breeding season in order to carry out display-courtship rituals.

Subclass Holocephali

The chondrichthyans so far described are known only from the Carboniferous and Permian periods, but another assemblage that radiated in the Carboniferous, the holocephalans, chimaeras or rat fish, still survive today. *Chondrenchelys* from the Early Carboniferous of Scotland (Fig. 6.2a) has a long tapering body with no tail fin and a small skull in which the palatoquadrate is firmly fused to the braincase (see p. 28). The pelvic fin is small and males have claspers. The iniopterygian *Iniopteryx* (Fig. 6.2b) has a large head, very long pectoral fins, and a rounded tail fin (Zangerl & Case 1973). The pectoral fins are attached to the pectoral girdle in a very high position, and they flapped up and down like the wings of a bird. The front of the fin bears a series of hook-like denticles.

Typical chimaeras of modern form appeared in the Jurassic, although there are some tantalizing chimaera-like fossils from the Carboniferous. *Ischyodus*, first known in the Middle Jurassic of Europe (Fig. 6.2c), is essentially the same in appearance as modern chimaeras. The skull is small and heavily fused, the gills lie beneath the braincase, and there are two pairs of tooth plates in the upper jaw and one pair in the lower. The tail is long and whip-like, the pectoral fins are large, and the tall spine in front of the dorsal fin may have borne a poison gland as in some modern forms.

Orders Xenacanthida and Ctenacanthiformes

The xenacanths, freshwater forms known from the Devonian to the Triassic, seem to be shark-like in their fin structure. *Xenacanthus* (Fig. 6.2d) has a small skull with a long spine just behind, large paired fins, and a tapering symmetrical narrow **diphycercal tail**, as in *Chondrenchelys*. The strange narrow long form of *Xenacanthus* may have allowed it to swim in and out of closely-growing lake vegetation.

The ctenacanths are also known from the Devonian to the Triassic, and they showed an even closer approach in their fins to the form of modern sharks than the groups so far described.

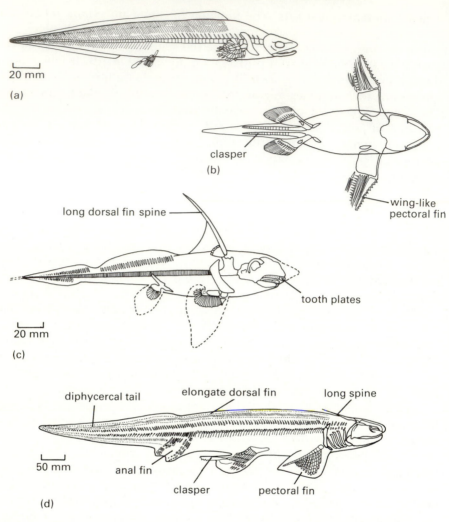

Figure 6.2 Early holocephalans (chimaeras) from (a) & (b) the Carboniferous, and (c) the Jurassic; and (d) a shark form the Permian: (a) *Chondrenchelys*; (b) *Iniopteryx* in ventral view; (c) *Ischyodus*; (d) *Xenacanthus*. (Fig. (a) after Moy-Thomas & Miles 1971; (b) after Zangerl & Case 1973; (c) after Patterson 1965; (d) after Schaeffer & Williams 1977.)

The age of sharks

Recent studies of Early Carboniferous localities in central Scotland (Dick *et al.* 1986) and Montana, USA (Lund 1985, Janvier 1985) have revealed a wealth of new fossils, often beautifully preserved. The Montana fauna, from the Bear Gulch limestone (Namurian, c. 325 Myr) is particularly striking – a world of sharks (Fig. 6.3)! The 'crested' *Stethacanthus* and *Falcatus* are the largest. *Harpagofutator*, a relative of *Chondrenchelys*, has forked appendages on the forehead, but only in the male. *Delphyodontos*, a possible early chimaera, seems to have no fins at all and a fat body covered with small denticles. Another chimaera, *Echinochimaera*, has spines and teeth in different parts of its body. The male also has pelvic claspers and forehead 'claspers', short spines over the eyes, as in some modern forms. The other Bear Gulch fishes include a 'telescoped' coelacanth, *Allenypterus*, a narrow eel-like bony fish, *Paratarassius*, and the oldest-known true lamprey, *Hardistiella*.

THE MODERN SHARKS

The xenacanths and ctenacanths of the Carboniferous looked rather like modern sharks. However, another group, the hybodonts (Carboniferous–Palaeocene), appear to come closest to the modern forms, the Neoselachii. The Mesozoic hybodonts (Maisey, 1982), such as *Hybodus* (Fig. 6.4a), were fast swimmers, with streamlined bodies and powerful muscles that

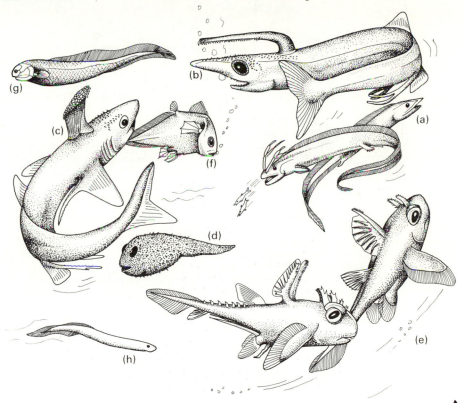

Figure 6.3 (a)–(e) Chondrichthyans and (f)–(h) other fishes of the mid-Carboniferous Bear Gulch limestone of Montana: (a) *Harpagofutator*; (b) *Falcatus*; (c) *Stethancanthus*; (d) *Delphyodontos*; (e) *Echinochimaera*; (f) the coelacanth *Allenypterus*; (g) the actinopterygian *Paratarassius*; (h) the lamprey *Hardistiella*. (After Janvier 1985.)

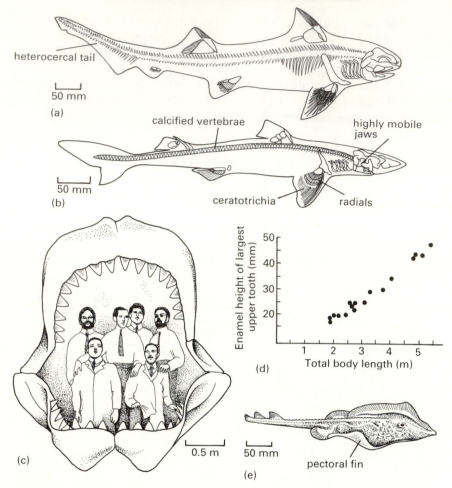

Figure 6.4 Sharks and rays: (a) the Jurassic shark *Hybodus*; (b) the modern shark *Squalus*; (c) the jaws of the giant Tertiary shark *Carcharodon*, reconstructed from isolated teeth, and probably too large; (d) regression of tooth size on actual body length for living *Carcharodon* suggests that the fossil examples reached only half the length implied by the reconstruction in (c); (e) the modern ray *Raja*. (Figs (a) & (b), after Schaeffer & Williams 1977; (c) based on Pough *et al.* 1989; (d) based on Randall 1973; (e) after J. Z. Young 1981.)

produced a swimming motion by bending the body from side to side. The paired fins were used for steering and stabilization. The tail is fully heterocercal, with the backbone bending upwards. Hybodonts have a number of tooth shapes, some high and pointed and others low, which suggests that they fed on a variety of prey types, ranging from fishes to bottom-living crustaceans.

The neoselachians, including all modern sharks and rays, arose in the Triassic, and radiated particularly during the Jurassic and Cretaceous to the modern diversity of 35 families. They are characterized by numerous derived characters (Maisey 1984). The jaws open more widely than in earlier forms because of greater mobility about the joint, and a highly kinetic palatoquadrate and hyomandibular (see pp. 28–9, Fig. 2.10c). The snout is usually longer than the lower jaw and this means that the mouth opens beneath the head rather than at the front. In larger sharks, this jaw apparatus, combined with large numbers of sharp teeth, is extremely effective at gouging flesh from large prey. The notochord is enclosed in calcified cartilage vertebrae, whereas the primitive chondrichthyans had only cartilage around it. The basal elements (the radials) in the paired fins are reduced and most of the fin is supported by flexible cartilaginous rods called ceratotrichia (Fig. 6.4b).

The early neoselachians were large near-shore hunters which probably radiated in response to the evolution of teleost fishes and squid, while later neoselachians became fast off-shore hunters. The neoselachians fall into three main groups (Compagno 1977). The first, the squalomorphs, such as *Squalus* (Fig. 6.4b), the spiny dogfish, generally live in deep cold waters. They retain spines in front of the dorsal fins.

The galeomorphs, the largest group, mainly inhabit shallow tropical and warm temperate seas, and include the dogfishes, grey shark, basking shark, whale shark, and great white shark. These feed on, respectively, crustaceans and molluscs, fishes, krill, and, on occasion, humans. The basking and whale sharks, up to 17 m long, are the largest living sharks, but they feed on krill, small floating crustaceans which they strain from the water. An even larger shark has been reported, *Carcharodon*, known only from triangular teeth up to 150 mm long which are found in sediments dating from Cretaceous to Pleistocene times. Early reconstructions of its jaws, based on these large teeth (Fig. 6.4c) suggested that it had a 3 m gape and a total body length of 18–30 m. However, a comparative study of its teeth (Randall 1973) has suggested that *Carcharodon* was a mere 13 m long! The teeth are very like those of the living (but much smaller) species of *Carcharodon*. Randall (1973) plotted the tooth size against body length for living examples and was able to extrapolate the body size of the fossil giants from his graph (Fig. 6.4d).

The third modern group, the batoids, or skates and rays, are specialized for life on the sea-floor, and have flattened bodies with broad flap-like pectoral fins at the sides and long whip-like tails. The eyes have shifted to the top of the head, while the mouth and gill slits are underneath. The batoids swim (Fig. 6.4e) by undulating the pectoral fins up and down in waves. The teeth are usually flattened, arrayed in pavements, and adapted for crushing hard-shelled molluscs.

The living chondrichthyans fall into a small number of quite well-defined groups, but the diversity of Carboniferous forms and their poor preservation has made classification difficult. Phylogenetic analyses have been attempted by several authors (e.g. Compagno 1977, Schaeffer & Williams 1977, Maisey 1982, 1984, 1986, Thies & Reif 1985), and these suggest a cladogram (Fig. 6.5) in which there are a number of primitive generalized chondrichthyans below the split into holocephalans (chimaeras) and true sharks.

THE EARLY BONY FISHES

The ray-finned bony fishes, subclass Actinopterygii, arose in the Devonian period, when forms like *Cheirolepis* were abundant (see pp. 37–9). The bony fishes underwent three major radiations:

(a) chondrostean radiation, Carboniferous–Triassic;
(b) holostean radiation, Triassic–Jurassic;
(c) teleost radiation, Jurassic–present.

The chondrostean radiation

The majority of chondrosteans fall in the order 'Palaeonisciformes' (Gardiner 1984b), known from the Devonian to the Cretaceous, but rare after the Triassic. The early forms like *Cheirolepis* (Fig. 2.18) were heavily built and covered with large bony scales. Most palaeonisciforms look roughly like *Cheirolepis*, but some later forms such as *Cheirodus* from the Carboniferous (Fig. 6.6a) were deep-bodied and very flat from side to side. The dorsal and anal fins are very long and the paired fins tiny. *Cheirodus* has flattened teeth and tooth plates which were presumably used for crushing hard-shelled prey.

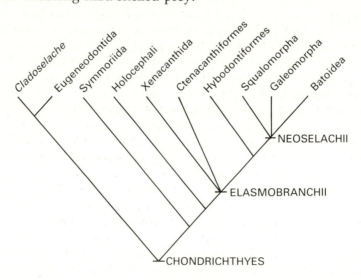

Figure 6.5 Cladogram showing the postulated relationships of the main groups of chondrichthyan fishes.

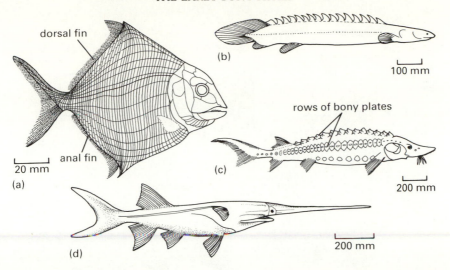

dorsal fin

(b)

100 mm

rows of bony plates

(c)

200 mm

20 mm   anal fin

(a)

200 mm

(d)

200 mm

Figure 6.6 'Chondrostean' fishes: (a) the palaeonisciform *Cheirodus*; (b) the living bichir *Polypterus*; (c) the living sturgeon *Acipenser*; (d) the living paddlefish *Polyodon*. (Fig. (a) after Woodward 1891–1901; (b) & (c) after Carroll 1987; (d) after Stahl 1974.)

The chondrosteans have survived to the present, but at the very low diversity of six genera in three families. The bichirs, family Polypteridae, are small heavily armoured fishes (Fig. 6.6b) which live in the streams and lakes of tropical Africa. Their dorsal fins run the whole length of the body, and are divided into segments, each with a spine at the front. The sturgeons, family Acipenseridae, are large fishes, 1–6 m long, which live in northern waters, and supply commercial caviar, their eggs. Sturgeons have very poorly ossified skeletons and the scales are reduced to five rows of large bony plates (Fig. 6.6c). The paddlefishes, family Polyodontidae, have long flat snouts which are about one-third of the total length of 2 m, and they feed by straining plankton and small fishes out of the water (Fig. 6.6d). Some fossil 'chondrosteans' of the Triassic, in particular, show advances over the living forms, groups such as the Redfieldiiformes and Perleidiformes, which appear to fall on the line to more advanced actinopterygians.

The holostean radiation

The holosteans, like the chondrosteans, are a grade group, or paraphyletic group that may represent a particular adaptive stage in evolution, but is not a clade. There are three main groups of holosteans, the gars, the semionotids and other extinct forms, and the bowfins (Patterson 1973).

The gars, family Lepisosteidae (order Ginglymodi), were for a long time classified as chondrosteans. *Lepisosteus* (Fig. 6.7a), a 1–4 m predatory fish, lives in warm-temperate fresh and brackish waters of North America. It

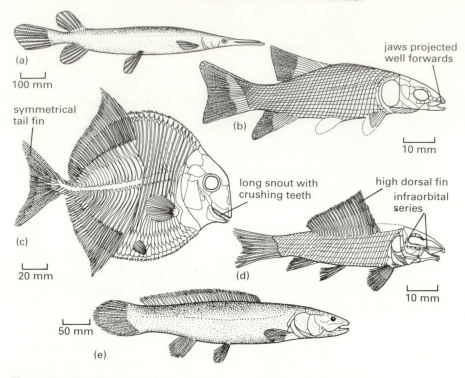

Figure 6.7 The diversity of 'holostean' fishes: (a) the living gar *Lepisosteus*; (b) the Triassic *Semionotus*; (c) the Jurassic *Proscinetes*; (d) the Jurassic *Macrosemius*; (e) the living bowfin *Amia*. (Figs (a) & (e) after Goode & Bean 1895; (b) after Schaeffer & Dunkle 1950; (c) after Woodward 1916; (d) after Bartram 1977.)

has long jaws and captures its prey by lunging and grasping them with its long needle-like teeth. There are seven living species of gars, and the genus *Lepisosteus* has been traced back to the Cretaceous, so it would seem to be a good example of a **living fossil**, an apparently slowly evolving lineage that has always remained at low diversity.

Five fairly diverse orders of holosteans arose in the Triassic, and radiated in the Jurassic in particular, but only one has survived to the present. The Semionotidae include about 25 genera of small, actively swimming fishes, such as *Semionotus* (Fig. 6.7b), that have nearly symmetrical tails and large dorsal and ventral fins. The tooth-bearing elements, the maxilla and dentary, are projected well forwards and lined with small sharp teeth. Semionotids occur in great diversity in some areas, such as the Newark Group (Late Triassic and Early Jurassic) lakes of the eastern seaboard of North America. These were inhabited by semionotids of all shapes and sizes, often as many as 10–20 similar species together in the same lake. Whole faunas were wiped out by catastrophic drying episodes, and replaced by new species flocks which evolved rapidly when the lakes became re-established. Modern parallels exist today in central African lakes where the small cichlid fishes have achieved great diversity by rapid diversification.

The pycnodonts of the Jurassic and Cretaceous are mostly deep-bodied forms with long dorsal and anal fins, and a symmetrical (homocercal) tail fin. *Proscinetes* (Fig. 6.7c) has an elongated snout and a pavement of crushing teeth on the upper and lower jaws which were presumably used to crush molluscs or echinoderms. The macrosemiids, also of the Jurassic and Cretaceous, were small fishes (Fig. 6.7d) often with a long high dorsal fin. They have some unusual bones in the skull, a series of seven rolled little bones beneath the orbit (the infraorbitals), and two tubular infraorbitals behind it.

The bowfins, subdivision Halecomorphi, were also generally regarded as holosteans. The modern bowfin, *Amia* (Fig. 6.7e), lives in freshwaters of North America, where it is an active predator on a wide variety of organisms ranging in size up to their own body length of 0.5–1 m. The halecomorphs include several families that date back to the Triassic, and all are characterized by a specialized jaw joint involving the **symplectic** and the quadrate (see box).

JAWS AND FEEDING

Amia illustrates an intermediate kind of jaw apparatus between that of the palaeonisciforms and the advanced teleosts. The skull of *Amia* (Fig. 6.8a) shows how the jaws are relatively shorter than in the chondrosteans (cf. Fig. 2.18e). The maxilla is highly mobile and a new element, the **supramaxilla**, is attached to it. This mobile maxilla hinges at the front and can swing out some way to the side. This is associated with changes to the main jaw joint between the lower jaw and an internal unit composed of the hyomandibular,

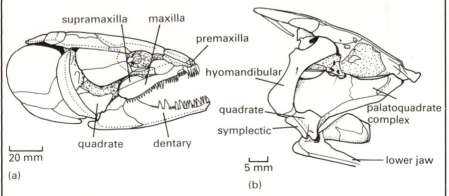

Figure 6.8 The jaws of non-teleost neopterygians: (a) skull of *Amia* showing the major jaw elements; (b) detailed view of jaw joint elements in the early neopterygian *Watsonulus*, reconstructed with the outer skull elements removed. (Fig. (a) after Patterson 1973; (b) after Olsen 1984.)

symplectic, and palatoquadrate, termed the jaw **suspensorium** (Fig. 6.8b). When the jaws of a holostean or teleost open, the cheek region of the skull expands a long way sideways which gives a sucking effect, useful for drawing in small particles of food or prey animals.

The heads of bony fishes of chondrostean, holostean, and teleost grades show three rather different sets of jaw opening adaptations (Fig. 6.9). Palaeonisciforms open their jaws in a wide 'grin' which is suitable for grabbing large prey, while most holosteans and teleosts protrude their jaws forwards and the open mouth is roughly circular (Schaeffer & Rosen 1961). This protrusion is most marked in teleosts where the sudden opening of the mouth produces a marked suction effect. The jaw-closing action is equally important. When the tube-like teleost mouth is closed by pulling the lower jaw and maxilla back, the food is retained, whereas simple closure by raising the lower jaw would blow some of the food out again.

Advanced teleosts, the Neoteleostei (see pp. 138–43), show a further modification of the jaw apparatus (Alexander 1967). The maxilla loses its role as the main tooth-bearing element in the upper jaw, and the premaxilla takes over, while the maxilla acts as a lever, pushing the premaxilla forwards as the jaws open (Fig. 6.10a). The maxilla is attached to the lower jaw (at B) and to the suspensorium (at A). As the mouth opens, an anterior adductor muscle (Fig. 6.10b) pulls the top of the maxilla back, and the lower jaw is pushed forward. The maxilla also rotates slightly about its long axis, and a process on the top of the maxilla, which interlocks with one on the premaxilla, causes the premaxilla to be protruded.

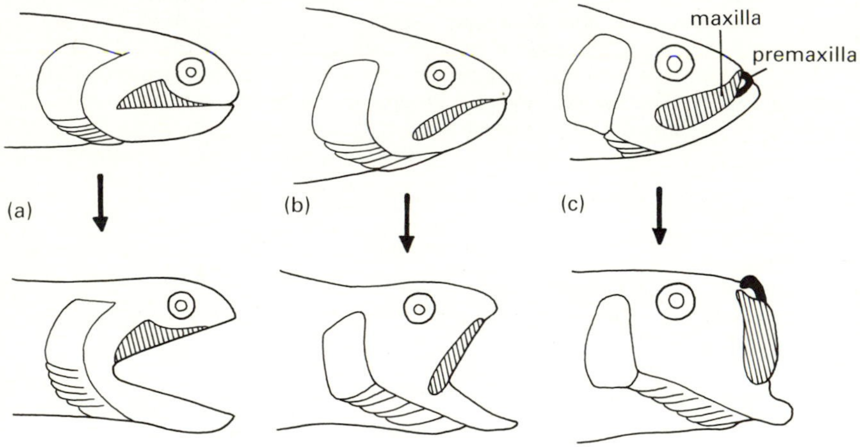

Figure 6.9 Sketches of the heads of (a) a palaeoniscid ('chondrostean'), (b) a primitive neopterygian ('holostean') or early teleost, and (c) a herring (typical teleost), showing the jaws closed (top) and open (bottom). (After Alexander 1975, courtesy of Cambridge University Press.)

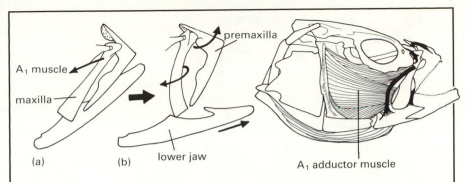

Figure 6.10 The jaw action and musculature of acanthopterygian teleosts: (a) lateral diagrammatic view of the major jaw elements with the mouth closed (left) and open (right), showing the relative movements and rotations of the bones; (b) jaw musculature of *Epinephelus*, showing the muscles and bones indicated in (a). (Fig. (a) after Alexander 1975, courtesy of Cambridge University Press; (b) after Schaeffer & Rosen 1961.)

RADIATION OF THE TELEOSTS

The teleosts are an extremely diverse group of fishes, with 20 000 living species which are classified in 40 orders (Nelson 1984). This enormous diversity is clearly impossible to survey in detail, and only the main groups can be mentioned. The teleosts are characterized by modifications to the tail which give it a symmetrical (**homocercal**) appearance, but with the vertebral column not running into the upper lobe. In addition, teleosts all have a mobile premaxilla, not seen in holosteans, and some modifications to the jaw musculature (Lauder & Liem 1983). Living teleosts may be divided into four main groups, the Osteoglossomorpha, the Elopomorpha, the Clupeomorpha, and the Euteleostei. In addition, a series of extinct forms fall between the holosteans and these living teleost groups.

Early teleosts

The most primitive teleosts, the pachycormids and aspidorhynchids (Fig. 6.11a) of the Jurassic and Cretaceous, have long bodies covered with heavy scales, and a long pointed snout. The pholidophorids of the Late Triassic and Early Jurassic were small hunting fishes which show advances in the jaws. The leptolepids of the Jurassic and Cretaceous were also small (Fig. 6.11b), often as little as 50 mm long, and they may have fed on plankton. These fishes have fully ossified vertebrae and the scales are **cycloid** (circular in shape, and of light construction).

Another important extinct group, the ichthyodectiforms (Fig. 6.11c), of the Jurassic and Cretaceous, were mostly large predaceous fishes

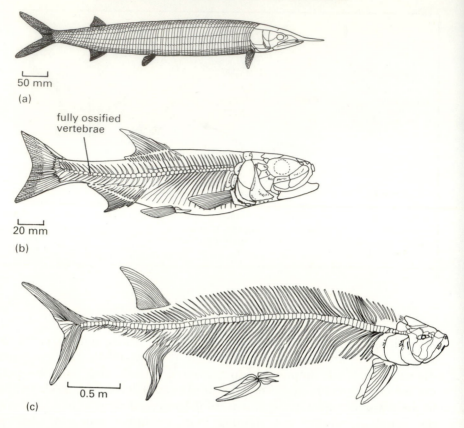

50 mm

(a)

fully ossified
vertebrae

20 mm

(b)

0.5 m

(c)

Figure 6.11 Primitive teleosts of (a) & (b) the Jurassic and (c) the Cretaceous: (a) the aspidorhynchid *Aspidorhynchus*; (b) the leptolepid *Varasiichthys*; (c) the ichthyodectiform *Xiphactinus*. (Fig. (a) after Woodward 1891–1901; (b) after Arratia 1981; (c) after Osborn 1904.)

(Patterson & Rosen 1977). They generally swallowed their prey head-first, as is shown by fossils with preserved stomach contents. A specimen of *Xiphactinus*, 4.2 m long, was found with a 1.6 m ichthyodectiform in its stomach area, and smaller relatives have been found with as many as ten recognizable fish skeletons preserved inside them.

Subdivision Osteoglossomorpha

The osteoglossomorphs, a relatively small group of about 150 species (Fig. 6.12a) which live in freshwaters mainly of the southern hemisphere, arose possibly in the Late Jurassic. They are characterized by features of the feeding system (Lauder & Liem 1983) in which a bony element in the tongue bears large teeth which bite against teeth in the roof of the mouth (hence the name osteoglossomorph which means 'bony-tongue-form').

136

Subdivision Elopomorpha

The elopomorphs (literally 'eel forms') include about 650 species of eels, tarpons, and bonefishes, and the group is known from the Early Cretaceous. The tarpon, *Megalops* (Fig. 6.12b), is typically 'fish-shaped', and it seems hard to see how it can be regarded as a close relative of the eel, *Anguilla* (Fig. 6.12c). However, all elopomorphs are characterized by the possession of a specialized larval stage, the leptocephalus (Fig. 6.12d) which is thin and leaf-shaped. The leptocephalus larvae can migrate long distances before they metamorphose. Eels have many skeletal modifications including the loss of the pelvic girdle, overall elongation of the body, and fusion of elements in the upper jaw. The deep-sea eels called saccopharyngoids, are even more modified, having lost their scales, ribs, tail fin, and many skull bones. Indeed the skull (Fig. 6.12e) is really just a huge pair of jaws with a tiny cranium set in front. These fishes sit quietly on the deep dark ocean floors and lever their huge mouths open to seize prey animals many times their own size.

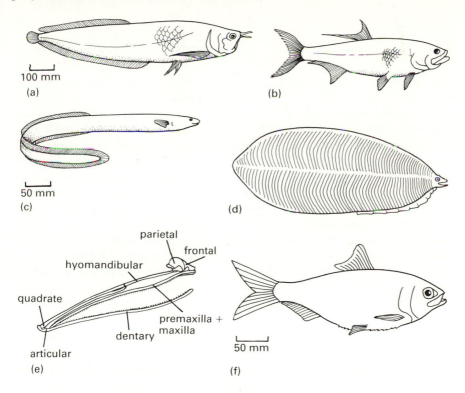

Figure 6.12 The (a) osteoglossomorph, (b)–(e) elopomorph, and (f) clupeomorph teleosts: (a) *Osteoglossum*; (b) the tarpon *Megalops*; (c) the eel *Anguilla*; (d) the leptocephalus larva of an elopomorph; (e) the skull of the saccopharyngoid eel *Gastrostomus*; (f) the herring *Clupea*. (Figs (a)–(d) & (f) after Greenwood *et al.* 1966; (e) after Gregory 1933.)

Subdivision Clupeomorpha

The clupeomorphs contain over 300 species of extant herring-like fishes and over 150 fossil species which date back to the Early Cretaceous (Grande 1985). They are generally small silvery marine fishes, some of which, like the herring (Fig. 6.12f) and anchovy, occur in huge shoals, and feed on plankton. Clupeomorph characters include a peculiar type of abdominal scute, an unusual arrangement of the bony plates at the base of the tail, and a specialized air sac within the exoccipital and prootic bones in the braincase. Most teleosts have a sausage-shaped air sac called the **swim bladder** in the body cavity which is used to achieve neutral buoyancy. Gas is pumped into the bladder, or removed, via the bloodstream in order to match the 'weight' of the fish to the pressure that acts at whatever depth it finds itself. In clupeomorphs, the swim bladder has a unique extension into the braincase and is also concerned with hearing.

Subdivision Euteleostei

The largest teleost group, the euteleosts, consists of 17 000 species in 375 families. These may be divided into three main subgroupings, the salmoniforms, the ostariophysans, and the neoteleosts (Greenwood *et al.* 1966, Patterson & Rosen 1977, Lauder & Liem 1983). The most primitive are generally considered to be the salmon and trout group, the order 'Salmoniformes'. Early representatives include the tiny *Gaudryella* from the mid Cretaceous (Fig. 6.13a). True salmon appeared only later.

The second euteleostean group, the superorder Ostariophysi, contains carp, goldfish, minnows, catfish, and indeed most freshwater fishes (Fink & Fink 1981). They are characterized by several features, including a specialized hearing system composed of modified cervical vertebrae, ribs, and neural arches, called the Weberian ossicles (Fig. 6.13b). There are five key bony elements that articulate, and provide a link between the anterior swim bladder and the ear. The os suspensorium and the tripus rest on the taut surface of the swim bladder. When sound waves reach the fish, the swim bladder vibrates and effectively amplifies the sound. The two bones in contact pivot, and the vibrations pass through the intercalarium, scaphium, and claustrum to the inner ear.

The Neoteleostei (see box), containing the remaining 10 000 species of advanced teleosts, are characterized by a specialized muscle in the upper throat region that operates small pharyngeal toothplates in the roof of the mouth, an important adaptation for manipulating prey before swallowing.

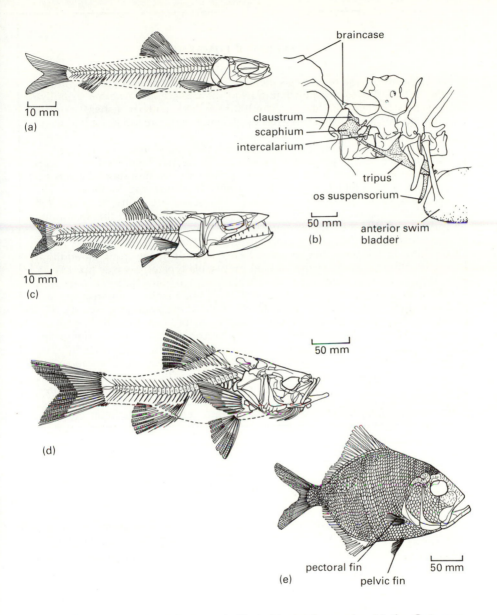

Figure 6.13 (a)–(c) basal euteleosts and (d) & (e) acanthomorphs: (a) the Cretaceous 'salmoniform' *Gaudryella*; (b) the Weberian ossicles that transmit vibrations from the swim bladder to the inner ear in ostariophysian fishes (ossicles are shaded and named); (c) the aulopiform *Eurypholis*; (d) the Cretaceous paracanthopterygian *Mcconichthys*; (e) the percomorph *Berycopsis*. (Fig. (a) after Patterson 1970; (b) after Goody 1969; (c) after Fink & Fink 1981; (d) after Grande 1988; (e) after Patterson 1964.)

NEOTELEOST DIVERSITY

The primitive living neoteleosts include the stomiiforms and the aulopiforms (Fig. 6.13c), deep-sea fishes dating from the Cretaceous. Another primitive neoteleostean group, the myctophiforms, the lantern fishes, possess photophores, light-producing structures which show up when they descend into deep waters.

The remaining neoteleosts are termed the spiny teleosts, or acanthomorphs, because they bear stiff fin spines. These may be moved by muscles at the base, and they have a defensive function since they can be erected if the fish is threatened, effectively stopping it from being swallowed! The acanthomorphs show other advances. The scales are spiny, the so-called **ctenoid** (i.e. 'comb') type. The body is short and rigid, and the swimming thrust is produced by rapid movements of the tail fin instead of by bending the whole body. This allows great speeds to be achieved – as much as 70 km per hour in the tuna, compared to 5 km per hour in the trout. In addition the acanthomorphs have the toothed premaxilla type of jaw (see pp. 134–5).

Within the Acanthomorpha, the paracanthopterygians (Rosen & Patterson 1969) include some 250 genera of cod, haddock, anglerfishes, and clingfishes, and they date back to the Late Cretaceous (Fig. 6.13d). The acanthopterygians contain 8000 species which fall into two major groups, the atherinomorphs (830 species) and the percomorphs. The atherinomorphs (killifishes, flying fishes, guppies) date from the Eocene to the present. The percomorphs include a tremendous range of forms, from seahorses to flatfishes, and from tunas to porcupine fishes. The group as a whole is known from the Late Cretaceous, but most families have a very limited fossil record, often confined to the past 20 Myr or so. The beryciforms (Fig. 6.13e) have deep bodies, and the pelvic fin is below the pectoral fin.

RELATIONSHIPS

The actinopterygians are a vast and diverse group, and there is still much confusion over their relationships. However, most studies (e.g. Greenwood *et al.* 1966, Patterson 1973, 1975, Patterson & Rosen 1977, Lauder & Liem 1983, Gardiner 1984b) have agreed on a few points (Fig. 6.14). The chondrosteans and holosteans are seen as grade groups, and others such as the 'Palaeonisciformes', 'Semionotidae', and 'Salmoniformes' are almost certainly polyphyletic. The Aspidorhynchidae, Leptolepidae, Euteleostei, Paracanthopterygii, and Percomorpha may be monophyletic, but they are hard to characterize (Lauder & Liem 1983).

The evolution of modern fishes (Fig. 6.15) shows roughly parallel patterns between the chondrichthyans and the actinopterygians. The 'palaeonisciform' radiation is matched by the Carboniferous–Permian shark groups. The 'semionotids' and others of the Triassic and Early

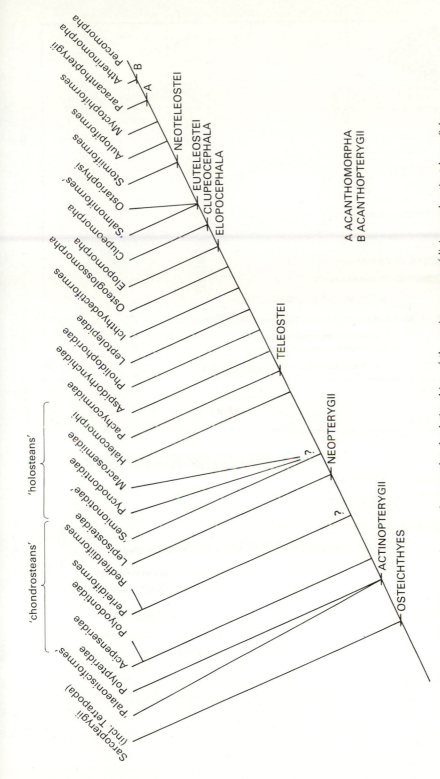

Figure 6.14 Cladogram showing the postulated relationships of the major groups of living and extinct bony fishes.

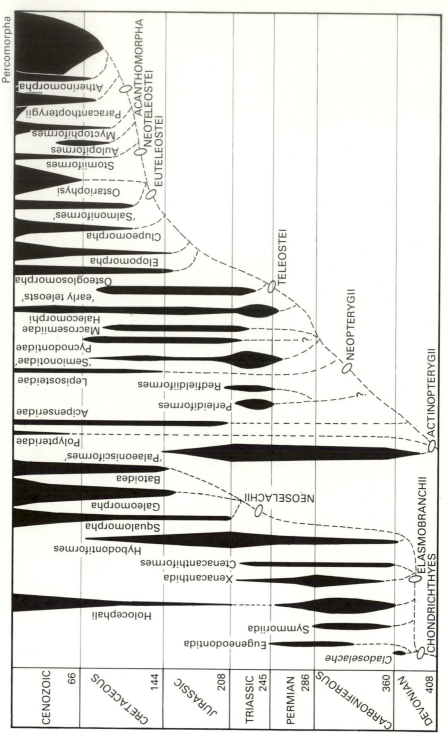

Figure 6.15 Phylogenetic tree showing the radiations of the cartilaginous fishes (left) and the bony fishes (right). Postulated relationships are shown by dashed lines, the known fossil record by solid shading in the vertical dimension (time scale on the left), and relative abundance by the width of the 'balloons'.

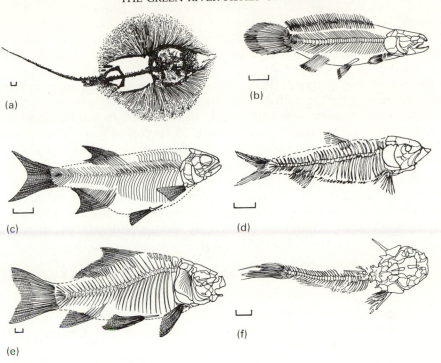

Figure 6.16 Typical fishes of the Eocene Green River Formation of Wyoming: (a) the ray *Heliobatis*; (b) the bowfin *Amia*; and the teleosts (c) *Eohiodon*; (d) *Knightia*; (e) *Amyzon*; (f) *Astephus*. Scale bar is 10 mm. (Figs (a)–(d) after Grande 1984; (e) after Grande, Eastman & Cavender 1982; (f) after Grande & Lundberg 1988.)

Jurassic had their heyday at the same time as the hybodonts, and the greatest radiations of all, of teleosts and neoselachians seem to go in parallel. The radiation of modern sharks began rather earlier, in the Early Jurassic, than did that of the teleosts. The fossil record of fishes was apparently little affected by the well-known mass extinction events that wiped out hoards of marine invertebrates and tetrapods (Benton 1989a).

THE GREEN RIVER FISHES OF WYOMING

Some of the best fossil fishes have been found in deposits that contain thousands of excellently preserved specimens as a result of unusually good conditions of preservation. An example is the Green River Formation of southwestern Wyoming, with deposits varying in age from Late Palaeocene to Late Eocene in some areas (57–38 Myr). Many of the specimens are found in finely layered buff-coloured limestones which were laid down in three large inland lakes (Lake Gosiute, Lake Uinta, Fossil Lake; Grande 1989), somewhat like the Devonian fish beds of

Scotland (see pp. 35–7). Lake Gosiute and Lake Uinta are interpreted as playa lakes, ephemeral saltwater bodies in a generally hot arid setting, while Fossil Lake seems to have been a more stable body of freshwater (Grande 1989).

The fish faunas (Grande 1984) consist mainly of teleosts, such as the small herring *Knightia*, catfish, suckers, perch, as well as the garfish *Lepisosteus*, the bowfin *Amia*, the paddle fish *Crossopholis*, and a stingray *Heliobatis* (Fig. 6.16). Grande (1984) records 25 species of bony fishes from the three main Green River lakes. Their distribution generally differs from lake to lake. For example, the gar *Lepisosteus* is very common at some localities (25–50% of all specimens), but extremely rare at others. Only *Knightia* is common or extremely common (5–50% or more of all specimens) at most localities. Other taxa are present in only one lake (e.g. suckers and catfishes from Lake Gosiute (48 Myr), hiodontids, paddlefish, and stingrays from Fossil Lake (52 Myr)).

CHAPTER SEVEN

The age of dinosaurs

The most famous fossil vertebrates, the dinosaurs (literally 'terrible lizards') arose in the Late Triassic, about 230 Myr ago (see pp. 119–22), and dominated terrestrial faunas for the next 165 Myr until their extinction at the end of the Cretaceous (66 Myr). The earliest dinosaurs were moderate-sized bipedal carnivores, but large quadrupedal herbivores had come on the scene by the end of the Triassic. During the Jurassic and Cretaceous, the dinosaurs diversified into a great panoply of carnivores large and small, massive herbivores, small fast-moving specialized plant-eaters, and forms armoured with great bone plates, horns, carapaces, and clubs. The pterosaurs, relatives of the dinosaurs, filled the skies, and the birds, descendants of the carnivorous dinosaurs, also came on the scene. Other land animals included the more familiar lizards, snakes, turtles, crocodilians, and mammals. The seas were populated by ichthyosaurs and plesiosaurs, great marine reptiles that preyed on fishes, squid, and on each other. The diversity of these reptiles, and their biology, will be considered in this chapter. There are numerous books on dinosaurs and their contemporaries, and four recent good ones are Norman (1986a), Bakker (1986), Benton (1989b), and Weishampel *et al.* (1990).

BIOLOGY OF *PLATEOSAURUS*

The oldest-known dinosaurs are dated as Carnian in age (Late Triassic, 230–225 Myr) and they include forms such as *Herrerasaurus* from Argentina (see p. 199) and *Coelophysis* from North America. One of the best-known Late Triassic dinosaurs is *Plateosaurus*. The first specimens were found in southwestern Germany in 1837, and since then dozens of skeletons have been collected from over 50 localities in Germany, Switzerland and France, dated as mid Norian (c. 215 Myr). The best locality was Trossingen, south of Stuttgart, where 35 skeletons and fragments of 70 more were excavated (Weishampel 1984a). Initially, this deposit was interpreted as a mass grave in which a herd had perished while migrating across an arid desert plain in search of plant food.

However, it seems more likely that the animals died because of a variety of factors over a long period of time and they have been entombed in water-laid mudstones. The skeletons are of young and old animals, and many have been broken up by scavengers and by water movement.

Plateosaurus (Fig. 7.1) is about 7 m long and could have adopted either a bipedal or a quadrupedal posture. The body proportions are typical of early dinosaurs: a long tail, long hindlimbs about twice as long as the arms, and a long neck, but the skull is small and the limbs are heavily built because of its large size. *Plateosaurus* shows advanced dinosaurian characters of the limbs and vertebrae (Galton 1984): upright posture, slender pelvic rays, and distinctive vertebrae in the neck, trunk, and the lower back.

The diet of the plateosaurs has been controversial recently. They have generally been regarded as herbivores because of their size, their great abundance, and their weak leaf-shaped teeth (Fig. 7.1c & e). However, some skeletons were found in association with dagger-like teeth and these suggested a diet of meat. Galton (1985) had countered these arguments with a strong defence of herbivory. The blade-like teeth were almost certainly those of rauisuchians and others which were scavenging on plateosaur carcasses and shed their teeth when biting on bones, a common enough phenomenon among sharks and crocodilians today. The teeth of *Plateosaurus* have serrated edges, but these are more like the teeth of herbivorous lizards which cut up tough plants than the steak-knife teeth of true carnivores. The jaw joint in *Plateosaurus* is set low (Fig. 7.1c), an adaptation seen in herbivorous mammal-like reptiles (see p. 90) and other dinosaurs (see p. 159) which gives a sustained and evenly spread bite along the tooth row, useful in dealing with tough plant stems.

Plateosaurus swallowed its plant food whole and could not chew it as modern mammals do since sideways jaw movements were not possible. However, it avoided indigestion by the use of a gastric mill. Just as chickens today swallow grit which lodges in the gizzard, above the stomach, and grinds the food up, the plateosaurs swallowed pebbles. This is shown by finds of gizzard stones, or **gastroliths**, inside the upper rib cage of plateosaur skeletons. A herd of feeding plateosaurs must have rattled, grunted and burped furiously as their rough plant diet was reduced to a digestible state!

THE JURASSIC AND CRETACEOUS WORLD

During the Triassic and Jurassic, the super-continent Pangaea was at its most extensive, with continuous land stretching from North America to Europe, and South America to Africa, Antarctica, Australia and India (Fig. 7.2). Jurassic climates were moister than in the Triassic (see p. 103) and warm conditions prevailed right to the polar regions (Hallam 1985).

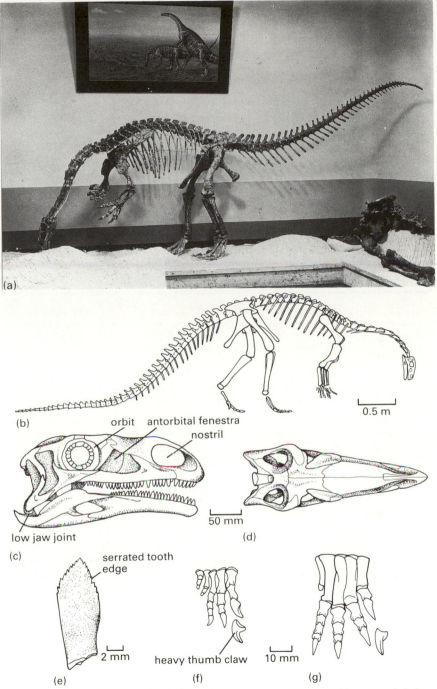

Figure 7.1 The plant-eating prosauropod *Plateosaurus* from Germany: (a) mounted skeleton in the Tübingen museum; (b) reconstructed skeleton; (c) & (d) skull in lateral and dorsal views; (e) tooth; (f) hand in anterior view, with lateral view of heavy thumb claw; (g) foot in anterior view, with lateral view of heavy claw on digit 1. (Fig. (a) courtesy of Dr D. B. Weishampel; (b) after Galton 1971; (c) & (d) after Galton *in* Carroll 1987; (e)–(g) after Galton 1985.)

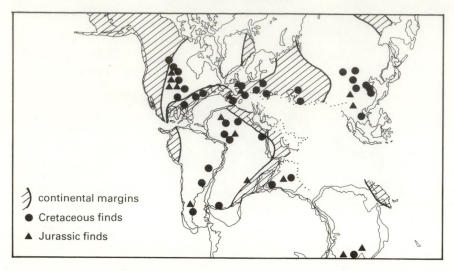

Figure 7.2 Map of the Jurassic/Cretaceous world, showing the distribution of land and sea at the time (ancient coastlines shown with heavy lines, and shallow seas cross-hatched), and localities of major dinosaur finds in the Jurassic and Cretaceous. (Map based on the Late Jurassic (150 Myr), from several sources.)

Ferns and conifers of subtropical varieties have been found as far north as 60° palaeolatitude, and rich floras are known from Greenland and Antarctica. Cretaceous floras show similar patterns. The polar regions had warm-temperate climates and the boundary between the subtropical and temperate floras was 15° closer to the poles than it is today. Thus most of the United States, Europe as far north as Denmark, and most of South America and Africa enjoyed tropical climates.

A major change took place in the world's floras during the Cretaceous. Triassic and Jurassic landscapes contained low ferns, horsetails, and cycads, and tree-sized club mosses, seed ferns and conifers. In the Early Cretaceous, the first flowering plants (angiosperms) appeared, and they radiated progressively during the Late Cretaceous until they reached modern levels of dominance. The earliest angiosperms include magnolia, beech, fig, willow, palm, and other such familiar flowering shrubs and trees.

DIVERSITY OF SAURISCHIAN DINOSAURS

The dinosaurs have traditionally been divided into two groups, the Saurischia and the Ornithischia on the basis of their radically different pelvic regions, the so-called 'lizard hip' and 'bird hip' respectively. The more primitive structure is seen in the saurischians in which the pubis points forwards and the ischium back (Fig. 7.3a), as in all basal

148

archosaurs of the Triassic (e.g. Figs 5.3, 5.6, 5.12). In ornithischians, on the other hand, the pubis runs back in parallel with the ischium, and an additional **prepubic process** develops in front (Fig. 7.3b).

Many of the dinosaurian characters of the hindlimbs are related to their upright posture (see p. 120). The acetabulum is fully open, and the pubis and ischium are long slender elements. The legs are brought in close to the vertical midline of the body (Fig. 7.3c) by a shift of the hip-thigh joint from the top of the femur to a distinct ball on its inside. The reorientation of the limbs from a partial sideways sprawl has also changed the angle of the knee and ankle joints to simple hinges. The fibula is reduced, often to a thin splint, and the tibia has a 90° twist so that its proximal head is broadest from back to front, and its distal end from side to side. The AM ankle (Fig. 5.5) is dominated by a wide astragalus with a distinctive vertical process that wraps round the front of the tibia, and the calcaneum is a small block-like element. In the foot, the dinosaur stands up on its toes (the **digitigrade stance**) rather than on the flat of its whole foot (the **plantigrade stance**), as most thecodontians (and humans) do. Toes 1 and 5 are much reduced, and the dinosaur really uses only numbers 2–4.

The saurischian dinosaurs, carnivorous theropods, and herbivorous sauropodomorphs like *Plateosaurus* and its descendants will be reviewed here. The ornithischians are considered later (see pp. 157–71).

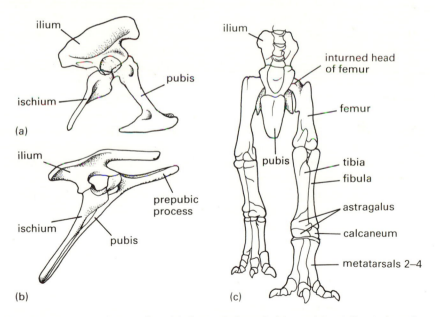

Figure 7.3 Dinosaurian pelvic girdles: (a) the typical saurischian pelvic girdle, in lateral view, in *Tyrannosaurus*; (b) the typical ornithischian pelvic girdle, in lateral view, in *Thescelosaurus*; (c) anterior view of the hindlimbs of *Tyrannosaurus* to show the fully upright posture. (Figs (a) & (c) after Osborn 1916; (b) after Romer 1956, courtesy of the University of Chicago Press.)

Suborder Theropoda

The basal theropods, the ceratosaurs, include forms like *Dilophosaurus* (Fig. 7.4a) and *Ceratosaurus* (Fig. 7.4b) from the Jurassic, which reach lengths of 5–7 m. *Dilophosaurus* (Welles 1984) has an unusual skull with two flat-sided crests over the skull roof, which must have had a signalling function of some kind. *Ceratosaurus* (Fig. 7.4b) has a stronger skull with heavier teeth than in *Dilophosaurus*, a low eyebrow ridge, and a pair of 'horns' on the nasal bones.

The giant meat-eating carnosaurs are more advanced than ceratosaurs in possessing a large opening in the maxilla, the **maxillary fenestra** (Fig. 7.4c), which is seen in all other theropods and primitive birds. *Allosaurus*, a Late Jurassic carnosaur from North America (Madsen 1976), has a short skull which is narrow from side to side (Fig. 7.4c). The orbit is high and smaller than the antorbital fenestra, there are heavy crests over the orbits, and the mandibular fenestra is much reduced. *Tyrannosaurus*, the largest terrestrial carnivore of all time at 14 m long, is one of a group of Late Cretaceous theropods which radiated in North America (Russell 1970) and central Asia. *Tyrannosaurus* has a large head (Fig. 7.4d) with an extra joint in the lower jaw between the dentary and the elements at the back. This allowed *Tyrannosaurus* to increase its gape for biting large prey. The tiny forelimbs of *Tyrannosaurus* would seem to be quite useless since they cannot even reach the mouth. They may have been used to help *Tyrannosaurus* get up from a lying position, by providing a push while the head was thrown back and the legs straightened.

The ornithomimids of the Late Jurassic to Late Cretaceous (Russell 1972) are highly specialized theropods (Fig. 7.5a) with a slender ostrich-like body and long arms and legs. The hands have three powerful fingers which may have been used for grasping prey items. The lightly built body indicates that *Struthiomimus* could have run fast, and speeds of 35–60 km/h have been estimated. The skull is completely toothless in later forms, and their diet may have included small prey animals such as lizards or mammals, or even plants.

The remaining theropods have been placed in the major clade Maniraptora by Gauthier (1986) since they share a large number of derived characters with each other and with the birds. These include *Compsognathus*, the smallest adult dinosaur, at 0.7–1.4 m long, the curious egg-eating *Oviraptor*, and the dromaeosaurids, such as *Deinonychus* (see pp. 155–7), and the troodontids from the Late Cretaceous of North America and Mongolia. *Saurornithoides* has a long slender skull (Fig. 7.5b) with orbits facing partly forwards so that it may have had binocular vision (Russell 1969). The braincase is bulbous and relatively large, which has led to the interpretation of the troodontids as the most intelligent (or least stupid?) dinosaurs. The foot (Fig. 7.5c) is also advanced, having three functional toes, of which number 4 is the longest. The metatarsals are

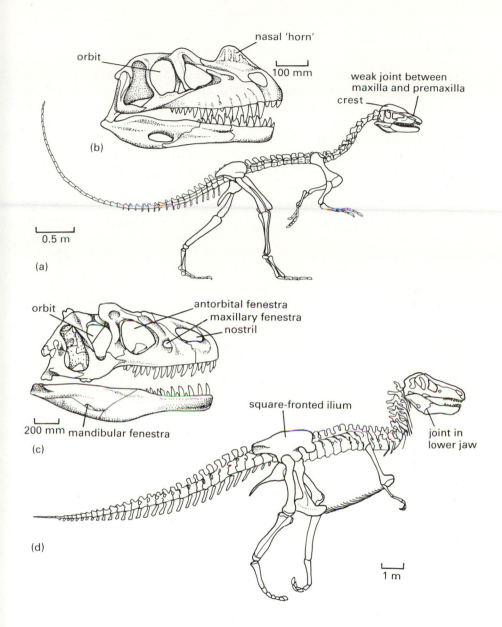

Figure 7.4 (a) & (b) Ceratosaurs, and (c) & (d) carnosaurs: (a) *Dilophosaurus*; (b) *Ceratosaurus* skull; (c) *Allosaurus* skull; (d) *Tyrannosaurus*. (Fig. (a) after Welles 1984; (b) after Marsh 1884; (c) after Madsen 1976; (d) after Newman 1970.)

tightly bunched, with numbers 2 and 4 meeting in front of number 3 which is 'squeezed' tightly at its upper end.

Suborder Sauropodomorpha

The Sauropodomorpha, the second major saurischian clade, first appears in the record in the Late Triassic a little later than the Theropoda, and the early forms of the Late Triassic and Early Jurassic are generally referred to as prosauropods, although this is a paraphyletic group. *Thecodontosaurus* from England (Fig. 7.6a), a 1.25 m long lightly-built herbivore (Kermack 1984), shows all of the basic sauropodomorph hallmarks: a small skull (c. 5% of body length), a downwards curve to the tip of the dentary (Fig. 7.6b), lanceolate teeth with serrated crowns (Fig. 7.6c), a long neck with ten or more cervical vertebrae, a huge thumb claw and no claws on fingers 4 and 5 (Fig. 7.1f), and a short blade on the ilium. More advanced

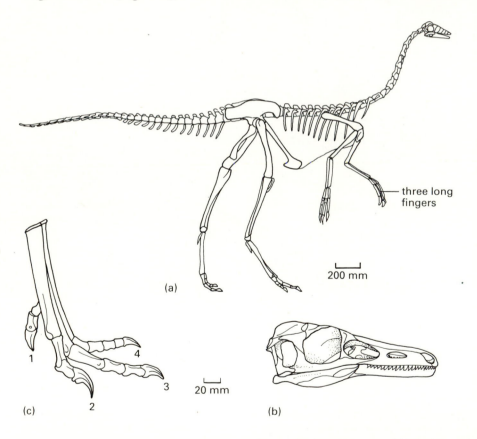

Figure 7.5 (a) Ornithomimids, and (b) & (c) troodontids: (a) skeleton of *Struthiomimus*; (b) skull of *Saurornithoides*; (c) foot of *Troodon*. (Fig. (a) after Russell 1979; (b) after Russell 1969; (c) after Sternberg 1932.)

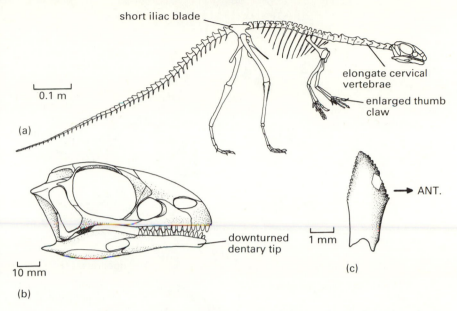

short iliac blade

0.1 m

elongate cervical vertebrae

enlarged thumb claw

(a)

ANT.

downturned dentary tip

1 mm

10 mm

(b)

(c)

Figure 7.6 The prosauropod *Thecodontosaurus*; (a) skeleton; (b) skull in lateral view; (c) tooth (ANT., anterior). (After D. A. Kermack 1984.)

prosauropods include *Plateosaurus* (see pp. 145–7) and some obligatory quadrupeds up to 10 m long.

The prosauropods appear to have evolved directly into the sauropods, and it is hard to draw the line between the two groups. The Middle Jurassic sauropod *Cetiosaurus* (Fig. 7.7a) from England, and *Shunosaurus* (Fig. 7.7b) from China, show some advances. The part of the skull behind the eye is much shortened, the nostrils and nasal bones are well back, and the lower temporal fenestra lies almost below the orbit instead of behind it. The external mandibular fenestra in the lower jaw has been lost, and the teeth are peg-like. The neck contains 12 or more cervical vertebrae, and there are five or six sacral vertebrae. The limb bones are short and massive, and the hands and feet very short and broad. The legs are rather pillar-like, as in elephants, an obvious adaptation for weight-supporting. The cetiosaurids are thought to be a paraphyletic group that includes the ancestors of the four main sauropod lineages, the camarasaurids, brachiosaurids, diplodocids, and titanosaurids.

The camarasaurids and brachiosaurids share a skull pattern (Fig. 7.7c & d) in which there is an arched internarial bar formed by very narrow premaxillae between the nostrils, and a clearly defined snout. The diplodocids and titanosaurids share several characters of the skull (Fig. 7.7e & f) such as the steeply sloping quadrate, and a long broad snout with a small number of cylindrical pencil-like teeth at the front. The jutting teeth may have been used in a pincer-like fashion to crop

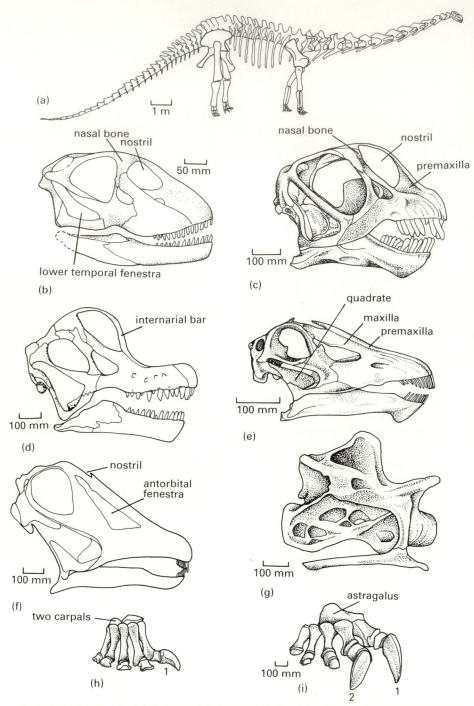

Figure 7.7 Sauropods: (a) skeleton of *Cetiosaurus*; skulls of (b) *Shunosaurus*, (c) *Camarasaurus*, (d) *Brachiosaurus*, (e) *Diplodocus*, and (f) *Antarctosaurus*; (g) cervical vertebra of *Diplodocus*; (h) & (i) hand and foot of *Diplodocus*. (Fig. (a) after Crowther & Martin 1976, (b) after Dong & Tang 1984; (c) after Osborn & Mook 1921; (d) after Lapparent & Lavocat 1955; (e) after Ostrom & McIntosh 1966; (f) after Huene 1929; (g) after Hatcher 1901; (h) & (i) after Coombs 1975.)

vegetation. The vertebrae are very large but lightweight since they contain a complex of cavities in the sides, supported by a latticework of narrow bone struts (Fig. 7.7g). The limbs of *Diplodocus* seem relatively slender, but the hands and feet (Fig. 7.7h & i) are shortened weight-supporting structures. The wrist contains only two carpals, and the ankle only the astragalus, all other elements having been lost or present only as cartilaginous masses. The first finger and the first two toes bear long claws which may have been used in digging, but the other digits bear only small hoof-like nubbins of bone.

BIOLOGY OF *DEINONYCHUS*

Deinonychus, a theropod from the Early Cretaceous Cloverly Formation of Montana, USA (Ostrom 1969) is one of the most remarkable dinosaur finds of recent years. The reconstructions (Fig. 7.8a) show that it was a small animal about 3 m long, 1 m tall, and weighing 60–75 kg. The skull (Fig. 7.8b) is incompletely known. The curved sharp teeth have serrated edges, as in all other theropods, which were presumably as effective in cutting flesh as a steak knife.

Deinonychus held its backbone roughly horizontal when it was moving about (Fig. 7.8a). At one time, bipedal dinosaurs were reconstructed in kangaroo mode with the backbone sloping or close to the vertical. However, there are three lines of evidence that *Deinonychus* and others adopted the posture shown here: (a) it allows the body weight to balance correctly with the centre of gravity over the hips; (b) the shapes of the joints between the cervical vertebrae shows that the neck curved up in a swan-like S-shape rather than acting simply as a pillar beneath the head (Fig. 7.8d); and (c) the dorsal vertebrae bear very distinctive scars on the front and back of the neural spines (Fig. 7.8e) which Ostrom (1969) argues are exactly like the ligament scars seen in flightless birds such as the ostrich; tough short interspinous ligaments, which are fused into the bone by the long fibres, prevent the flexing that would occur if the back were not held horizontal.

The tail is also rather exceptional in *Deinonychus* since it is stiffened by long bony rods which invest it on all sides in the middle and posterior sections. These bony rods are parts of the individual caudal vertebrae that have become unusually elongated (Fig. 7.8f). The **prezygapophyses**, normally a pair of short processes in front of the neural spine which interlock with the **postzygapophyses** of the vertebra in front, have become as long as nine vertebrae. They run forwards and intertwine with the prezygapophyseal rods of other vertebrae. Beneath the caudal vertebrae are the **chevrons** which are separate elements that normally run back and down a short distance and provide attachment sites for the tail muscles. In *Deinonychus*, the chevrons have unusually long anterior rods,

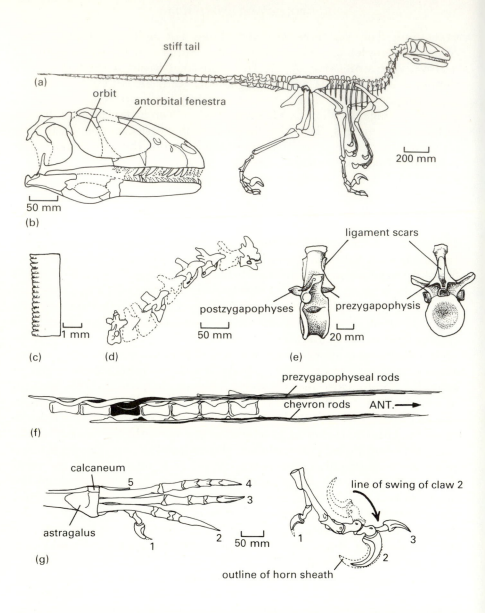

Figure 7.8 The dromaeosaurid *Deinonychus*: (a) skeleton in running pose; (b) lateral view of the skull; (c) posterior edge of a maxillary tooth, showing the serration; (d) reconstructed neck, showing the curvature; (e) a dorsal vertebra in lateral and posterior views; (f) outline of a series of caudal vertebrae, with one vertebra, and its elongate prezygapophyseal rods (above) and chevron rods (below), shaded black (ANT., anterior); (g) left foot in anterior view; (h) left foot in lateral view, showing the swing of the scythe claw. (After Ostrom 1969.)

probably formed from ossified tendons which intertwine beneath the vertebrae. The rods did not entirely immobilize the tail since they remain separate and could slide across each other to some extent. However, the tail of *Deinonychus* was largely a stiff rod which acted as a single unit, and was almost certainly used as a balancing organ.

The arms are strong, and the hands armed with deep claws on the three long fingers. Indeed, the hand is nearly half the length of the arm, an advanced feature seen in few other theropods that is similar to birds. Further, the wrists of *Deinonychus* were unusually mobile, and the hands could be turned in towards each other. The hand was clearly used for grasping prey and the claws for tearing at flesh.

The hindlimbs have long bird-like proportions: a short femur, long tibia and fibula, long metatarsals, three functional toes, and a small backwardly-pointing first toe (Fig. 7.8g). The astragulus has a high process that wraps around the tibia, and the calcaneum is a small block of bone firmly attached to it. The key feature of the foot is the elongate toe 2 which is armed with a vast sickle-shaped claw up to 120 mm long, which could be bent right back, and then swung down, but the whole toe could bend only a short way below horizontal. This foot claw would have got in the way during walking, so it must have been held in the upright position most of the time. Ostrom's (1969) functional interpretation was based on his insight that *Deinonychus* was an active biped like a modern flightless bird which could balance readily on a single foot. The toe claw is ideal for disembowelling prey. *Deinonychus* (literally 'terrible claw') ran up to its victim with the claw held up to keep it from scraping on the ground, raised one foot, balanced, and slashed with a backwards kick at its flanks causing a deep gash up to 1 m long. The most likely prey for *Deinonychus* seems to be *Tenontosaurus*, a fairly abundant relative of *Hypsilophodon* (see pp. 161–2), that reached 6–7 m in length. *Deinonychus* may have hunted in packs like certain wild dogs today, which would have enabled it to harry and weaken much larger prey animals before killing them with fatal slashes to the belly region.

DIVERSITY OF ORNITHISCHIAN DINOSAURS

The Ornithischia are the second major dinosaurian group, and they are relatively easy to define. They have a pubis that points backwards (Fig. 7.3b) as well as over 30 other derived characters of the skull and skeleton (Norman 1984, Sereno 1986).

The ornithischians presumably arose at the same time as the saurischians, possibly during the Carnian (Late Triassic, 227 Myr), but fossils are clearly known only at the beginning of the Jurassic (208 Myr). The ornithischians were all herbivorous and they divide into two main groups, the Cerapoda (the bipedal ornithopods, bone-headed pachycephalosaurs, and horned ceratopsians) and the Thyreophora (the armoured ankylosaurs and stegosaurs).

Pisanosaurus – the first possible ornithischian

Pisanosaurus from the Late Triassic of Argentina is known from only its jaws, neck, and a few limb elements (Bonaparte 1976). The cheek teeth (Fig. 7.9a) have low triangular crowns with a well-developed narrow neck beneath, and they are set over to the inside of the jaws, leaving a broad shelf on the outside. This indicates that *Pisanosaurus* had cheeks, pouches of skin that lay on either side of the tooth rows, which could retain unchewed plant material while other food was being processed. Cheeks are typical of ornithischians and other reptiles in which the skin of the face is firmly attached to the jaw margins just below the tooth rows. If *Pisanosaurus* is an ornithischian, and many regard it simply as an undifferentiated 'early dinosaur', it is followed by a long gap in their evolution. The next ornithischians appear after 20 Myr.

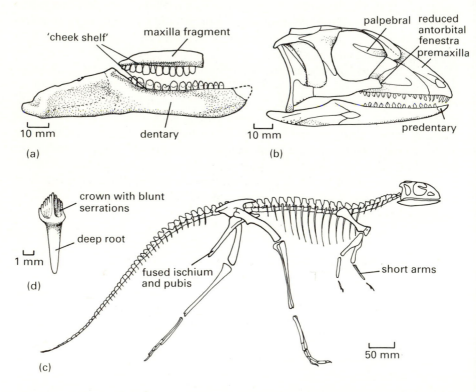

Figure 7.9 Early ornithischians: (a) *Pisanosaurus*, maxilla fragment and partial lower jaw in lateral view; (b)–(d) *Lesothosaurus*, skeleton, skull, and tooth. (Fig. (a) after Bonaparte 1976; (b)–(d) after Galton 1978.)

158

Lesothosaurus

Small ornithischians often called fabrosaurids have been reported from the Early Jurassic of several parts of the world (Galton 1978), but only *Lesothosaurus* (sometimes called *Fabrosaurus*) from southern Africa is reasonably complete. It is a lightly built 0.9 m long animal with long hindlimbs and short arms (Fig. 7.9c). It has the typical ornithischian pelvis, an ilium with a narrow anterior process, and fusion of the ischia and pubes at their tips. The skull (Fig. 7.9b) shows even more ornithischian characters. The tip of the premaxilla is toothless and roughened and it is matched by an entirely new bone in the lower jaw, the **predentary**. The orbit also contains a new bone, the **palpebral**. The teeth (Fig. 7.9d) are more typically ornithischian than those of *Pisanosaurus* since they have a bulbous base to the crown and rounded denticles on the edges. The wear facets lie symmetrically on either side of the pointed tip of the crown, and this suggests an up and down jaw action with no possibility of back and forwards or side to side chewing.

Infraorder Ornithopoda

The ornithopods were the largest and most successful ornithischian group, containing only about 55 genera, but achieving very high abundance in Cretaceous faunas in particular. There are four main groups, the heterodontosaurids, hypsilophodontids, 'iguanodontids', and hadrosaurids.

The heterodontosaurids lived at the same time as the fabrosaurs. *Heterodontosaurus*, from southern Africa (Charig & Crompton 1974, Santa Luca 1980), just over 1 m long, is similar to *Lesothosaurus* in many ways. The bodily proportions (Fig. 7.10a) differ only in the slightly longer arms and the shorter body. The skull (Fig. 7.10b) shows the most remarkable features. *Heterodontosaurus* (literally 'different tooth reptile') has differentiated teeth, two incisors, a canine, and about 12 cheek teeth. The canines are long and the lower one fits into a deep notch in the upper jaw. One specimen has no tusks, and it has been suggested that their presence may be a secondary sexual character of males. If so, the canine tusks may have been used for defence and for social display, as in modern herbivorous mammals with tusks, like certain pigs and the musk deer.

The jaw joint in *Heterodontosaurus* is set well below the level of the tooth rows, a common feature of other herbivorous dinosaurs and mammal-like reptiles (see pp. 90, 146) that increases the duration and force of the bite. The cheek teeth wear against the opposite teeth of the lower jaw forming a straight line at the crest of the teeth (Fig. 7.10c). The outer surfaces of the lower teeth fit inside the upper teeth and wear them from the inside. Unlike *Lesothosaurus*, *Heterodontosaurus* was capable of a small amount of sideways chewing by rotation of the lower jaw about its

159

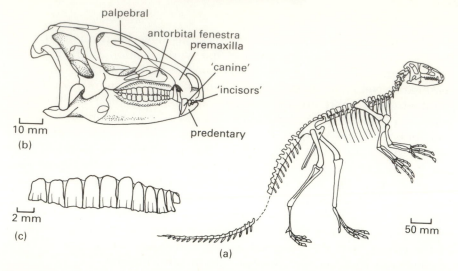

Figure 7.10 The heterodontosaurid *Heterodontosaurus*: (a) skeleton; (b) skull in lateral view; (c) maxillary tooth row. (Fig. (a) after Santa Luca 1980; (b) & (c) after Charig & Crompton 1974.)

long axis. The articular-quadrate joint at the back, and a special ball and socket joint at the front between the dentary and predentary, allowed rotation as the jaws opened and closed (Crompton & Attridge 1986). This mode of jaw rotation provided one solution to the problem of creating an efficient shearing scissor-like cutting movement between the cheek teeth (Fig. 7.11a).

All later ornithopods adopted the other option, of rotating the maxilla, in order to achieve lateral shearing, and this adaptation is said to lie at the root of the tremendous success of the ornithopods in the Cretaceous (Norman & Weishampel 1985). The hypsilophodontids, 'iguanodontids', and hadrosaurids have essentially fixed lower jaws that simply move up

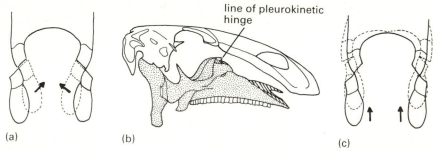

Figure 7.11 Ornithopod jaw mechanics: the lower jaws of *Heterodontosaurus* (a) slide outwards as they close, hence producing a kind of 'chewing', while later ornithopods have a pleurokinetic hinge, which allows the cheek portion of the skull and the maxillary teeth, shown stippled in (b), to move outwards as the jaws close (c). (Figs (a) & (c) after Crompton & Attridge 1986; (b) after Norman 1984, courtesy of the Zoological Society of London.)

160

and down without distortion during chewing, while the side of the skull (maxilla, lacrimal, jugal, quadratojugal, quadrate), as well as attached palatal elements (ectopterygoid, palatine, pterygoid) flap in and out. This specialized **pleurokinetic hinge** (Fig. 7.11b) produces the same lateral shearing effect (Fig. 7.11c) as did the rotating mandible of the hetero-dontosaurids.

The hypsilophodontids, typified by *Hypsilophodon* from the Early Cretaceous of England (Galton 1974), but known from the Late Jurassic to the Late Cretaceous, ranged in length from 3–5 m or so. The bodily proportions (Fig. 7.12a) and the skull (Fig. 7.12b & c) are similar in broad outlines to those of *Heterodontosaurus*, except that the skull lacks tusks, and is narrower in the midline. The ventral view (Fig. 7.12c) shows the extent of the cheeks very clearly, represented by the broad area of the maxilla lying outside the tooth rows.

An early view of *Hypsilophodon* was that it lived in trees, grasping the branches with its feet, but the foot (Fig. 7.12d) was incapable of grasping, being a typical elongate running foot with hoof-like 'claws'. Further, the end of the tail is sheathed in ossified tendons which stiffened it and caused it to act as a stabilizer during running, as in *Deinonychus* (Fig. 7.8a). The overall limb proportions of *Hypsilophodon* are similar to those of a fast-moving gazelle, especially the very long shin and foot.

The well-preserved specimens of *Hypsilophodon* have allowed Galton (1969) to make a detailed restoration of the muscles of the hindlimbs. He analyzed the occurrence of muscle scars and processes on the bones, and compared these with dissections of birds and alligators in order to arrive at a convincng restoration of the muscles (Fig. 7.12e). These muscles have complex names that record the bones to which they attach at each end. They fall into four groups that define their functions in walking:

(a) muscles that pull the femur forwards and up: ilio-femoris, pubo-ischio-femoralis internus (upper part);
(b) muscles that pull the femur back: pubo-ischio-femoralis internus (lower part), caudi-femoralis longus and brevis, adductor femoralis;
(c) muscles that extend the lower leg: ilio-tibialis, femoro-tibialis;
(d) muscles that pull the lower leg back: ilio-fibularis, flexor-tibialis-internus.

During a single step all of these muscles came into play. As the leg swung forwards, muscles of group (a) pulled the femur forwards and upwards and the muscles of group (c) extended the lower leg. The foot touched the ground, and the power stroke in which the body moves forward was achieved by muscles of groups (b) and (d) which pulled the femur and lower leg back respectively.

The 'iguanodontids' are a paraphyletic group representing stages of the acquisition of advanced hadrosaurian characters (Sereno 1986). *Iguanodon*

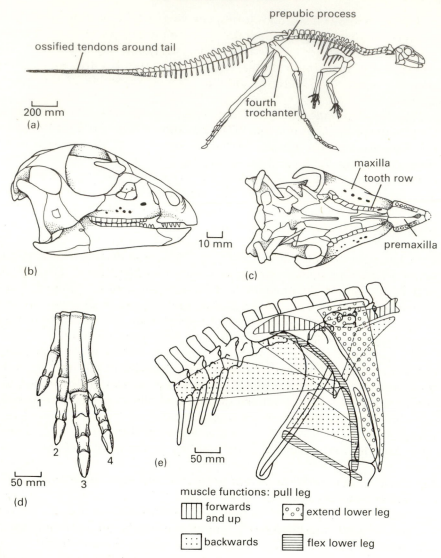

Figure 7.12 The ornithopod *Hypsilophodon*: (a) skeleton in running pose; (b) & (c) skull in lateral and ventral views; (d) foot in anterior view; (e) restoration of the muscles of the pelvis and hindlimb, coded according to their functions. (After Galton 1974.)

from the Early Cretaceous of Europe (Norman 1980, 1986b) has a horse-like skull (Fig. 7.13a). In the skeleton (Fig. 7.13b), the prepubic process is expanded, the postpubic process is very short, and there is a complex lattice of ossified tendons over the neural spines of all vertebrae of the trunk and tail. The most remarkable modifications are seen in the hand (Fig. 7.13c), in which the carpals and metacarpal 1 are fused to form a single block in the wrist, digit 1 is reduced to a thumb spike, digits 2–4

162

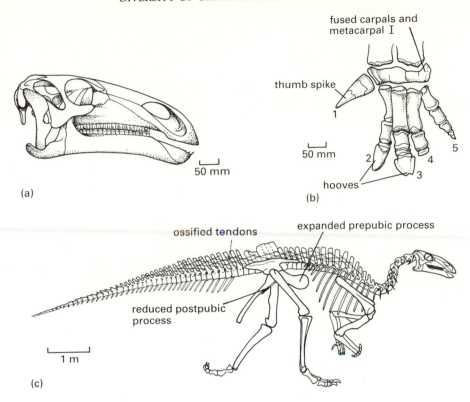

Figure 7.13 Ornithopod anatomy: (a) skull of *Iguanodon*; (b) skeleton of *Iguanodon* in running pose; (c) hand of *Iguanodon* in anterior view. (Figs (a) & (b) after Norman 1986b; (c) after Norman 1980.)

form a bunch, and digits 2 and 3 have small hooves. This hand was clearly used in walking (hooves) and in defence or display (thumb spike). *Iguanodon* could walk on all fours, or equally well on its hindlegs alone with the tail and the backbone extended horizontally.

The most diverse, and most successful, ornithopod clade were the hadrosaurs or 'duck-billed' dinosaurs of the Late Cretaceous. They are especially well-known from North America (Ostrom 1961), Central Asia, and China, where hundreds of specimens have been found. Frequently, three or four distinct hadrosaurian species are found side by side in the same geological formation, and it seems evident that large mixed groups roamed over the lush lowlands rather as closely related antelope do today in Africa.

Biology of the hadrosaurs

The hadrosaurs are famous for their expanded duck-like bills (Fig. 7.14a & b) in which both the premaxillae and maxillae are flattened and spread

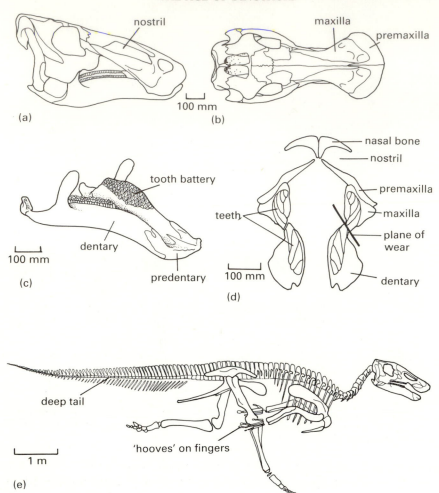

Figure 7.14 Hadrosaur anatomy: (a) & (b) skull of *Edmontosaurus* in lateral and dorsal views; (c) lower jaw of *Kritosaurus* seen at an angle to show the dental batteries; (d) cross-section through the snout of a hadrosaur to show patterns of tooth replacement; (e) skeleton of *Anatosaurus* in running pose. (Figs (a) & (b) after Norman 1984, courtesy of the Zoological Society of London; (c) & (d) after Ostrom 1961; (e) after Galton 1970a.)

out to the sides. The nostrils are long and low, and the orbit and lower temporal fenestra are pushed back. The teeth of hadrosaurs consist of long rows of grinding cheek teeth set well back from the front of the mouth, and arranged in closely-packed batteries within the jaws (Fig. 7.14c). There may be as many as five or six rows, each containing 45 or 60 teeth which are 'budded' at the bottom and move up progressively to the jaw margin where they come into wear. The wear surfaces can be seen in a cross-section through a hadrosaur skull (Fig. 7.14d) as sloping downwards and outwards. As the lower jaw closes, the cheek region of

164

the skull moves outwards on the pleurokinetic hinge, and the plant food is ground with a strong sideways shearing movement. In addition, the jaws move back and forwards a little, giving a further grinding action. Only the top rows of teeth are in use at any time, but they must have worn down quite rapidly since there are so many back-up teeth below ready for use.

This advanced and evidently powerful plant-grinding jaw system (Weishampel 1984b) may be one reason for the success of the hadrosaurs. But what did they eat? Some hadrosaur specimens have been 'mummified', preserved with their skin and some internal parts intact. These include stomach contents such as conifer needles and twigs, as well as remains of other land plants, which suggests that the hadrosaurs were terrestrial browsers that stripped trees of their foliage by stretching up on their hindlegs.

The hadrosaur skeleton (Fig. 7.14e) is interpreted (Galton 1970a) as adapted for efficient running with the body held horizontally as in other bipedal dinosaurs. The hands bear small hooves on the fingers, so they could also be used in slower locomotion. However, certain skeletal features were used by earlier authors as evidence that the hadrosaurs actually lived like ducks in the water most of the time: the deep tail could be used in swimming, the hands and feet are paddle-like in shape, and may have been webbed, and the duck-like beak was used in dabbling for soft plant food. These ideas are largely speculative and contradicted by other evidence, as already noted.

Hadrosaurs all have essentially the same skeletons and skulls (Fig. 7.14a & e), but some have an impressive array of headgear. The premaxillae and nasal bones extend up and backwards to form in some a high flat-sided 'helmet', either low or high, square or semicircular, in others a long 'tube', spike, or forwards-directed rod (Fig. 7.15a).

It has long been realized that the nasal cavities extended from the nostrils and into these crests, but it was once assumed that they acted as 'snorkels', especially in *Parasaurolophus*. However, this is impossible since there is no opening at the top of the crest. There are four separate air passages within the crest (Fig. 7.15b), two running up from the nostrils, and two running back down to the throat region. Air breathed in or out through the nose had to travel round this complex passage system.

The function of the crests has obviously given rise to a great deal of palaeobiological speculation: snorkels, salt glands, olfaction enhancers, foliage deflectors? The most likely explanation (Hopson 1975) is that the crests were used as visual species and sexual signalling devices. Just as modern birds use colourful and often elaborate patterns of feathers to recognize potential mates, and to signal their position in dominance hierarchies, so too did the hadrosaurs, but with cranial crests. Seemingly, males and females of the same species had rather different crests (Fig. 7.14c & d). Further, Weishampel (1981) has shown that the hadrosaurs

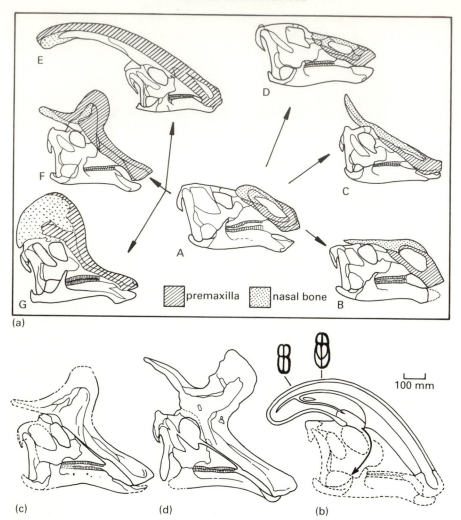

Figure 7.15 Hadrosaur skull evolution: (a) several lines of hadrosaur evolution from a crestless form (A, *Kritosaurus*) to crested genera (B, *Brachylophosaurus*; C, *Saurolophus*; D, *Edmontosaurus*; E, *Parasaurolophus*; F, *Lambeosaurus*; G, *Corythosaurus*); (b) internal structure of the crest of a possible female *Parasaurolophus*, showing the complex passages within the premaxillae and nasals, the passage of air (arrow) and two cross-sections through the crest; (c) & (d) sexual dimorphism in hadrosaurs; probable female (left) and male (right) *Lambeosaurus*. (Modified from Hopson, 1975.)

augmented their visual display with an auditory one too. The shapes of the air passages within the crests, particularly those of forms like *Parasaurolophus* (Fig. 7.14b), are very like musical wind instruments. A powerful snort would create a low resonating note, and the shape of the air passages in males and females, and in juveniles, would give a different note. Species differences would have been even more marked. We can imagine the Late Cretaceous plains of Canada and Mongolia reverberating to deep growls and blaring squawks as the hadrosaurs went about their business!

Infraorder Pachycephalosauria

The pachycephalosaurs, a small group of mainly Late Cretaceous herbivores from North America and central Asia (Maryánska & Osmólska 1974), are characterized by their remarkably thick skull roofs (Fig. 7.16a). The parietal and frontal bones are fused into a great dome in some forms with the bone up to 0.22 m thick in a skull that is 0.62 m long! This great thickened mass of bone is ringed by the normal skull roof elements as well as two supplementary supraorbital elements. Several of the skull bones are further ornamented by lines of bony knobs. Galton (1970b) speculated that the pachycephalosaurs used their thickened heads in butting contests when seeking mates, as is seen today among wild sheep and goats. The pachycephalosaur, a biped, adopted a horizontal-backbone posture during the charge (Fig. 17.6b) so that the force of the impact ran straight round the skull margins and down the neck to the shoulders and hindlimbs. This system of force dissipation was paralleled in the dinocephalian mammal-like reptiles (see p. 88). Confirming evidence for this theory is that males have thicker skulls than females.

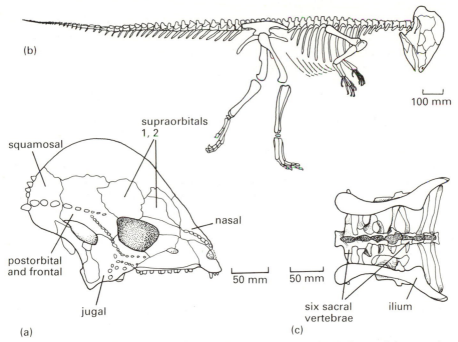

Figure 7.16 The pachycephalosaurs: (a) skull of *Prenocephale*: (b) skeleton of *Stegoceras* in butting position; (c) pelvis of *Homalocephale* in dorsal view. (Figs (a) & (c) after Maryánska & Osmólska 1974; (b) modified from Galton 1970b.)

Pachycephalosaurs are also characterized by an unusually broad pelvis (Fig. 7.16c) with gently curved iliac blades that contact the ribs of up to eight sacral vertebrae. This firm attachment of the pelvis may relate to the need to dissipate the forces of head-butting. Pachycephalosaur relationships are hard to establish, but they seem to be allied to the horned ceratopsians (Sereno 1986).

Infraorder Ceratopsia

The Ceratopsia (literally 'horned faces') are a larger group of about 20 genera known mainly from the Late Cretaceous of North America (Ostrom 1966), but with primitive representatives in the mid Cretaceous of Mongolia. All are characterized by a triangular-shaped skull when viewed from above (Fig. 7.17b), an additional beak-like **rostral bone** in the midline at the tip of the snout, a high snout, and broad parietals at the back. Some early ceratopsians were bipeds, while *Protoceratops* was a quadruped with the beginnings of a nose horn, a thickened bump in front of the orbit (Fig. 7.17a). It also shows the second major ceratopsian characteristic, a bony frill formed from the parietals and squamosals (Fig. 7.17b). The frill probably formed the origin of part of the jaw adductor muscles, the posterior adductor mandibularis muscle, which would have produced a strong biting force.

The later ceratopsians have a skeleton with adaptations for galloping (long limbs, digitigrade posture) (Fig. 7.17c). The vertebrae of neck and trunk have high neural spines for the attachment of powerful muscles to hold the head up, and there are bundles of ossified tendons over the hips. The real variation is seen in the skulls: some forms like *Centrosaurus* (Fig. 7.17c) have a simple horn formed by fused nasal bones, while others have this and a pair of 'horns' on the jugals. The frill may be short or long, and indeed *Torosaurus* had a 2.6 m long skull in which the frill is longer than the skull itself, the largest skull known from any land animal. The frills and horns may have been used in defence and as visual species-signalling structures as well as in threat displays. Male ceratopsians may have engaged in head wrestling with the horns interlocked, just as deer do today.

Infraorder Stegosauria

The thyreophorans, the truly armoured ornithischians, radiated in the Middle Jurassic, and include stegosaurs and ankylosaurs. Typical stegosaurs, such as *Stegosaurus* from the Late Jurassic of North America (Fig. 7.18), have low, almost tubular skulls. The hindlimbs are much longer than the forelimbs (evidence of a bipedal ancestry), and the massive arched backbone supports large triangular bone plates which sit either in a double row (as shown here) or in a single row. The correct

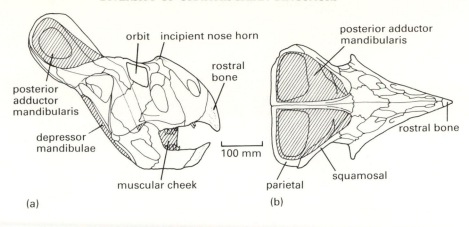

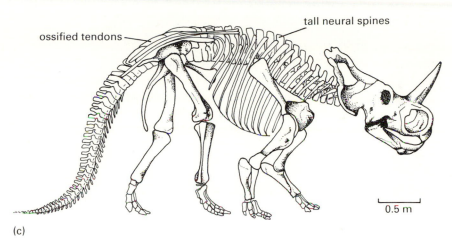

Figure 7.17 The ceratopsians: (a) & (b) skull of *Protoceratops* in lateral and dorsal views, with the cheek and major muscles restored; (c) skeleton of *Centrosaurus*. (Figs (a) & (b) modified from Ostrom 1966; (c) after Brown 1917.)

arrangement is unknown since these bony plates developed independently within the skin and did not meet the bones of the skeleton at all. They must have simply been rooted in the muscles of the back and held firm by massive ligaments.

What were these plates used for? The plate surface is covered by branching grooves which probably housed blood vessels in life, and they must have been covered by skin. Postulated functions for the plates include: (a) armour, (b) sexual display structures, (c) deterrent display structures, (d) thermoregulatory devices. Farlow *et al.* (1976) noted that the arrangement of the plates fitted engineering design models for heat-dissipation structures. As with the sails of the pelycosaurs (see pp. 85–6), *Stegosaurus* could have modified its body temperature by

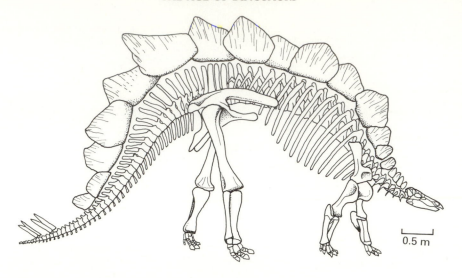

Figure 7.18   The stegosaur *Stegosaurus*. (After Gilmore 1914.)

adjusting the blood flow to the fins and its orientation to the wind. An overheated animal could cool down rapidly by pumping high volumes of blood over the plates and standing broadside on to the prevailing wind.

Infraorder Ankylosauria

Like the stegosaurs, the ankylosaurs (Coombs 1978) arose in the Middle Jurassic, but they are not well known until the Early Cretaceous. *Polacanthus*, a nodosaurid from southern England (Blows 1987), is a typical early form with a mixture of spiny plates along the flanks and a fused mass of smaller plates over the hips (Fig. 7.19a). The later ankylosaurids such as *Euoplocephalus* and *Ankylosaurus* (Fig. 7.19b–d) have broad armoured skulls and a body armour of plates rather than spines covering the neck, trunk, and tail. Ankylosaurids also have massive bony bosses at the ends of their tails, formed by the fusion of the last caudal vertebrae and the incorporation of bony plates from the skin (Fig. 7.19c). A blow from this club would readily disable *Tyrannosaurus* or any other contemporary predator!

The ankylosaur skull (Fig. 7.19d) is a heavy box-like structure with massive overgrowths of the normal bones of the skull roof by a mosaic of new bone plates generated within the skin over the head. These cover the upper temporal fenestra in all genera, and the lower one in most. Only a small orbit and nostril remain, and even they are heavily overgrown.

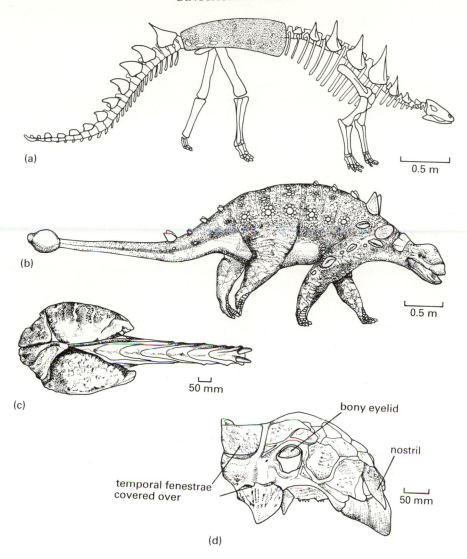

Figure 7.19 The ankylosaurs: (a) *Polacanthus*; (b) *Euoplocephalus* body restoration; (c) *Ankylosaurus* tail club; (d) *Euoplocephalus* skull in lateral view. (Fig. (a) after Blows 1987; (b) after Carpenter 1982; (c) & (d) after Coombs 1978.)

DINOSAURIAN BIOLOGY – WARM-BLOODED OR NOT?

The dinosaurs were clearly a successful and diverse group. How can that success be explained? One view could be that the Mesozoic was an era during which there was little competition and the dinosaurs, heavy and lumbering as many were, managed well enough living their slow reptilian lives. The other view, accepted by many Victorians, was that the dinosaurs were a vital warm-blooded lot, active and powerful, even if inferior, in the long-run, to the mammals. This debate continues today, and dozens of papers have been published in the past 20 years debating the likely thermoregulatory physiology of the dinosaurs (a representative collection of papers may be found in Thomas & Olson 1980).

Robert Bakker in particular has argued strongly (Bakker 1972, 1975, 1980, 1986) that all dinosaurs were fully warm-blooded, just like living birds and mammals, and that this explains their success. He was arguing that the dinosaurs were **endotherms**, animals that control their body temperature internally, rather than **ectotherms**, which rely only on external sources of heat. His arguments are set out briefly here, with critical comments presented by other authors.

The evidence
Erect gait and high speeds
Dinosaurs had an erect stance and advanced gait compared with most of their predecessors. Among living animals, only endotherms (birds, mammals) have erect gait, and Bakker suggested that this, and the supposed ability of dinosaurs to achieve fast speeds, indicated endothermy. However, there is no demonstrable causal link between endothermy and erect gait and the data on dinosaur running speeds are also equivocal. Estimates of speeds, based on fossilized trackways and limb dimensions, range from 6–60 km/h (1.5–7 m/s: Alexander 1976, Thulborn 1982). However, the higher speeds apply only to the smaller bipedal dinosaurs (35–60 km/h), and 40 km/h may be a more likely maximum. Larger dinosaurs were probably restricted to walking or slow trotting gaits and speeds of 10–20 km/h.

Haemodynamics
The long-necked sauropods must have had problems in pumping blood up their necks to supply the brain and face. It has been suggested that these dinosurs probably had to have a powerful four-chambered heart, a feature seen only in living birds and mammals, and that dinosaurs were thus endothermic. This correlation is not certain, not least since crocodilians have a four-chambered heart.

Activity levels
Many of the small lightly-built theropods were probably agile and fast-moving. However, living (ectothermic) lizards and snakes are capable of rapid movement, in short bursts at least.

Palaeoclimatology and distribution
Finds of dinosaurs within the Cretaceous Arctic Circle have been thought to indicate endothermy, since a typical reptile could not survive in cold

polar conditions (Bakker 1975). However, there is little evidence of glaciation in the Mesozoic, and climates at high latitudes were much warmer than today.

Predatory–prey ratios

Herbivores (whether endothermic or ectothermic) can support about 5% of their biomass of endothermic predators, and for ectothermic (reptile) carnivores, this predator–prey ratio is apparently nearer 30–50%. Bakker (1972, 1975, 1980) showed that predator–prey ratios for fossil populations dropped from 50–60% in the Early Permian to 10% in the Late Permian, and to 2–3% in Late Triassic, Jurassic, and Cretaceous dinosaur faunas (Fig. 7.20a). He interpreted this as strong evidence for dinosaur endothermy. There are many practical problems in calculating such ratios, and the ratios for large ectothermic predators closely approach those for endothermic predators (Fig. 7.20b). The ratios seem to vary with the size of the animals involved rather than simply with their thermoregulatory state.

Bone histology

Early work on the bone histology of dinosaurs showed that they had highly vascular bone, apparently very like that of mammals, but quite unlike the bone of lizards and other living reptiles. Many specimens of dinosaur bone show a vascular primary structure, and extensive secondary remodelling with the development of true Haversian systems

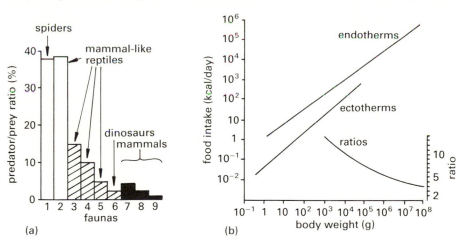

Figure 7.20 Thermal physiology: (a) predator–prey ratios for various invertebrate and vertebrate faunas; faunas 1 and 2 have ectothermic ratios, faunas 5–9 endothermic ratios, and faunas 3 and 4 intermediate values; (b) body weight versus food intake (which is equivalent to metabolic rate) for a variety of living animals; regression lines of ectotherms and endotherms are shown; the values converge at large body weights (declining ratio curve). (Fig. (a) based on Bakker 1975; (b) based on McNab 1978.)

173

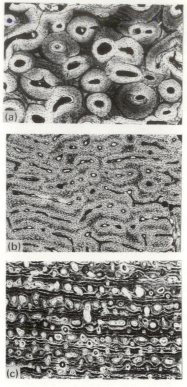

Figure 7.21 Dinosaur bone at high magnification: (a) Haversian bone tissue, showing secondary remodelling; (b) fibrolamellar bone; (c) lamellar-zonal bone, showing growth rings running vertically. (Photographs courtesy of Mr R. E. H. Reid.)

(Fig. 7.21). This was interpreted by Bakker (1972, 1980) as evidence for mammal-like endothermy in dinosaurs. However, true Haversian bone can occur in modern ectothermic reptiles, as well as in endotherms (Benton 1979, Reid 1984), and many small mammals and birds have no Haversian systems, despite having the highest metabolic rates found in endotherms. A second histological argument for dinosaur endothermy is based on the presence of fibrolamellar bone in many dinosaurs (Fig. 7.21b; Ricqlès 1980). This is a type of primary compact bone which grows quickly, without formation of growth rings, and it is now found in large fast-growing mammals (e.g. cattle) and some birds (e.g. ostriches). Modern reptiles instead have lamellar-zonal bone, which grows slowly and often intermittently, producing growth rings which are known to be annual in, for example, crocodilians. Fibrolamellar bone implies only fast growth rates, and not necessarily endothermy, but dinosaurs commonly also show lamellar-zonal bone (Fig. 7.21c), so a mixed thermoregulatory regime is suggested with a combination of fast and episodic growth rates.

Brain size

Most dinosaurs had relatively small 'lizard-like' brains, but the ornitho-mimids, dromaeosaurids, and troodontids had large 'bird-like' brains. This has been interpreted as evidence for endothermy in these dinosaurs at least (Bakker 1975), but the large brains are associated with good eyesight and balance, and they do not necessarily imply advanced mammal-like intelligence (Hopson 1977).

Birds and dinosaurs

Birds arose from theropod dinosaurs. *Archaeopteryx* had feathers, and was almost certainly an endotherm, and this has been taken to add weight to the argument that dinosaurs too were endothermic (Bakker 1975, 1980). However, this is merely suggestive and by no means conclusive evidence.

Inertial homoeothermy?

Most authors seem to have come to the view that the typical larger dinosaurs were neither 'reptilian' nor 'mammalian' in their physiology, but that they had some intermediate condition. Experiments on large living reptiles have shown that rates of internal temperature change are very slow during normal subtropical daily air temperature fluctuations. In living reptiles over 30 kg body weight, the rate of heat loss (thermal conductance) becomes equivalent to that of mammals (Fig. 7.22). By extrapolation, the body temperatures of medium- to large-sized dinosaurs living in similar climatic conditions would have remained constant to within 1 or 2°C inertially without internal heat production (Spotila *et al.*

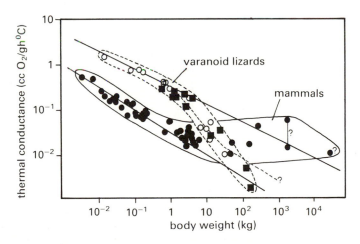

Figure 7.22 The effect of body size on thermal biology of varanid lizards (○), other reptiles (■), and mammals (●); as body size increases, thermal conductance values converge. (After McNab & Auffenberg 1976.)

1973). Dinosaurs probably achieved thermal constancy (homoeothermy) by large size, and their relatively high metabolic rates were probably produced by activity metabolism rather than by thermoregulation.

There are problems in taking this view too literally since the dinosaurs are such a varied group. Many were too small to have been inertial homoeotherms, and yet some of them, such as *Hypsilophodon*, are known to have had fibrolamellar bone which has generally been interpreted as an indicator of sustained high growth rates.

BABY DINOSAURS

Many reports of dinosaur eggs have been published, and it seems that all groups of dinosaurs laid eggs, from the small theropods to the giant sauropods. Most dinosaurs laid their eggs in nest-like structure that were dug in the sand or earth and then covered over for incubation, as in

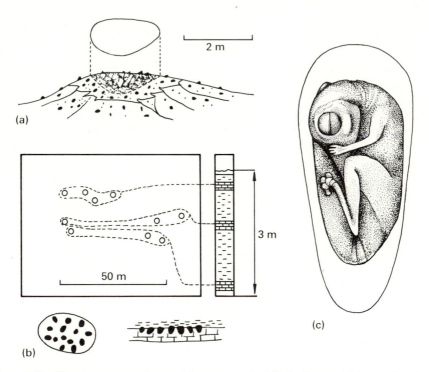

Figure 7.23 Dinosaur nests and eggs: (a) nest mound of the hadrosaur *Maiasaura* from Late Cretaceous sediments of Egg Mountain, Montana, in lateral view and plan view; (b) plan of ten neighbouring nest sites at three different stratigraphic levels, plan of a nest and section showing the eggs in black; (c) reconstruction of an embryo of the hypsilophodontid *Orodromeus* in a 170 mm long egg. (Fig. (a) after Horner & Makela 1979, copyright © 1979 Macmillan Magazines Ltd; (b) after Horner 1982, copyright © Macmillan Magazines Ltd; (c) based on a reconstruction by Matt Smith.)

present-day crocodilians. Recent work has suggested that dinosaurs may have exhibited parental care. Excavations of Late Cretaceous nests of the hadrosaur *Maiasaura* (Horner & Makela 1979, Horner 1982, 1984) have revealed a great deal about dinosaurian parental behaviour. Skeletons of 15 juvenile *Maiasaura*, each about 1 m long, were found around a nest mound which contained egg fragments (Fig. 7.23a). The juveniles have well-worn teeth, and are clearly not hatchlings, thus suggesting that they stayed together for some time after hatching.

A second nesting site nearby has yielded different eggs and small skeletons of hypsilophodontids. There are ten nests, all close together, but at different stratigraphic levels (Fig. 7.23b), thus suggesting site fidelity over many years (Horner 1982, 1984). The eggs are long and ellipsoid in shape, and they were laid upright with the narrow end downwards and partially buried. On hatching, the young left through the top portion of the eggs, leaving the lower halves intact within the sediment.

Some of the unhatched eggs from these sites have been dissected to reveal the tiny bones of embryonic dinosaurs (Horner & Weishampel 1988). The hypsilophodontid embryo just before hatching (Fig. 7.23c) would have been about 0.2 m long, while adults reached lengths of 2.5 m. The bones of the embryo hypsilophodontid seem to be well ossified, while those of *Maiasaura* have cartilaginous joint surfaces, which may indicate that the former was ready to fend for itself when it hatched, while the latter was not.

Juvenile dinosaurs have been reported more widely than embryos. Baby dinosaurs have big heads, short necks, and big feet, and the proportions change in a fairly regular way. One of the smallest baby dinosaurs, a young *Psittacosaurus* (Coombs 1982), is about 0.24 m long, compared to an adult length of 2 m (Fig. 7.24a). A sequence of juvenile to adult skulls (Fig. 7.24b) shows how the proportions changed and the especially characteristic ceratopsian features – the beak, high snout, small orbit, large lower temporal fenestra – progressively developed.

EVOLUTION OF THE PTEROSAURS

The pterosaurs (literally 'winged reptiles') existed for the same amount of time as the dinosaurs. They were important small piscivores in the Jurassic, and adopted a variety of ecological roles in the Cretaceous when some truly gigantic forms arose.

The first pterosaurs from the Late Triassic, such as *Eudimorphodon* from northern Italy (Wild 1978), show all the unique characters of the group (Fig. 7.25a): the short body, the reduced and fused hip bones, the five long toes (including a divergent toe 5), the long neck, the large head with pointed jaws, and the arm. The hand (Fig. 7.25b) has three short grasping

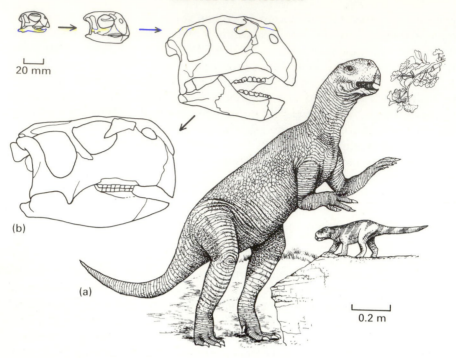

Figure 7.24 Development and growth of the early ceratopsian *Psittacosaurus*: (a) adult and juvenile reconstructed; (b) sequence of skulls from hatchling to adult, showing changes in proportion with growth. (Fig. (a) after restoration by Jonathan Sibbick in Norman 1986b; (b) after Coombs 1982.)

fingers with deep claws, and an elongate fourth finger which supports the wing membrane. In front of the wrist is a new element, the **pteroid**, a small pointed bone that supported a small anterior flight membrane which joined on to the short robust humerus (Fig. 7.25a). The pelvis (Fig. 7.25c) is a solid little structure with short blunt pubes and ischia. An additional element, the **prepubis**, is attached in front, and it may have had a function in supporting the guts. The tail is stiffened with ossified tendons as in many dinosaurs, and it may have been used as a rudder during flight.

The pterosaurs diversified remarkably in the Jurassic and Cretaceous (Wellnhofer 1978, Langston 1981), and much of this diversity can be appreciated by an examination of a selection of skulls (Fig. 7.26). Firstly the sizes vary remarkably from a length of 90 mm in *Eudimorphodon*, little larger than a pigeon, to 1.79 m in *Pteranodon*. These skulls also show some broad evolutionary changes (Fig. 7.26a–e): forward shift of the jaw joint to lie below the orbit, elongation of the skull, and fusion of the nostril and antorbital fenestra with reduction of the nasal bone. The long spaced teeth of *Rhamphorhynchus* (Fig. 7.26b) were probably used for piercing and holding fish which it caught by trawling its lower jaw

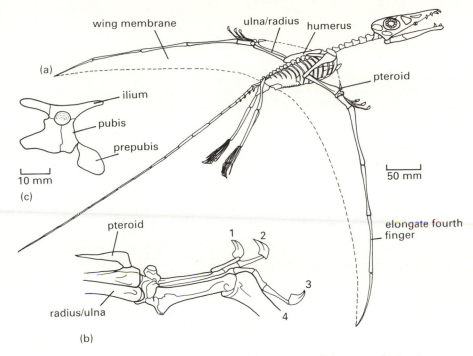

Figure 7.25 The first pterosaur *Eudimorphodon*: (a) skeleton in flying pose; (b) hand region of the right wing; (c) pelvis in lateral view. (After Wild 1978.)

through the seawater, whereas the shorter teeth of *Dimorphodon* (Fig. 7.26a) may have been used for insect-eating. *Pterodaustro* (Fig. 7.26c) has 400–500 flexible teeth in each jaw which were used to catch microscopic plankton. The teeth would have acted as a fine filter mesh in trapping thousands of small organisms which could be licked off and swallowed. The jaws of *Dsungaripterus* and *Pteranodon* (Fig. 7.26d & e) are deep and hatchet-shaped and bear very few, or no, teeth. These forms also probably fished by beak-trawling, and swallowed their catch so rapidly that no teeth were needed.

Pteranodon, one of the best-known and largest pterosaurs from the Niobrara formation of Kansas, USA (Eaton 1910), has a wingspan of 5–8 m. The skull is longer than the trunk (Fig. 7.26e), and its length is doubled by the pointed crest at the back which may have functioned like a weathercock to keep the head facing forwards during flight. Each massive cervical vertebra (Fig. 7.26f) has a pneumatic foramen in the side which led into open spaces inside, a weight-reducing feature. The dorsal vertebrae are nearly all involved in one or two heavily fused girder-like structures, the **notarium** and the **synsacrum** (Fig. 7.26g & h), which stabilize and support the shoulder girdle and pelvis. The shoulder girdle is attached to the side of the notarium above and to a large bony **sternum**

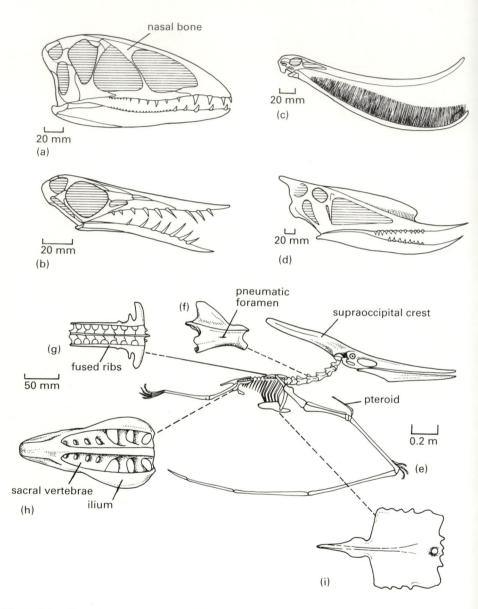

Figure 7.26 Diversity of pterosaurs (a) *Dimorphodon*; (b) *Rhamphorhynchus*; (c) *Pterodaustro*; (d) *Dsungaripterus*; (e)–(i) *Pteranodon*: (e) flying skeleton in lateral view; (f) cervical vertebra in lateral view; (g) notarium in dorsal view; (h) synsacrum in dorsal view; (i) sternum in ventral view. (Figs (a)–(d) after Wellnhofer 1978; (e)–(i) after Eaton 1910.)

(Fig. 7.26i) below, which holds the ribcage firm. The sternum bears a slight keel for the attachment of flight muscles. This massive stabilization of the shoulder girdle and pelvis is probably as much to do with the stresses of the landing impact as with flight stresses.

Pteranodon was not the largest pterosaur. That honour goes to *Quetzalcoatlus* from the late Cretaceous of Texas (where else?), which is represented by parts of a single wing, giving an estimated wing span of 11–15 m. *Quetzalcoatlus* was the largest flying animal of all time, three times the size of the biggest bird, and more like a small aeroplane in size than any familiar living animal.

PTEROSAUR FLIGHT

Pterosaurs have often been portrayed in the past as rather inefficient gliding animals that were incapable of flight. On the ground, their locomotion was supposed to be an awkward bat-like form of progression, consisting of staggering and tumbling on all fours. New studies (e.g. Wellnhofer 1978, Brower 1983, Padian 1984) have countered these views and presented a new picture of the pterosaurs being as well adapted to flapping flight as birds are. The first line of evidence is the possession of wings and all sorts of other aerodynamic and flight adaptations (hollow bones, reinforced landing gear, streamlined head). The second key aspect is that the pterosaurs were almost certainly endothermic, since they have hair, as is shown by several remarkably well-preserved specimens (Wellnhofer 1978). Only endotherms have insulation, and endothermy gave the pterosaurs the high sustained metabolic rates necessary for flight.

The wing is composed of skin that attached to the side of the body and along the entire length of the arm and of the elongated flight finger 4 (Figs 7.25a, 7.27a). Older reconstructions show a broad flight membrane, but well-preserved specimens have now shown that the pterosaur wing is a slender structure rather like that of a gull (Wellnhofer 1978, 1987), although in *Pterodactylus* at least the membrane also attaches to the femur (Fig. 7.27a). The wing membrane is reinforced with parallel stiff fibres or tight folds that indicate internal elastic fibres (Fig. 7.27b).

The pterosaur power stroke was directed down and forward, and the recovery stroke was an up and backward motion, so that the wing tip, viewed from the side, described a figure-of-eight shape. The downstroke was powered by the massive **pectoralis muscle**, and the upstroke by the **supracoracoideus muscle** (Fig. 7.27c & d) which ran from the sternum, over a pulley arrangement at the shoulder joint, to the dorsal face of the humerus. When it contracted, the supracoracoideus muscle, although placed below the wing, actually pulled it up, just as in birds (Padian 1984). Pterosaurs probably took off from trees or cliffs, or after a short run

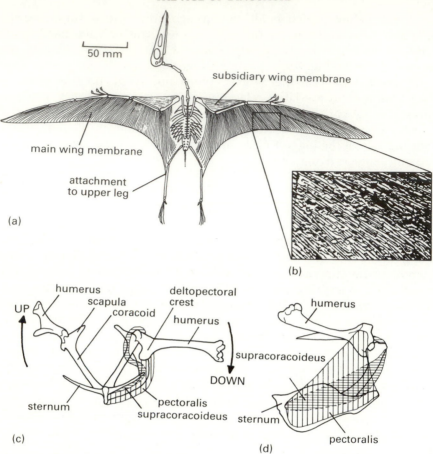

Figure 7.27 Pterosaur wings and flight: (a) skeleton of *Pterodactylus* with the wing membranes preserved, and showing partial attachment to the legs; (b) enlarged view of the fibres in the wing membrane; (c) & (d) anterior and lateral views of the shoulder girdle and humerus of a pterosaur showing the humerus in the upstroke and downstroke positions, and the main flight muscles (pectoralis, downstroke; supracoracoideus, upstroke). (Figs (a) & (b) after Wellnhofer 1987; (c) & (d) after Padian 1984.)

to pick up speed. Even in the larger pterosaurs, the take-off speed was low, possibly 4 m/s in *Pteranodon* (Brower 1983). Landing was probably awkward for the larger pterosaurs, just as it is for large birds, and the reinforced pelvis, sacrum and pectoral girdle would have had to withstand large impacts on occasion.

If there is relatively little controversy over the flying abilities of pterosaurs, there certainly is argument over how well they could walk. Padian (1984) argued that they could walk well on fully erect hindlimbs. He reconstructs the pelvic girdle of various pterosaurs as firmly fused beneath, and the limb motions just like those of a small bipedal dinosaur. The wings are held tucked horizontally beside the body during running.

Wellnhofer (1988a) and others have argued, on the other hand, that the pelvis is wide open at the bottom and that the hindlimbs point sideways in an awkward sprawling posture. During walking, the pterosaur would have to use the claws on its hands in a slow quadrupedal mode of locomotion with the wing tips sticking up on either side of the head.

EVOLUTION OF THE CROCODILIANS

Crocodilians are a small group today of eight genera of crocodiles, alligators, and gavials, that live in fresh and salt waters of the tropics. They have long snouts with the nostrils at the tip (Fig. 7.28a & b) so that they can breathe with only the nostril bump showing above water. There is a secondary palate formed from ingrowths of the maxillae, palatines, and pterygoids (Fig. 7.28c), which separates the air stream from the mouth cavity and allows the crocodilian to breathe with its mouth open underwater while feeding (Iordansky 1973). Crocodilians typically seize antelope and other mammals by a leg and drag them underwater until

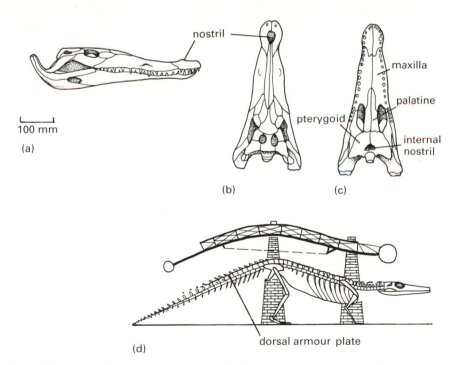

Figure 7.28 Crocodilian anatomy: (a)–(c) skull of the modern African crocodile *Crocodylus* in (a) lateral, (b) dorsal, and (c) ventral views; (d) mechanical analogy between the box-like girder structure of the crocodilian backbone and dorsal scutes, and a box-girder bridge. (Figs (a)–(c) based on Iordansky 1973; (d) after Frey 1984.)

they drown, and then tear off chunks of flesh by sinking their sharp teeth well into the flanks, and twisting with the whole body. In this way they are able to achieve much greater force for tearing at the meat than by simply twisting their heads from side to side (Taylor 1987).

On land crocodilians appear to be capable of four modes of locomotion:

(a) belly run, in which the body is pushed along like a toboggan by the hindlimbs only, for escape down river banks;
(b) sprawling, in slow locomotion, with the knees and elbows sticking out sideways;
(c) high walk, in which the limbs are tucked well under the body, for faster movement; and,
(d) galloping, the most unexpected mode, in which the forelimbs and hindlimbs act in pairs.

The skeleton of crocodilians does not seem to be well adapted for this last mode, galloping. It has been proposed that crocodilian backbones are braced in a way analogous to a box-girder bridge (Frey 1984). There is a double row of dorsal bony scutes in the skin which adhere closely to the backbone, and the vertebral column is braced by longitudinal muscle systems that attach to the dorsal armour over the back and tail (Fig. 7.28d).

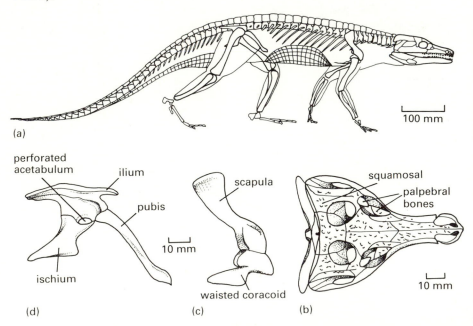

Figure 7.29 The early crocodilians (a), (c) & (d) *Protosuchus* and (b) *Orthosuchus*: (a) skeleton and armour plates; (b) skull in dorsal view; (c) shoulder girdle; (d) pelvic girdle. (Figs. (a), (c) & (d) after Colbert & Mook 1951; (b) after Nash 1975.)

The first crocodilomorphs such as *Terrestrisuchus* (see pp. 117–8) were lightly built and probably bipedal. The first true crocodilians, forms like *Protosuchus* (Colbert & Mook 1951) and *Orthosuchus*, appeared in the Early Jurassic. These small 1 m long animals are quadrupedal (Fig. 7.29a), but the hindlimbs are longer than the forelimbs, betraying the bipedal ancestry. The skull is ornamented with irregular pits in the bone surface (Fig. 7.29b), as in modern crocodilians, and the posterior part of the skull roof is square in outline because of the great overhang of the squamosals on either side. The squamosal bears a specialized ridge to which a fleshy 'ear lid' attached in life, a device to keep out the water during diving. There are additional palpebral bones in the eye socket, an independent evolution of bones also seen in some ornithischian dinosaurs (see p. 159). The whole posterior region of the skull is **pneumatic**, with complex air passages whose function is not clear. *Protosuchus* shows crocodilian characters in the skeleton as well: an elongate 'waisted' coracoid (Fig. 7.29c), a perforated acetabulum and reduced pubis (Fig. 7.29d), elongate wrist elements, and extensive armour covering. *Protosuchus* probably fed mainly on small terrestrial animals.

In the classification of the 150 or so genera of fossil crocodilians, most of those of the Jurassic and Cretaceous have classically been placed in a group called the 'Mesosuchia', mainly aquatic forms that lack the specializations of the living groups, the Eusuchia. The 'Mesosuchia' is however a paraphyletic group containing a great diversity of forms (Buffetaut 1982, Benton & Clark 1988; see box).

'MESOSUCHIAN' DIVERSITY

Geosaurus from the Late Jurassic of Europe (Fig. 7.30a & b) is 2–3 m long, and its skeleton is heavily modified for a wholly aquatic existence and swimming by powerful undulations of the body (Massare 1988). The caudal vertebrae bend down to support a tail fin, the limbs are paddle-like, and the body armour is lost which would improve the hydrodynamic efficiency of the body. It is likely that the geosaurs had difficulty in walking on land. Some long-snouted 'mesosuchians' lived on in marine and fresh waters through the Cretaceous and well into the Tertiary.

Other mesosuchian groups adopted a much more terrestrial lifestyle. In South America and Africa, several lineages became tiny and almost mammal-like in habits. For example, *Argentinosuchus* (Fig. 7.43c), less than 1 m in length, has differentiated teeth. The pointed teeth at the front may have been used in seizing prey, and the flatter 'cheek teeth' for cutting up the flesh. The Sebecidae, known from the Palaeocene to Miocene (60–10 Myr) of South America (Colbert 1946) have large skulls (Fig. 7.30d) with a high snout, no antorbital fenestra, and unusual flattened teeth. The sebecids were successful carnivores that probably preyed on mammals, but they were eventually replaced by mammalian carnivores in the later Tertiary.

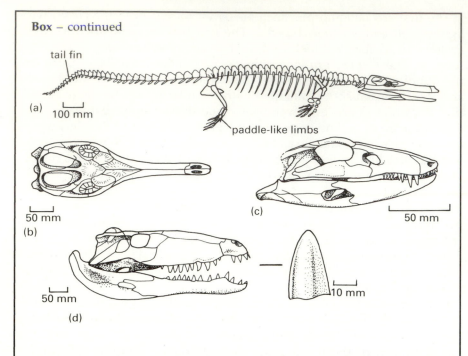

Box – continued

tail fin

(a)
100 mm

paddle-like limbs

50 mm
(b)

(c)

50 mm

50 mm
(d)

10 mm

Figure 7.30 'Mesosuchian' crocodilians: (a) & (b) the metriorhynchid *Geosaurus*, skeleton and skull in dorsal view; (c) the notosuchid *Argentinosuchus*; (d) the sebecid *Sebecus*, skull in lateral view and a characteristic flattened (ziphodont) tooth. (Figs (a) & (b) after Steel 1973; (c) based on Price 1959; (d) after Colbert 1946.)

The Eusuchia (literally 'true crocodilians') appeared in the Late Cretaceous, and most of the early representatives are very like modern forms. The group is distinguished from the 'protosuchians' and 'meso-suchians' by a full secondary palate formed from the maxillae, palatines, *and* pterygoids, as well as other characters (Benton & Clark 1988). In the Late Cretaceous and Tertiary, alligators, crocodiles and gavials were much more widespread than they are now, with dozens of species reported from Europe and North America as far north as Sweden and Canada, as well as all tropical regions and southern continents. The present array of crocodilians is a much reduced representation of their former glory.

HISTORY OF THE TURTLES

The turtles and tortoises, order Testudines or Chelonia, arose in the Late Triassic and later achieved a diversity of 25 families (Młynarski 1976). It seems that early on they hit on a successful design, the 'shell', and stayed with it.

The first turtles, *Proganochelys* and *Proterochersis* (Gaffney & Meeker 1983), show the key features that are common to all modern forms. The skull (Fig. 7.31a & b) is anapsid (no temporal fenestrae) and massively built, being firm and immovable. *Proganochelys* could no doubt have

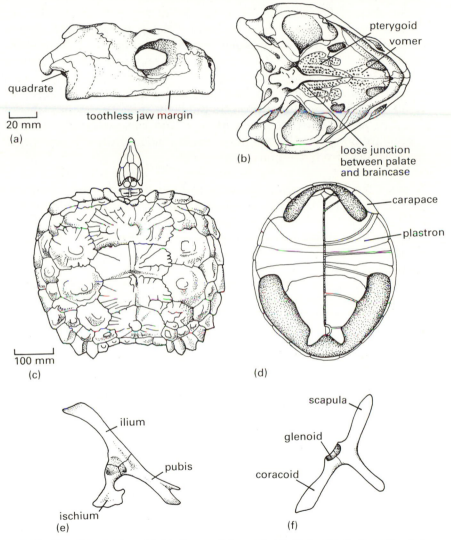

Figure 7.31 The first turtles, (a)–(c) *Proganochelys* and (d) *Proterochersis* from the Late Triassic of Germany: (a) & (b) skull in lateral and ventral views; (c) carapace and skull in dorsal view; (d) reconstructed plastron, showing the divisions between the bones (left) and between the horny covering scutes (right); (e) & (f) pectoral and pelvic girdles of a modern turtle. (Figs (a) & (b) after Gaffney and Meeker 1983; (c) after Jaekel 1915; (d) based on Młynarski 1976; (e) & (f) after Carroll 1987.)

187

survived a *Plateosaurus* playfully standing on its head! In side view (Fig. 7.31a), the skull shows two further turtle characters: toothless jaws and a deep curved embayment on the quadrate which supports a large ear drum. The palate (Fig. 7.31b) is primitive for turtles since it retains teeth on the vomer and pterygoid, and is linked only loosely to the braincase. In later forms, the teeth are lost, and the palate fuses firmly to the base of the braincase.

The shell of *Proganochelys* is composed of two portions, as in modern forms, a domed **carapace** on top and a flat **plastron** below, which are attached to each other at the sides, leaving broad openings at the front for the head and arms, and at the back for the legs and tail. The carapace is composed of bony plates that form within the skin, and these are covered by broad horny scutes in regular patterns (Fig. 7.31c). The plastron is a small unit (Fig. 7.31d) that protects the belly area. The main plates of the carapace are attached to the vertebrae and ribs, while the plastron is formed from expanded elements of the shoulder girdle and equivalents of the gastralia of other reptiles (see p. 77).

The shoulder girdle of modern turtles (Fig. 7.31e) is triradiate with two scapular heads, one facing upwards and one inwards, and a long narrow coracoid running back. The pelvis is smaller, but also three-pointed (Fig. 7.31f), with a narrow iliac blade running up and back, and a narrow pubis and ischium running forwards and backwards respectively. The limbs are short and held in a sprawling posture, and the hands and feet are large in swimming forms.

The remaining turtles from the Jurassic to the present day are classified into two major groups, the Pleurodira and the Cryptodira (Gaffney & Meylan 1988; see box).

PLEURODIRES AND CRYPTODIRES

Most turtles, unlike *Proganochelys*, can retract their heads under the carapace when they are threatened by danger, and the way in which they achieve this defines the two groups. The pleurodires pull the head in by making a sideways bend in the neck (Fig. 7.32a), while the cryptodires make a vertical bend (Fig. 7.32b).

Proterochersis from the Late Triassic of Germany (Fig. 7.31d) is classified as the first pleurodire because its pelvis is fixed to the carapace and plastron (Gaffney & Meylan 1988). Living pleurodires, the snake necks and matamatas, are freshwater in habitat and are limited to the southern continents. However, fossil forms are known from all continents and include terrestrial and possibly marine forms. The largest non-marine turtle, a 2.2 m long pleurodire from the Pliocene of Venezuela has been named *Stupendemys*.

The cryptodires date back to the Early Jurassic, but they radiated only

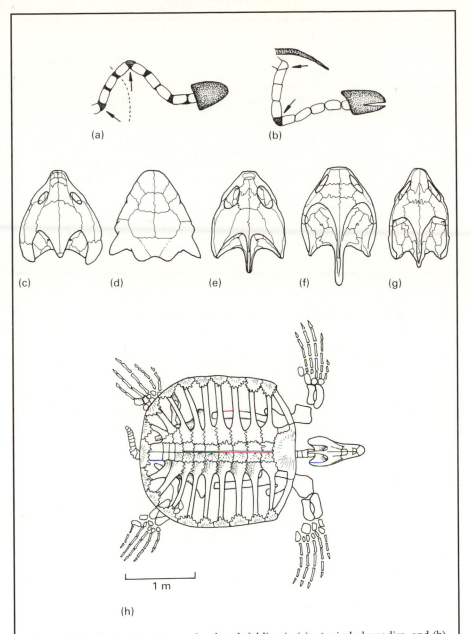

Figure 7.32 Turtle diversity: mode of neck folding in (a) a typical pleurodire, and (b) a cryptodire; (c)–(g) the diversity of turtle skulls, all in dorsal view: (c) *Eubaena*, a baenid; (d) *Meiolania*, a meiolanid; (e) *Toxochelys*, a chelonioid; (f) *Adocus*, a trionychoid; (g) *Mauremys*, a testudinoid; (h) the giant Cretaceous marine turtle *Archelon* in dorsal view. (Figs (a) & (b) after Młynarski 1976; (c)–(g) after Gaffney & Meylan 1988; (h) after Wieland 1909.)

Box – continued

after the Late Jurassic. They fall into five main clades, each characterized by features of the skull and shell (Młynarski 1976, Gaffney & Meylan 1988). The baenids (Fig. 7.32c) from the Late Jurassic to Eocene of North America and Europe have a narrow snout region. The meiolanids (Fig. 7.32d), an odd group mainly from the Pleistocene of Australia, have broad skulls up to 500 mm wide armoured with horns. The living cryptodires, the chelonioids (marine turtles), trionychoids (soft-shells), and testudinoids (tortoises), are distinguished from their extinct relatives by characters of the vertebrae and ribs. They also share a general skull outline (Fig. 7.32e–g) in which the parietals and supraoccipitals extend backwards as a vertical plate with a deep curved conch cut into the skull table on each side. The chelonioids have their four limbs modified as long paddles which they beat like wings to 'fly' through the water. Some, such as the leatherback, reach shell lengths of 2 m and weights of 500 kg, while *Archelon* from the Late Cretaceous of North America (Fig. 7.32h) is 4 m long.

SPHENODONTIANS – REPTILIAN 'LIVING FOSSILS'

Sphenodon, the living tuatara (Fig. 7.33a–c), is an unusual lizard-like animal known today only from some offshore islands in New Zealand. It reaches a length of 600 mm and it has nocturnal habits, feeding mainly on invertebrates. *Sphenodon* was originally classified as a lizard, but it is now regarded as the sister group of lizards and snakes (e.g. Carroll 1977, Evans 1984, 1988, Benton 1985a). *Sphenodon* is said to be a 'living fossil' since it lacks the special features of lizards and snakes (for example, the lower temporal bar is still complete and the skull is immobile) and since it is the single surviving member of a group known only much earlier in time.

The earliest sphenodontians are known from the Triassic when as many as eight or nine genera lived in Britain. These animals vary in body length from 150–350 mm, and their skulls and teeth vary, suggesting a variety of diets, including insectivory and herbivory. *Planocephalosaurus* (Fraser & Walkden 1984) is about 150 mm long (Fig. 7.33d), smaller than *Sphenodon*, and it has a blunt-snouted skull. The long slender limbs and body outline are very lizard-like, and indeed the Triassic sphenodontians show all the characters of the Lepidosauria – Sphenodontida + Squamata (i.e. lizards and snakes) – such as a broad opening in the pelvis between the pubis and ischium, the **thyroid fenestra**, a fused astragalus and calcaneum in the ankle, and a metatarsal 5 hooked in two planes (Fig. 7.33e). Later sphenodontians include bizarre Late Jurassic and Early Cretaceous forms from North America with broad grinding teeth, and probably also some aquatic forms.

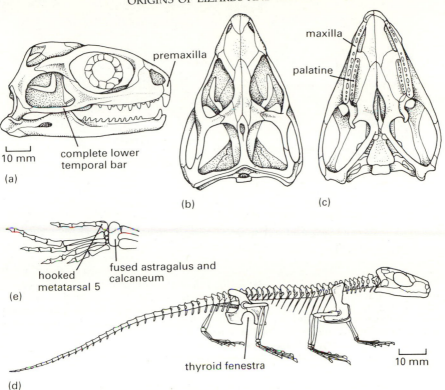

Figure 7.33 The sphenodontians: (a)–(c) skull of the living *Sphenodon* in lateral, dorsal, and ventral views; (d) skeleton of the Late Triassic sphenodontid *Planocephalosaurus*; (e) left foot and lower leg of the Jurassic sphenodontid *Homoeosaurus*. (Figs (a)–(c) after Carroll 1987; (d) after Fraser & Walkden 1984; (e) based on Cocude-Michel 1963.)

ORIGINS OF LIZARDS AND SNAKES

Early lizard fossils might be expected in rocks of Triassic or even Late Permian age. Indeed, a number of poorly preserved skeletons of small diapsid reptiles have been described as the first lizards (e.g. Carroll 1977, Estes 1983), but these have all turned out to lack sufficient clearcut characters of the Squamata (Evans 1984, 1988, Benton 1985a).

Lizards have an incomplete lower temporal bar which is associated with a high degree of skull kinesis, or mobility, of a special kind. Typical living lizards have three hinging systems (Fig. 7.34a & b).

(a) between the frontal and parietal in the skull roof and a matching joint in the palate, the **mesokinetic joints**;
(b) between the braincase and the skull (parietal, supratemporal, and quadrate), the **metakinetic joints**; and

191

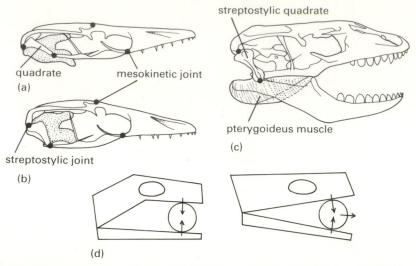

Figure 7.34 Lizard jaw mechanics: (a) & (b) skull of *Varanus*, showing the skull flexed (a) up and (b) down; (c) lizard skull with the jaws open, and the streptostylic quadrate swung back so that the pterygoideus jaw muscles have their maximum effect; (d) diagrammatic lizard skulls showing the advantages of kinesis in holding a food particle (left) which would otherwise be forced out by the bite in a non-mobile skull (right). (Figs (a) & (b) after Alexander 1975, courtesy of the Cambridge University Press; (c) after Smith 1980, copyright © 1980 Macmillan Magazines Ltd; (d) after Frazzetta 1986.)

(c) between the quadrate and squamosal at the top, and the pterygoid and lower jaw at the bottom, the **streptostylic joints**.

When the jaws open (Fig. 7.34a), the snout tips up and the quadrate is nearly horizontal. When the jaws close, the snout tips down and the quadrate becomes more vertical. This kinetic system has important adaptive advantages (Smith 1980, Frazzetta 1986). The pterygoideus muscle, which runs from the pterygoid to the outside of the lower jaw (Fig. 7.34c) is able to deliver a strong closing force to the kinetic lizard skull because of the rotations. The lizard jaws both effectively close on a food item at the same time, exerting equal perpendicular forces on it (Fig. 7.34d). With akinetic jaws there is a risk of losing a food item because the forces are not perpendicular and there is a force directed out of the mouth (Fig. 7.34e).

The oldest-known squamates are Late Jurassic in age, such as *Ardeosaurus* from Germany (Mateer 1982), which reached a total length of only 120–40 mm. The skeleton (Fig. 7.35a) is like that of most modern lizards, with a slender flexible body, long tail, and short sprawling limbs. The skull (Fig. 7.35b) shows a number of squamate derived characters: the parietals are fused and they meet the frontals on a broad transverse suture which can hinge up and down, the lacrimal and quadratojugal bones have been lost, and the quadrate is streptostylic. Another Late

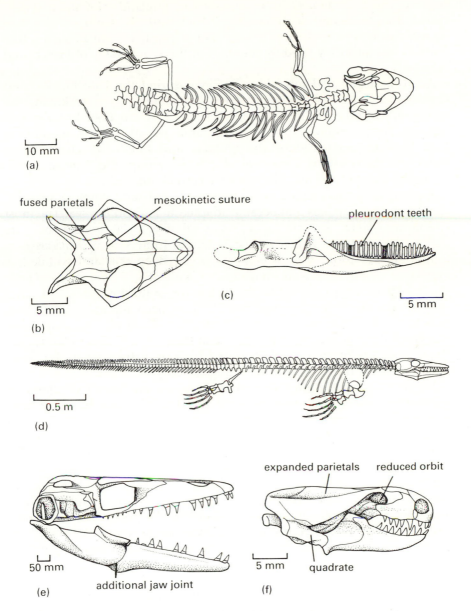

Figure 7.35 Fossil lizards: (a) & (b) skeleton and skull of the Late Jurassic gekkotan *Ardeosaurus* in dorsal view; (c) lower jaw of the Late Jurassic scincomorph *Paramacellodus* in internal view; (d) & (e) the Late Cretaceous mosasaur *Platecarpus*, skeleton in swimming pose and skull in lateral view; (f) the living amphisbaenid *Amphisbaena*. (Fig. (a) after Estes 1983; (b) after Mateer 1982; (c) after Hoffstetter 1967; (d) & (e) after Russell 1967; (f) after Romer 1956, courtesy of the University of Chicago Press.)

Jurassic lizard, *Paramacellodus* from southern England (Hoffstetter 1967), shows the **pleurodont** dentition (Fig. 7.35c), with the teeth set in a 'half groove', as is typical of most lizards. These peg-like teeth in such a tiny animal were probably used in penetrating the tough skins of insects and centipedes.

The lizards radiated into six main lines during the Late Jurassic and Cretaceous (Estes 1983, Rieppel 1988). *Ardeosaurus* is a gekkotan, the group which today includes the tiny geckos that can cling to walls and ceilings. The second main group, the Iguania, was also represented in the Late Jurassic, and includes the iguanas and tree-living chamaeleons today. *Paramacellodus* is a member of a larger group, the Scincomorpha, including today the skinks, European lacertids, and others.

The anguimorphs include living monitor lizards and limbless anguids, as well as three families that became adapted to a life in the sea. The most spectacular were the mosasaurs (Russell 1967), 20 genera of Late Cretaceous predators which ranged in length from 3–10 m. *Platecarpus*, a typical smaller form, has an elongate body, deep tail, and paddle-like limbs (Fig. 7.35d). Mosasaurs have heavy skulls (Fig. 7.35e), and the jaws are lined with sharp conical teeth, clearly for capturing fishes and other marine animals. Some ammonite shells have been reported which bear puncture holes that exactly match the tooth spacing of a mosasaur that has bitten them across, but failed to crush them. In addition to the typical lizard flexibility of the skull, mosasaurs have an extra joint in the lower jaw to increase the gape and the biting force.

The fifth squamate group, the amphisbaenisns, are heavily modified for a life of burrowing, with their heads reduced to miniature battering rams with which they force a passage through the soil. The front of the skull is tipped downwards and the whole structure is reinforced (Fig. 7.35e). The orbit is reduced, and the temporal bar has disappeared so that the back of the skull is largely the parietal fixed to an enlarged braincase and palate.

The sixth squamate group, the snakes (Serpentes or Ophidia), are believed to have arisen from 'lizard' ancestors, but the nature of those ancestors is a mystery (Gauthier *et al.* 1988c, Rieppel 1988). The main characters of snakes of course include limblessness (living boas still have a small remnant of a hindlimb), a greatly increased number of vertebrae (120–500), venom in certain advanced forms, and a great increase in skull kinesis.

The snake skull (Fig. 7.36a & b) is of very light construction, with several points of flexure. On opening, the palate moves forward, the fangs are erected, and the supratemporal-quadrate system enlarges the jaw joint two or three times. The snake then strikes at its prey, and passes it down its throat by moving the lines of backwardly pointing teeth on its maxillae, palatines, and pterygoids. These can be moved independently so that the prey is virtually stuffed down the throat and has no chance of escape. In advanced venomous snakes, the action of

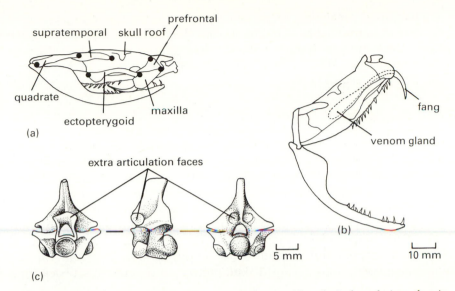

Figure 7.36 Snake anatomy: (a) & (b) the skull of a viperid snake in lateral view, showing the jaws closed and open; (c) mid-trunk vertebra of the living *Python* in posterior, lateral, and anterior views. (Figs (a) & (b) after Alexander 1975, courtesy of the Cambridge University Press; (c) after Rage 1984.)

striking at the prey squeezes a poison sac above the palate, and venom is squirted down a groove in the main fang.

Snakes are known from the Early Cretaceous onwards, and they radiated greatly during the Tertiary (Rage 1984) in line with the radiation of the mammals on which they preyed. Most of the early forms killed their prey by suffocation, as boas and pythons do today: they coil tightly around the ribcage of the victim, and tighten up when it breathes out. Death is by asphyxiation rather than crushing of the body, as is often assumed. The poisonous forms appeared first in the Late Eocene. Snakes range in length up to 6–7 m in a large python, but a huge vertebra from the Palaeocene of North Africa indicates a 9 m monster. Snake vertebrae have a complex shape (Fig. 7.36c) with extra processes on the sides of the neural arches which control the sideways and vertical bending of the body and give the snake considerable flexibility.

SEA DRAGONS

Jurassic and Cretaceous seas were filled with 'holostean' and teleost fishes and neoselachian sharks which preyed on them (Chapter 6). A broad range of predatory reptiles also hunted fishes, ammonites, belemnites, and other marine life. Pterosaurs and crocodilians seized fishes near the surface (see pp. 177–86), while mosasaurs (see p. 194) were important carnivores in the Late Cretaceous. Certain groups of birds also fed on marine fishes (Chapter 8), but the main sea reptiles were the ichthyosaurs and plesiosaurs, both of which groups had appeared in the Triassic (see pp. 113–7). After early finds, they came to be known collectively as 'sea dragons'.

Order Plesiosauria

The first true plesiosaurs are known from the very end of the Triassic, and they are believed to be closely related to the nothosaurs. Plesiosaurs were generally larger, ranging typically from 2–14 m in total body length, and they were highly adapted for submarine locomotion, with powerful paddle-like limbs and heavily reinforced limb girdles. There were three or four main groups (Brown 1981), some with long and some with short necks. The long-necked cryptoclidids (Fig. 7.37a & b) have over 30 cervical vertebrae, and a skull with a long snout, a single (upper) temporal fenestra, the euryapsid skull pattern, and a nostril set back from

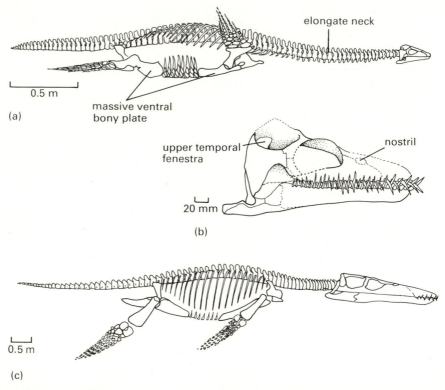

Figure 7.37 The plesiosaurs: (a) & (b) the Late Jurassic *Cryptoclidus*, skeleton in swimming pose and skull in lateral view; (c) the Late Jurassic pliosaur *Liopleurodon*. (Figs (a) & (b) after Brown 1981; (c) after Robinson 1975.)

the tip of the snout. The jaw joint is set below the level of the tooth row which shifts the strongest biting point forwards. The long pointed conical teeth interlock when the jaws are shut, an adaptation to retain slippery fishes in the mouth as the jaws close. The pliosauroids (Fig. 7.37c) have a long heavy skull and (usually) a relatively short neck. The massive pliosaurs, up to 12 m long, may have fed on other smaller plesiosaurs and ichthyosaurs.

Plesiosaur locomotion has been a subject of interesting speculation recently. Plesiosaurs have relatively small tails, but large and powerful paddles, so it has been assumed that the latter were used in creating thrust. Earlier scientists thought that plesiosaurs 'rowed' by simply beating the paddles back and forwards (Fig. 7.38a). The problem with this mode of locomotion would be that the backstroke could create a counter thrust that would cancel the forward motion to some extent since the paddles could be feathered (i.e. tipped to the horizontal), but not removed from the water. Robinson (1975) showed that the plesiosaur limb was more closely designed for underwater 'flying' (Fig. 7.37a, 7.38b), as in living sea turtles and penguins. The paddle is flat with an aerofoil cross-section like a bird's wing. If such a paddle pushes down and backwards at an angle above the horizontal, it generates lift and forwards

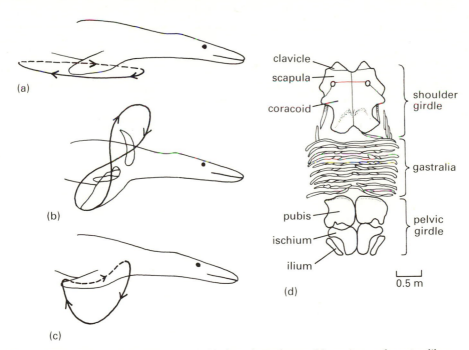

Figure 7.38 Plesiosaur locomotion: (a)–(c) three hypotheses: (a) rowing underwater like a duck, (b) flying underwater like a penguin, and (c) an intermediate style like a sealion; (d) ventral view of the heavy bony covering of the plesiosaur belly. (Figs (a)–(c) after Taylor 1986, copyright © 1986 Macmillan Magazines Ltd; (d) after Robinson 1985.)

thrust. In the upstroke, the paddle is tipped the other way, and a further smaller forwards lift force is generated. The paddle tip describes a figure-of-eight pattern and each stage of the cycle produces forwards movement.

Godfrey (1984) has proposed a modified version of the flying model, in which the paddle tip describes a crescent-shaped path (Fig. 7.38c) as in the sealion. His argument is based on the anatomy of the plesiosaur skeleton. Plesiosaurs could not have moved their paddles up and down in a figure-of-eight because the pectoral and pelvic girdles are both flattened heavy units of bone that form an immovable ventral bony plate with the gastralia between the limb girdles (Fig. 7.38d), and the limb girdles are too weak for strong vertical movements.

The short-necked plesiosaurs probably hunted their prey by pursuit since they could achieve quite fast speeds (Massare 1988), but the longer-necked forms were slower. They may have hunted by 'ambushing' fishes.

Order Ichthyopterygia

The ichthyosaurs of the Jurassic and Cretaceous are superficially very like those of the Triassic (see pp. 116–7). There is considerable variation in size, with body lengths of 1–16 m, but the dolphin-like body shape, long snout, and large eyes remain common features (Fig. 7.39a).

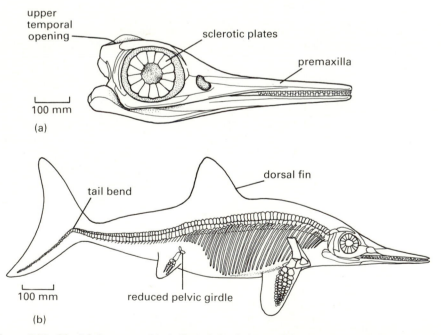

Figure 7.39 The ichthyosaurs: (a) skull, and (b) skeleton of the Early Jurassic *Ichthyosaurus*; the body outline is based on skin impressions preserved with some European material. (Fig. (a) after Andrews 1910; (b) after Stromer 1912.)

198

The body outline of ichthyosaurs (Fig. 7.39b) is well known because of the exquisite preservation of specimens especially in the Early Jurassic of southern Germany where they may show in some cases a black 'ghost' of the skin outline. This shows that the paddles were extended by skin and connective tissue, that the tail fin was roughly symmetrical, even though the vertebral column bends down, and that there was a high dorsal fin made entirely from soft tissues. Stomach contents include tiny hooklets from the arms of the squid and fish scales, but seemingly no belemnites or ammonites possibly because of their hard shells.

Ichthyosaurs no doubt swam efficiently (Massare 1988), and their adaptations to life in the sea were even more extreme than those of the plesiosaurs. The ichthyosaurs swam by beating their tails from side to side, and used their paddles to change direction and to control roll and pitch, as in fishes. The reversed heterocercal tail of ichthyosaurs (i.e. with a larger lower lobe) would have produced a forwards and slightly upwards propulsive force (Taylor 1987) by analogy with the present understanding of the heterocercal tail fin of sharks (see p. 128). The slight upwards force would counteract any sinking effect when the ichthyosaur was swimming at depth. Near the surface, movements of the forefins were used to counteract the effect of positive buoyancy, and slight adjustments to the flexibility of the tail and curve of the body would allow the ichthyosaur to dive rapidly.

The weakness of the limb girdles of ichthyosaurs, and their overall 'fishy' body shape suggests that they could not venture onto land. Marine turtles, and probably plesiosaurs, which spend most of their time at sea, do creep out onto a beach to lay their eggs. Ichthyosaurs, however, bore live young underwater as dolphins and whales do. Remarkable specimens from the Early Jurassic of Germany show two or three embryos within the ribcages of some specimens, and one or two actually show the young in the process of being born, tail first. The mother must have died while giving birth.

RELATIONSHIPS AND EVOLUTION OF MESOZOIC REPTILES

The terrestrial reptiles of the Triassic, Jurassic, and Cretaceous periods (Chapters 5 and 7) were mainly diapsids. In this account, the phylogeny of the Dinosauria will be presented first, and the other groups will then be added to a broader tree.

The Dinosauria (Fig. 7.40) consist of two main clades (Gauthier 1986, Benton 1990a), the Saurischia and Ornithischia, with a few primitive forms such as *Herrerasaurus* (see p. 120) at the base. The Saurischia fall into two main clades, the Theropoda and Sauropodomorpha. Within the Theropoda, a sequence of groups leads from the basal ceratosaurs to the birds (Aves), whose closest relatives are either the dromaeosaurids or the

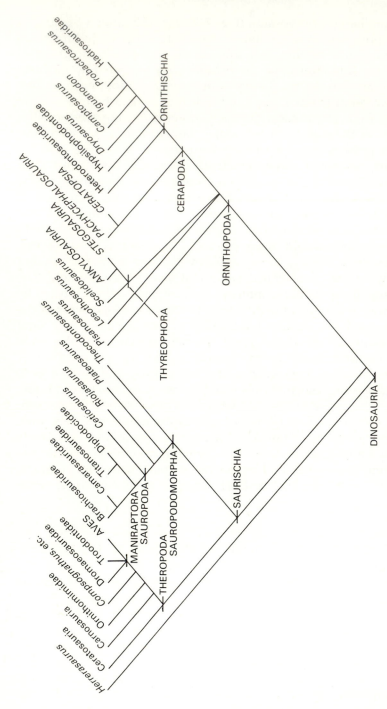

Figure 7.40 Cladogram showing the postulated phylogenetic relationships of the main groups of dinosaurs.

troodontids (Gauthier 1986). The Sauropodomorpha include a sequence of Triassic and Jurassic prosauropods and the four major families of giant sauropods.

The Ornithischia (Norman 1984, Sereno 1986) have a possible primitive member, *Pisanosaurus*, and two main clades, the Cerapoda and the armoured Thyreophora. The Early Jurassic *Lesothosaurus* could be related to either of these. The Cerapoda include a series of unarmoured bipedal ornithopods of the Jurassic and Cretaceous, leading to the hadrosaurs, as well as the horned ceratopsians and bone-headed pachycephalosaurs, which together make up the Marginocephalia. The Thyreophora consists essentially of the Stegosauria and the Ankylosauria with *Scelidosaurus*, an Early Jurassic form, as a primitive representative.

The dinosaurs appear to be closely related to the pterosaurs among other archosaurs, forming part of an ornithosuchian lineage (Fig. 7.41). Other archosaurs include the crocodilomorphs and the various Triassic thecodontians (Fig. 5.4). The archosaur branch of the Diapsida is part of a larger archosauromorph assemblage which branched off from the lepidosauromorphs in the Permian (Fig. 5.8). The main lepidosauromorph clade, the Lepidosauria, includes the sphenodontians and the squamates (five 'lizard' clades plus the snakes).

The four major groups of marine reptiles of the Mesozoic, the placodonts, ichthyosaurs, nothosaurs, and plesiosaurs are now generally reckoned to be diapsids, but their exact relationships are still uncertain (see p. 117).

The turtles and tortoises, Testudines, form a well-characterized clade that is separate from the Diapsida, and probably was separate since the Carboniferous (Fig. 4.21).

The evolution of all of these reptilian groups in the Mesozoic and Cenozoic (Fig. 7.42), shows two main phases of expansion, the radiation of the dinosaurs in the Late Triassic and Jurassic, and the radiation of the 'modern' groups (lizards, snakes, turtles) from the Late Cretaceous onwards. Mass extinctions among tetrapods have been postulated in the Late Triassic (see pp. 120–2) and, of course, at the end of the Cretaceous. Others, at the end of the Jurassic and in the mid-Cretaceous are not so clearly shown (Benton 1988, 1989a).

THE GREAT MASS EXTINCTION

As is well known, the dinosaurs died out 66 Myr ago at the Cretaceous–Tertiary (K–T) boundary. Over the years, hundreds of theories for this disappearance have been proposed, and yet none has gained general acceptance. There are several key problems in accounting for this extinction episode, not least of which concern its timing and the exact patterns of the disappearance of the dinosaurs and the other groups that also died out.

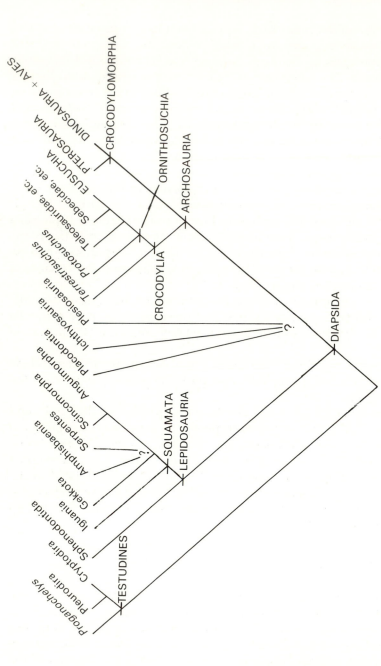

Figure 7.41 Cladogram showing the postulated phylogenetic relationships of the main groups of non-synapsid reptiles.

What died out?

Among terrestrial tetrapods, the dinosaurs and pterosaurs disappeared, as well as several families of birds and marsupial mammals, a total of 28 out of 89 families that existed in the latest Cretaceous (Benton 1988, 1989a). In the sea, plesiosaurs, mosasaurs, and a few families of teleost fishes disappeared. The ichthyosaurs had died out 30 Myr earlier. Among non-vertebrates, many important Mesozoic groups disappeared at about the same time, such as the ammonites, belemnites, rudist and trigoniid bivalves, and various plankton groups. Most plants and many animals, however, were apparently unaffected, such as gastropods, most bivalves, fishes, amphibians, turtles, lizards, and placental mammals.

It is hard to separate the survivors and non-survivors into simple ecological categories. Most of the land animals that survived were small, except for certain crocodilians. Most of the marine forms that died out were free-swimming or surface forms (plankton, ammonites, belemnites), but what about the fishes? Among forms that lived on the sea-bed, it was mainly the filter-feeders like corals, bryozoans, and crinoids that were heavily affected (possibly by loss of plankton food?), while forms that fed on detritus were little affected (Van Valen 1984). In general, however, there is little evidence for selectivity across the board at the K–T event, so that any explanation has to account for an essentially random set of disappearances.

How long did it take?

Some geologists assume that all major extinctions occurred essentially instantaneously, in as little as one week or one year. Others still posit a 'sudden' event, but allow several thousands or tens of thousands of years. At the level of discrimination that is possible, there is of course no way of distinguishing such time spans since both appear to be the same in the geological record. It is not possible to correlate marine and terrestrial rocks of the Late Cretaceous with an accuracy of less than 1–2 Myr.

The K–T boundary on land is marked by the disappearance of the last bones of dinosaurs, and by some pollen changes. In the sea, it is marked by changes in plankton and invertebrate fossils. Of course, these sorts of stratigraphic definitions lead to circular arguments when they are used to identify a mass extinction.

Other non-biological dating techniques can be used. Radiometric dating of certain rocks, particularly volcanic lavas, gives dates in Myr with uncertainties of ±0.5–4 Myr at that time. Another technique is to measure the polarity of magnetization of rocks. Every few Myr, the north and south poles flip over, and all iron-bearing minerals in rocks that are just being formed acquire the relevant magnetization. In the latest Cretaceous, the Earth's polarity changed eleven times, the K–T boundary lying in polarity band 29R (i.e. reversed), which lasted as little as 0.3 Myr.

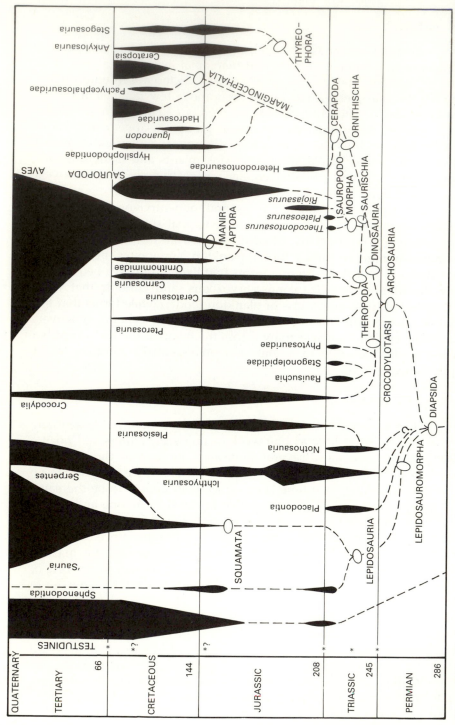

Figure 7.42 Phylogenetic tree of the diapsid reptiles and turtles, showing postulated phylogenetic relationships of the main groups (dashed lines), known fossil record of each (vertical time dimension), and their relative abundance through time (horizontal axis). Mass extinctions, and possible mass extinctions, are noted with asterisks on the left.

Magnetostratigraphic techniques can identify the likely age of particular geological formations, but the precision is still too poor for a decision on the exact duration of Late Cretaceous events.

The pattern of extinction

The ichthyosaurs were gone well before the end of the period, while the belemnites, ammonites, plesiosaurs, and pterosaurs had dwindled to very low numbers. The ammonites, so dominant throughout the Jurassic and Cretaceous, were represented by only nine genera at the end. What of the fossil reptiles?

For the dinosaurs, Sloan *et al.* (1986) argued that there was a long-term decline over 5–10 Myr (7.43). However, this study was based only on the dinosaur-bearing beds of Montana, USA and was not extrapolated to a global level, admittedly a task fraught with difficulties of stratigraphic correlation.

Sullivan (1987) found that 44 families of reptiles were in existence before the K–T boundary (Fig. 7.43), of which 22 survived it, 9 died out apparently at the boundary, and the other 13 died out well before. This represents a family extinction rate of 9/44 (20%) at the K–T boundary, which is a typical figure for vertebrates as a whole, or for marine animals during this and other postulated mass extinctions.

Nineteen 'families' of dinosaur are known from the last 20 Myr of the Cretaceous, and they had all disappeared by the K–T boundary. However, four of these 'families' are presently known only by single species from single localities. Of the remaining 15 families, two died out before the Maastrichtian, three in the early Maastrichtian (c. 72 Myr), two in the middle Maastrichtian (c. 70 Myr), and eight at the K–T boundary. These eight latest-surviving dinosaur families were represented by only 12 species in the late Maastrichtian (c. 70–66.4 Myr), the last 3 Myr of the age of the dinosaurs. These are (excluding odd teeth): the theropods *Albertosaurus*, *Avisaurus*, and *Tyrannosaurus*, the ornithopods *Edmontosaurus* and *Triceratops*, the pachycephalosaur *Pachycephalosaurus*, and the ankylosaurs *Ankylosaurus* and *Edmontonia*. This total of ten genera (12 species) is an unusually low figure, and contrasts with known totals of 40–80 species for equivalent time units earlier in the Late Cretaceous. Perhaps a significant decline of dinosaurs had taken place globally during the 10 Myr before the K–T boundary.

Theories of extinction

Over the years, hundreds of scientific papers have been written giving reasons for the extinction of the dinosaurs (Benton 1990b). A common view in the latter half of the nineteenth century and in the first three decades of the twentieth was that the dinosaurs simply died out because their time had come – they were described by many palaeontologists as

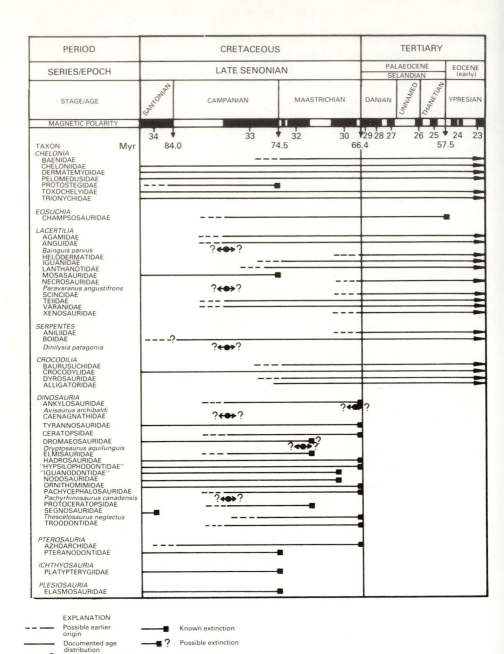

Figure 7.43 Range chart of reptilian families, and other taxa, that lived in the ten million years before the Cretaceous-Tertiary event, and just after it. The time-scale (top) shows major divisions, exact ages, and magnetic reversal patterns. The known fossil record of each group is shown by a solid line, doubtful intervals with a dashed line, and extinctions by a solid box. Solid circles indicate isolated occurrences, and arrows indicate the continuation of taxa after the Early Eocene. (Based on Sullivan 1987.)

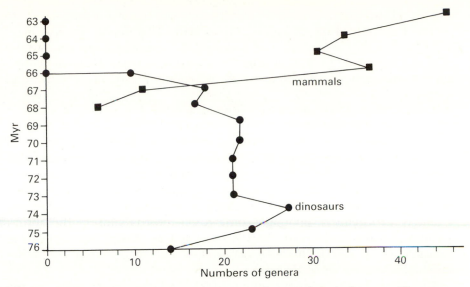

Figure 7.44 The relative abundance of dinosaurs and mammals in the last ten million years of the Cretaceous in the Hell Creek Basin of Montana. Total numbers of genera are plotted, and show a long-term decline of dinosaur diversity, and a rise in mammal diversity. (Based on data in Sloan *et al.* 1986.)

prime examples of racial senility – their genetic potential was exhausted, they exhibited giantism (if not acromegaly), excessive spinosity, and a loss of the ability to adapt. From about 1920, dozens of hypotheses were put forward, ranging from the physiological (slipped discs, excessive hormone production) to the ecological (competition with mammals, change in plant food), from the climatic (too hot, too cold, too wet) to the terrestrial catastrophic (vulcanism, magnetic reversal), from the topographic (marine regression, mountain building) to the extraterrestrial (sunspots, cometary impact). Many of these explanations were little more than whims, and most were hard to couch in terms that would allow them to be tested. Present hypotheses are more 'scientific'.

The two main current scenarios to explain the K–T event are the 'catastrophist' extraterrestrial model (Alvarez 1987) and the 'gradualist' ecological succession model (Van Valen 1984). The first model explains the disappearance of the dinosaurs as a result of the after-effects of a major extraterrestrial impact on the Earth and the evidence is essentially geochemical and astrophysical. The gradualist model sees a decline caused by long-term climatic changes in which the subtropical lush dinosaurian habitats gave way to the strongly seasonal temperate conifer-dominated mammalian habitats. The evidence for this hypothesis is mainly palaeontological and stratigraphic. A 'catastrophist' would envisage that the main extinction event lasted less than say a year, while a gradualist would regard the time-span as more than say 1000 years.

The catastrophic extraterrestrial scenarios postulate the impact of an asteroid, or a shower of comets, on Earth. The impact caused mass extinctions either by throwing up a vast dust cloud which blocked out the sun and prevented photosynthesis, by releasing cyanide, by flash heating of the atmosphere on entry, or by releasing poisonous arsenic and osmium.

There are two key pieces of evidence for the impact hypothesis, an iridium anomaly worldwide at the K–T boundary, and associated shocked quartz. Iridium is an element that reaches the Earth from space in meteorites, at a low average rate of accretion. At the K–T boundary, that rate increased dramatically, giving an iridium spike. Further, several sections have also yielded shocked quartz, grains of quartz bearing criss-crossing lines produced by the pressure of an impact. Other evidence is similarly geochemical, although some fossil records support the abrupt pattern. A catastrophic extinction is indicated by abrupt shifts in pollen ratios at some K–T boundaries, as well as abrupt plankton and other extinctions in certain sections.

The gradual ecosystem evolution model has been largely based on the occurrence of a mammal community (the *Protungulatum* community) of distinctive Palaeocene aspect up to 10 000–40 000 years before the K–T boundary in Montana (Sloan *et al.* 1986). This mammal community is devoid of dinosaurs, but dinosaurs occur in neighbouring sediments. The *Protungulatum* community spread gradually, and it seems to be correlated with the spread of temperate forest conditions in place of the lusher subtropical dinosaur habitats (Fig. 7.44). This model has been challenged on the basis of problems in exact correlation of the isolated mammal faunas. The gradualist scenario has been extended to cover all aspects of the K–T events on land and in the sea (e.g. Hallam 1987) especially in view of the gradual declines of so many groups throughout the Late Cretaceous. The climatic changes on land are linked to changes in sea-level and in the area of warm shallow-water seas.

Thus, while the geochemical data such as the iridium anomaly and the shocked quartz, are best explained at present by extraterrestrial causes, the patterns of extinction do not generally support an instantaneous catastrophe. Whether the two models can be combined so that the long-term declines are explained by gradual changes in sea-level and climate and the final disappearances at the K–T boundary were the result of impact-induced stresses is hard to tell. Although no plausible proximate causes of death have yet been proposed by the impacters to explain the known patterns from the fossil record, the timing is intriguing. No gradualist has yet been able to explain why the last stragglers of Dinosauria, Pterosauria, Ammonitida, Belemnitida, and others, should have finally given up the ghost apparently at the same time, unless some sharp catastrophe is invoked.

CHAPTER EIGHT

The birds

Birds are a large group of highly successful flying vertebrates, with nearly 9000 living species. The oldest bird, *Archaeopteryx* from the Late Jurassic, is known in some detail on the basis of several well-preserved specimens, but much of the later history of bird evolution is known only patchily. The present diversity of 155 or so families is not represented even in the fossil record of the last few million years: 47 modern families have no known fossil representatives. Nevertheless, some remarkable extinct bird groups are known, and the broad outlines of avian evolution are becoming clearer. A general account of bird evolution is given by Feduccia (1980), and papers in Hecht *et al.* (1985) give a good impression of the present debates on the origin of birds.

ARCHAEOPTERYX

Specimens and environment

Six skeletons of *Archaeopteryx* are now known, as well as a single feather impression (Fig. 8.1), all collected from the Late Jurassic limestones of Solnhofen, Bavaria. The first specimen to be found, a single feather, was collected in 1860, and the first skeleton with clear feather impressions, was named *Archaeopteryx lithographica* in 1861. The most famous example, the Berlin specimen, was found in 1877. It is a virtually complete skeleton, with the limbs and head in articulation, and the feathers of the wing and tail well preserved. Four more skeletons were collected in 1855 (recognized 1970), 1951, 1955, and 1987. The history of these specimens is described by Ostrom (1985), and several detailed monographs on their anatomy have been published (e.g. Wellnhofer 1974, 1988b, Ostrom 1976).

The six skeletons of *Archaeopteryx* were found in the Solnhofen Lithographic limestone, a fine sediment consisting of alternating layers of pure limestone and marly limestone containing clay. The pure limestones were quarried for the manufacture of printing blocks, hence their description as lithographic limestones. The limestones were deposited in

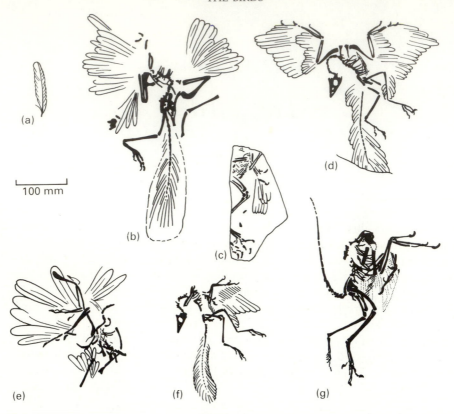

Figure 8.1 The specimens of *Archaeopteryx*, all drawn to the same scale, with the bones shown in black, and the feathers in rough outline. The commonly used specimen names, and dates of discovery are as follows: (a) Berlin/München 1860; (b) London 1861; (c) Haarlem 1855 (1970); (d) Berlin 1877; (e) Maxberg 1955; (f) Eichstätt 1951; (g) Solnhofen 1987. (After Wellnhofer 1988b.)

a subtropical lagoon and the fossils include marine or brackish-water forms (plankton, jellyfish, ammonites, crinoids, starfish, crustaceans, fishes), as well as terrestrial plant remains, insects, pterosaurs, crocodilians, sphenodontians, rare dinosaurs (*Compsognathus*), and *Archaeopteryx*.

The carcasses of *Archaeopteryx* appear to have drifted for some time at the surface, buoyed up by the gases of decomposition. Eventually, the guts burst and the carcasses sank rapidly to the bottom, where they lay essentially undisturbed. Most specimens lie on their sides with all limbs and other elements in articulation. The neck is always bent firmly back as a result of the contraction of strong muscles and ligaments.

Anatomy

Archaeopteryx is a medium-sized bird, 300–500 mm long from the tip of its snout to the end of its tail (Fig. 8.2a), and it may have stood about

210

250 mm tall, about the size of a magpie. The skull (Fig. 8.2b & c) is lightly built, and it may have been kinetic, with a movable quadrate (steptostyly), a bird feature paralleling that seen in lizards (see pp. 191–2). It is not certain whether the skull of *Archaeopteryx* was as kinetic as that of living birds, which can also move their beaks up and down relative to the rest of the skull (**prokinesis**). The lower jaw is narrow and delicate, and both jaws bear several small widely-spaced sharp teeth set in sockets. *Archaeopteryx* had large eyes and a bird-like brain with large optic lobes, which indicates that sight was a key sensory system.

Archaeopteryx has a curved neck, a short back, and a long straight tail with 22–3 caudal vertebrae. The shoulder girdle is lightly built, with a long narrow scapular blade and a short subrectangular coracoid. There are three fingers on the hand, and these are greatly elongated and bear long curved claws.

The pelvis is theropodan, but there has been some controversy over its

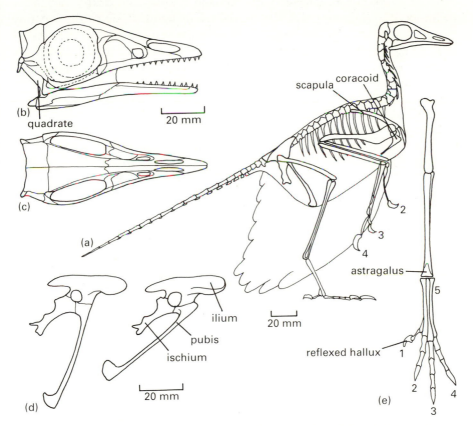

Figure 8.2 The anatomy of *Archaeopteryx*: (a) a skeleton in lateral view; (b) & (c) skull in lateral and dorsal views; (d) two reconstructions of the pelvis, the 'saurischian' (left) and the 'avian' (right); (e) the hindlimb in anterior view. (Fig. (a) after Yalden 1984; (b)–(d) after Wellnhofer 1974; (e) after Wellnhofer 1986; (f) after Wellnhofer 1988b.)

reconstruction, and in particular over the orientation of the pubis which may have run essentially vertically, as in some theropods, or backwards as in living birds (Fig. 8.2d). The hindlimb (Fig. 8.23e) is also like that of small theropods: the femur is short and slightly arched, the tibia is straight and the fibula very thin, the astragalus and calcaneum are firmly fused to the ends of the tibia and fibula, there appears to be an ascending process associated with the astragalus, the outer (5) toe is virtually lost, and the inner (1) toe is short and lies at the rear of the foot, the 'reflexed hallux' typical of many birds.

Relationships

The phylogenetic position of *Archaeopteryx* has been the subject of much controversy recently (Hecht *et al*. 1985). It is generally accepted that *Archaeopteryx* is the oldest-known bird, and that it is a form with a mixture of characters intermediate between reptiles and birds, the classic example of a 'missing link'. The reptile characters include teeth, separate fingers with claws in the hand, no ossified sternum, and the long bony tail, while the bird characters include feathers and a **furcula**, the fused clavicles, better known as the 'wishbone'. Recent reports of an older fossil bird from the Late Triassic of Texas (Chatterjee 1988) have yet to be confirmed, and indeed seem rather unlikely. There are four current hypothesis of bird origins:

(a) *The crocodilian-bird hypothesis*. Martin (1985) and others have argued that birds are most closely related to the early crocodilomorphs (see pp. 117–8) on the basis of shared characters such as streptostyly, air spaces in certain skull bones, and braincase features. However, most of these are common to other archosaurs or are hard to homologize precisely (Ostrom 1976).

(b) *The thecodontian-bird hypothesis*. Tarsitano & Hecht (1980) place the ancestry of birds among Triassic ornithosuchian thecodontians (see pp. 118–9). Some of these forms have generally bird-like proportions, but so far synapomorphies shared by them with birds have not been discovered.

(c) *The mammal-bird hypothesis*. Birds and mammals are endothermic, both groups have four-chambered hearts, advanced brains, and insulating skin coverings (feathers and hair) made from the protein keratin. Gardiner (1982) has enumerated these and 25 more shared characters to support his view that birds are not archosaurs, or even diapsids, but the closest living sister-group to the mammals. However, most of the shared characters are hard to defend when they are analyzed in detail (Gauthier *et al*. 1988b, Kemp 1988b).

(d) *The dinosaur-bird hypothesis*. Ostrom (1976) catalogued dozens of similarities in the skull and skeleton of *Archaeopteryx* and advanced theropod dinosaurs such as *Deinonychus*. Subsequently cladistic analyses (e.g. Gauthier 1986) fully support this view, and establish beyond all reasonable doubt that the birds are derived theropod dinosaurs (see p. 200).

FEATHERS AND THE ORIGIN OF FLIGHT

Archaeopteryx has feathers and wings, so it was clearly a flyer. However, Ostrom (1976) argued that *Archaeopteryx* could hardly fly at all since it lacks two bony elements that seem to be essential for flight in modern birds: an ossified keeled sternum for the attachment of the pectoralis and supracoracoideus muscles, used to produce the downstroke and the upstroke respectively (see p. 181), and the **acrocoracoid process** on the coracoid which provides a pulley round which the supracoracoideus operates, as also postulated in pterosaurs (see p. 181).

Two lines of evidence have been presented to show that *Archaeopteryx* was probably a good flyer. The pectoralis muscle could readily have originated from its robust furcula and the supracoracoideus muscle is not necessary for the recovery stroke of the wing. Bats, which are good powered fliers, have no keel on the sternum, and they also lack the acrocoracoid process. Secondly, *Archaeopteryx* has asymmetrical vanes on its feathers as in modern flying birds (flightless birds have symmetrical feathers), and the feathers are curved (Feduccia & Tordoff 1979). The asymmetry and the curve are necessary to allow the feathers to adjust aerodynamically to all stages of the wing beat.

The origins of bird flight must be entirely speculative. Numerous ideas have been aired, but there are two main current models, leaping 'from the ground up' or gliding 'from the trees down' (Fig. 8.3).

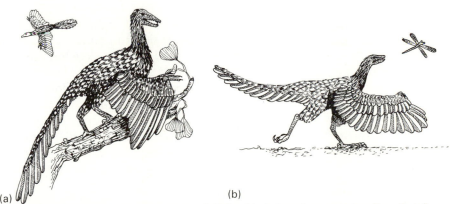

(a)   (b)

Figure 8.3 Two models for the origins of flight: (a) *Archaeopteryx* as a tree-liver that flew from branch to branch, and (b) as a ground-dweller that leapt up to catch insects. (Based on drawings in Rayner 1988.)

The leaping-up hypothesis has been developed from Ostrom's idea that *Archaeopteryx* was essentially a small running theropod dinosaur which used its feathered wings and tail as a kind of insect-catching system. Caple *et al.* (1983) presented an aerodynamic model in which the bird ancestors leapt into the air in pursuit of insect prey. Feathers and wings assisted and extended their leaps until eventually true powered flight evolved.

The gliding-down hypothesis is based on the older idea that *Archaeopteryx* was an arboreal form that could climb trees using the claws on its hands and feet. An ancestor initially leapt between branches, and then evolved to be able to leap and parachute between trees, or from the trees to the ground. In the end, gliding flight evolved into powered flapping flight. *Archaeopteryx* has hand claws that are well adapted for trunk climbing, but Rayner (1988) and others have presented aerodynamic arguments that gliding flight is pre-adapted for the evolution of flapping flight.

THE FLIGHT APPARATUS OF MODERN BIRDS

The Cretaceous record of fossil birds after *Archaeopteryx* is patchy, but there are enough good specimens to show that their flight apparatus advanced considerably towards the modern condition. In the forelimbs of modern birds (Fig. 8.4), the hand and wrist elements are greatly reduced, leaving essentially a single bony crank system which supports the feathers and forms the leading edge of the wing. Whereas *Archaeopteryx* has digits 2, 3 and 4 present and bearing claws, the modern bird has lost the claws, and retains only a splint-like remnant of metacarpal 2, metacarpal 4 is reduced to a slender element fixed to metacarpal 3, and digit 3 is the only one to retain phalanges. The carpals are fused to the metacarpals to form a simple hinge joint. The humerus has clearly defined processes at each end for the attachment of flight muscles, and a pneumatic foramen leading to an air space inside the bone.

The most dramatic modifications of the modern bird skeleton are seen in the shoulder girdle and sternum. There is a deep sternal keel which provides extensive areas of origin for the pectoralis muscle (downstroke) and the supracoracoideus muscle (upstroke) which insert on the lower and upper faces of the humerus respectively. The supracoracoideus runs over the acrocoracoid process on the coracoid which acts like a pulley. The sternum is a key element in the flight apparatus of modern birds, and it is stabilized by a long strut-like coracoid, very different from the small squarish element in *Archaeopteryx* (Fig. 8.2a).

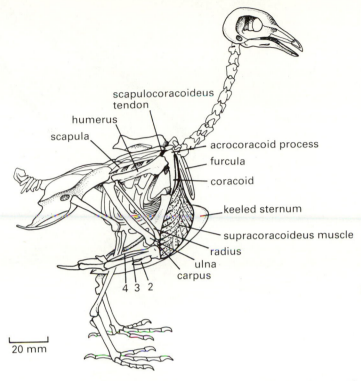

Figure 8.4 The skeleton of a typical modern bird, the pigeon *Columba*, showing the wing, and the supracoracoideus muscle that raises the wing by acting over the acrocoracoid process, a pulley-like system. (Based on Young 1981, and other sources.)

TOOTHED BIRDS OF THE CRETACEOUS

Some forms of toothed birds from the Late Cretaceous of North America are well represented by fossils, and these will be considered first. Other Cretaceous birds are less well known, and include supposed relatives of modern albatrosses and gulls (Olson 1985) as well as tantalizing primitive forms.

Order Hesperornithiformes

Hesperornis (Fig. 8.5a) is more than 1 m tall, and has a long neck, reduced tail, and long powerful legs. The forelimb is much reduced, being represented by only a pointed humerus that looks like a hat-pin. The remains of *Hesperornis*, and the related smaller *Baptornis* (Fig. 8.5b; Martin & Tate 1976), have been found abundantly in the Late Cretaceous Niobrara Chalk Formation of Kansas, USA which was deposited in the

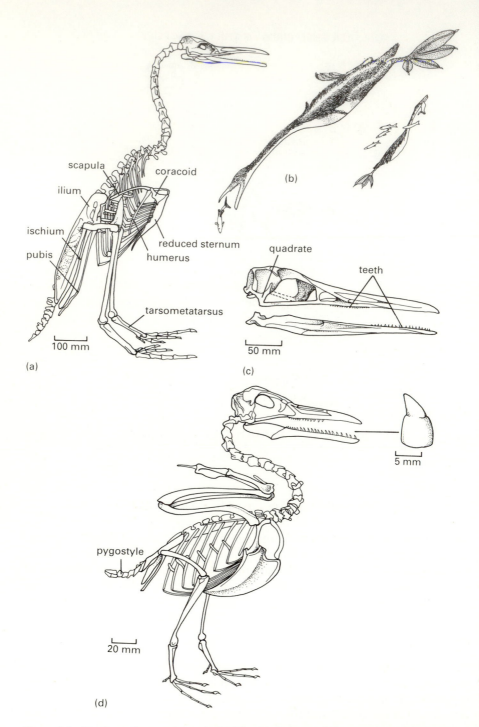

Figure 8.5 The Late Cretaceous toothed birds: (a) skeleton of *Hesperornis* in standing pose; (b) restoration of *Baptornis* swimming; (c) skull of *Hesperornis*; (d) *Ichthyornis* skeleton and tooth. (Figs (a) & (d) after Marsh 1880; (b) used by permission of the Smithsonian Institution Press from Martin & Tate 1976; (c) after Martin *in* Carroll 1987.)

shallow warm waters of the great sea channel that ran from north to south through North America at the time.

The hesperornithiforms were clearly flightless, and they are interpreted as foot-propelled divers that swam rapidly by kicking their feet. The toes are long and could spread widely. In life, they were probably linked by webs of skin or at least bore lobes to increase the surface area for swimming (Fig. 8.5b). The tiny wing stumps may have had a modest function in steering. The hesperornithiforms ate sea fishes, as is shown by coprolites, and parts of the jaws are lined with small pointed teeth.

Order Ichthyornithiformes

Ichthyornis, also from the Niobrara Chalk Formation of Kansas, is smaller than *Hesperornis*, being the size of a small gull (Fig. 8.5d). The wings are fully developed, and there is a deeply keeled ossified sternum, as in modern birds. The tail is more reduced than in *Hesperornis*, and the body is deeper. The head is large, and the massive jaws are lined with short pointed teeth set into a groove as in *Hesperornis*. *Ichthyornis* presumably caught fishes in the Niobrara sea by diving into the water from the wing, as terns do.

Relationships

Archaeopteryx is the basal bird, the sister group of the Ornithurae (Cracraft 1986, 1988) which includes all other birds (Fig. 8.6). These share characters not seen in *Archaeopteryx* such as the movable quadrate-quadratojugal joint (Fig. 8.5c), the reduced tail with the last few vertebrae fused together to form a **pygostyle**, an ossified sternum, fused pelvic bones with all three elements lying almost parallel (Fig. 8.5a), a reduced fibula, and fusion of the tarsals and metatarsals to form a **tarsometatarsus** (Fig. 8.5a). The hesperornithiforms are more primitive than *Ichthyornis* and modern birds, the Carinatae, since they lack the deep keel on the sternum (Figs 8.4 & 8.5d), the strut-like coracoid that attaches to it, and the much-modifed wing elements. The modern birds, termed the Neornithes (Fig. 8.6) are distinguished from the extinct groups by further derived characters (Cracraft 1988), including loss of teeth and a pneumatic foramen in the humerus.

The modern birds fall into two clades, the Palaeognathae, flightless ratites and tinamous, and the Neognathae, all other flying birds (Cracraft 1986, 1988). The palaeognathous palate (Fig. 8.7a) shows a large vomer firmly attached to the pterygoid, no joint between the pterygoid and the palatine, and a movable joint between the pterygoid and the base of the braincase. The 'neognathous palate' (Fig. 8.7b) is more loosely constructed and more mobile. The vomers are reduced or lost completely, there is a

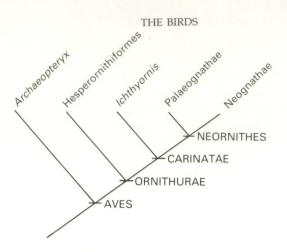

Figure 8.6 Cladogram showing the postulated relationships of the major groups of birds.

movable joint between the palatine and the pterygoid, and the pterygoid/braincase joint has been lost.

SUPERORDER PALAEOGNATHAE

The palaeognathous palate is primitive in most respects (Feduccia 1980, Olson 1985), but there are several synapomorphies (the extensive vomer-pterygoid joint, the elongate basipterygoid processes that meet the pterygoid) and in other parts of the skull (Cracraft 1988).

Palaeognath groups generally have short fossil records, extending back to the Miocene or Pliocene only, but a Late Cretaceous form has been reported from Argentina.

Modern palaeognaths fall into two groups, the tinamous, partridge-sized birds from South and Central America, and the ratites. The ratites include such well-known flightless birds as the rheas of South America, the cassowaries and emus of Australia, the kiwis of New Zealand, and

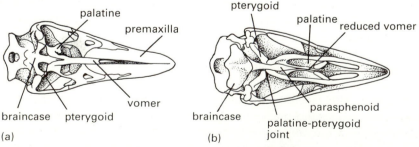

Figure 8.7 The (a) palaeognathus, and (b) neognathous palates, from a cassowary or rhea and a bronze turkey respectively. (Modified from various sources.)

Figure 8.8 Restoration of the giant flightless *Dinornis* from the subrecent of New Zealand. (Based on a Charles R. Knight painting.)

the ostriches of Africa. These all have reduced wings and they have lost the keel on the sternum, presumably having evolved from ancestors which could fly.

Some of the most spectacular ratites are now extinct, the elephantbird of Madagascar, and the moas of New Zealand. Both groups are known by subfossil bones and fossil bones no older than the Pleistocene. There were at least 13 species of moas (Fig. 8.8) which lived on both North and South Islands of New Zealand (Cracraft 1976), and these ranged in size from that of a turkey to heights of over 3 m. They fed on a variety of plants and, together with kiwis, flightless rails, ground parrots, geese, and others, formed unique communities in the absence of mammals. After the arrival of polynesian settlers 1000 years ago and the introduction of domestic mammals, the moas declined and eventually died out before 1775 at the latest.

SUPERORDER NEOGNATHAE

The neognaths, the majority of living birds, are characterized by features of the palate (Fig. 8.7b), as well as by a peculiar character of the ankle: the ascending process of the ankle bones which runs up in front of the tibia seems to have switched allegiance from the astragalus to the calcaneum.

The theropod ancestors of birds have a thin plate of bone which is attached to the enlarged astragalus and hugs the lower end of the tibia

219

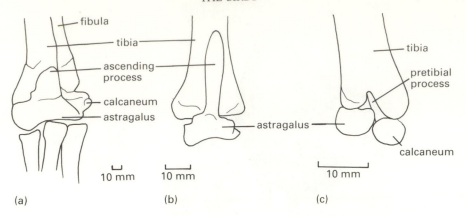

Figure 8.9 Anterior views of the tarsal regions of (a) the theropod dinosaur *Allosaurus*, (b) a juvenile ostrich, and (c) an embryonic chicken, to show the different origins of the ascending process. (Based on McGowan 1985, courtesy of the Zoological Society of London.)

(Fig. 8.9a). This process in ratites (Fig. 8.9b) has been interpreted as a new element called the **pretibial bone**. Hence, it could be argued that birds arose from some other source amongst the archosaurs, and that the ratites are degenerate neognaths. Embryological evidence (McGowan 1985), however, shows that the process in ratites is homologous with that of the theropod dinosaurs. The neognath ascending process (Fig. 8.9c) starts out as a small cartilaginous nubbin associated with the astragalus, but shifts to an attachment on the calcaneum before it ossifies.

The phlogeny of the 19 or so orders of modern birds has proved to be hard to establish. Cracraft (1988) has presented a tentative cladogram (Fig. 8.10) which will be followed here in outline (see box). These relationships are supported by studies of the anatomy of modern birds and, to some extent, by studies of the evolution of their protein and DNA structures (e.g. Sibley *et al.* 1988). Because of the diversity of the modern bird groups, and their generally poor fossil records, no attempt will be made here to catalogue them all (see Feduccia 1980, Olson 1985, Cracraft 1988 for fuller details).

NEOGNATH DIVERSITY

Ducks and swans date back to the Oligocene when some goose-like fossils appeared, while game birds date back to the Eocene. Feduccia (1980) and Olson (1985) regard the ducks as more closely related to the shorebirds on the basis of an Eocene fossil, *Presbyornis*, which supposedly has a duck's head and shorebird's body (Fig. 8.11a). However, *Presbyornis* shares no derived characters with either modern group, and Cracraft (1988) regards it as possibly a member of a completely extinct group. Unusual extinct relatives of the ducks and game birds include giant flightless plant-eaters from the Palaeocene and Eocene.

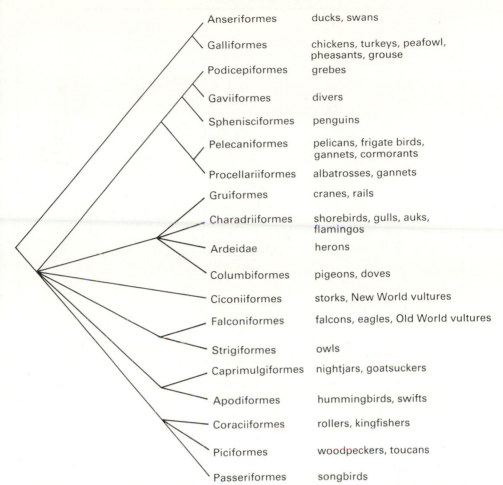

Anseriformes	ducks, swans
Galliformes	chickens, turkeys, peafowl, pheasants, grouse
Podicepiformes	grebes
Gaviiformes	divers
Sphenisciformes	penguins
Pelecaniformes	pelicans, frigate birds, gannets, cormorants
Procellariiformes	albatrosses, gannets
Gruiformes	cranes, rails
Charadriiformes	shorebirds, gulls, auks, flamingos
Ardeidae	herons
Columbiformes	pigeons, doves
Ciconiiformes	storks, New World vultures
Falconiformes	falcons, eagles, Old World vultures
Strigiformes	owls
Caprimulgiformes	nightjars, goatsuckers
Apodiformes	hummingbirds, swifts
Coraciiformes	rollers, kingfishers
Piciformes	woodpeckers, toucans
Passeriformes	songbirds

Figure 8.10 Cladogram showing the postulated relationships of the major groups of neognath birds. (Based on data in Cracraft 1988, and other sources.)

Box – continued

The grebes and divers are foot-propelled diving birds, superficially similar to *Hesperornis*, but still with the power of flight. The divers may date back to the Eocene, while the oldest grebe fossil is Miocene in age. The penguins have a rich fossil record with 25 genera dating back to the Eocene (Simpson 1975). They have completely lost the power of flight, but retain a deep keel and wings (Fig. 8.11b), which are used for 'flight' underwater. The diverse fossil penguins of the Oligocene and Miocene of New Zealand and Seymour Island, Antarctica, ranged in height from 0.3–1.5 m, but most are represented only by partial remains (Fig. 8.11c). Pelicans, frigate birds, gannets, and cormorants are all large fish-eating birds, with fossil representatives known from the Eocene. They have extensively webbed feet for swimming, and flexible throat pouches that allow them to hold large fishes. The related albatrosses and petrels include some very large birds,

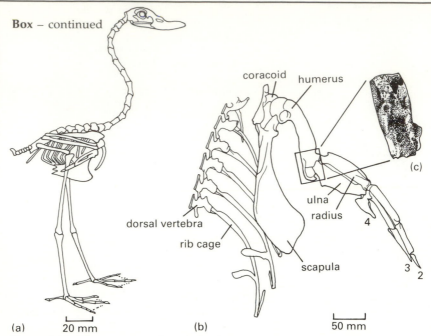

Figure 8.11 Diverse neognaths: (a) *Presbyronis*, a shorebird from the Palaeocene/ Eocene, of uncertain affinities; (b) the highly modified wing bones of a penguin in dorsal view; (c) distal end of the humerus of a giant fossil penguin. (Fig. (a) used by permission of the Smithsonian Institution Press from Olson & Feduccia 1980; (b) based on Van Tyne & Berger 1976; copyright © 1976 John Wiley & Sons Inc.; (c) based on Simpson 1975.)

with wingspans up to 3.5 m. Fossils are known from the Oligocene.

A third neognath clade is made up from a selection of wading birds and the pigeons. The cranes and rails date back to the Eocene, while the shorebirds, gulls, auks, and flamingos are said to have arisen in the Late Cretaceous (Feduccia 1980, Olson 1985). Most of the members of these two groups have long legs, and they seek food by wading in shallow water. The gulls are highly successful diving hunters, while the auks are wing-propelled divers, rather like penguins. The pigeons and doves seem very different, but they share some skull characters. Fossil pigeons date from the Eocene, but one of the most famous extinct forms is the dodo (Fig. 8.12a), a hefty flightless pigeon that was formerly abundant on the island of Mauritius. Sailors in the sixteenth century first discovered the tameness of these birds and they overcame their initial distaste for the 'hard and greasie' flesh. Specimens of this 'strange fowle' were exhibited in London in 1638, and a stuffed one was preserved in the Oxford University Museum. The last survivor was reported in 1681, and the Oxford specimen became so foul-smelling that it was burned in 1755.

Storks date back to the Oligocene. The oldest New World vultures are only Miocene in age, although ancestral forms from the Eocene and Oligocene of France have been reported. The largest flying bird, the condor, has a wingspan of 3 m, but the Pleistocene New World vulture *Teratornis* (Fig. 8.12b) has a wingspan of 4 m. Large mammals became trapped in the

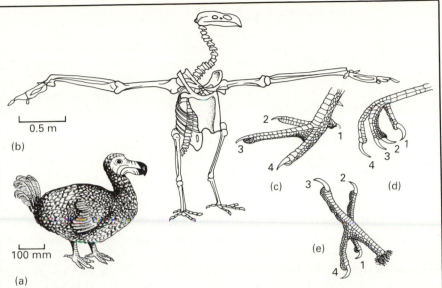

Figure 8.12 Diverse neognaths: (a) *Didus*, the dodo; (b) *Teratornis*, an extinct giant New World vulture; (c) the walking and scratch-digging foot of a pheasant; (d) the grasping foot of a sparrow; (e) the zygodactylous climbing foot of a woodpecker. (Figs (a), (c), (d) & (e) redrawn from various sources; (b) based on Van Tyne & Berger 1976, copyright © John Wiley & Sons.)

La Brea tar pits of California, and carrion-eating birds such as *Teratornis* and a dozen species of hawks and eagles were attracted to feed, some becoming trapped as well.

The main predatory land birds, falcons, eagles, Old World vultures, and owls, may form a clade. The oldest falcons are Miocene, while the oldest owl is Palaeocene in age.

The oldest known nightjars, goatsuckers, and swift-like birds are Eocene in age. All of these are basically insect hunters, the nightjars and goatsuckers being nocturnal and having large gaping mouths in which they engulf their prey. The hummingbirds, which feed on insects or nectar and include the smallest living birds (63 mm long), are unknown as fossils.

The kingfishers and rollers date back to the Eocene, the woodpeckers and toucans to the Miocene. The perching birds (passerines), consisting today of 5000 species of songbirds such as robins, thrushes, sparrows, crows, as well as flycatchers and antbirds, are known from the Oligocene, but most modern groups either lack a fossil record altogether, or are represented by sparse Miocene, Pliocene, or Pleistocene remains. These three groups are all perching birds in which the foot is specialized for grasping branches. In most birds there are three toes in front (numbers 2–4) and a small one (1) behind (Fig. 8.12c). This posterior toe is enlarged in passerines in order to help them grasp small branches (Fig. 8.12d). Some perching forms, such as the kingfishers and woodpeckers, also have the outer toe (4) pointing backwards as well to improve their grip, the **zygodactylous** condition (Fig. 8.12c).

GIANT HORSE-EATING BIRDS OF THE EOCENE

If tetrapod history had proceeded differently, the major carnivores on land today might have been giant birds. The radiation of mammals in the Palaeocene and Eocene after the extinction of the dinosaurs did not include any very large carnivores (see p. 265), and the phorusrhacids, probably related to the cranes and shorebirds, were able to achieve some diversity.

Phororhacos is 1.5 m tall (Fig. 8.13a), and its high-beaked skull shows bone-crushing adaptations. The phorusrhacids arose in Europe in the Eocene, and became well established in South America, where eight genera lived from Oligocene to Miocene times, some reaching heights of 3 m or so. *Andalgalornis* has been pictured (Fig. 8.13b) attacking a horse-like mammal of the Pliocene by seizing it with a huge clawed foot, and tearing the flesh with its powerful beak. The short wings and tail feathers would have helped it to balance.

These terror birds lived on in South America much longer than elsewhere, possibly feeding on larger prey than did the carnivorous

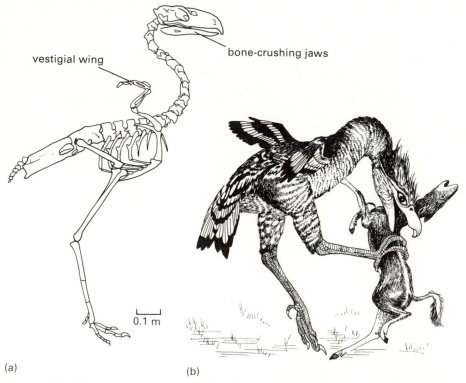

vestigial wing

bone-crushing jaws

0.1 m

(a)   (b)

Figure 8.13 Flightless carnivorous birds: (a) skeleton of *Phorusrhacus* (*Phororhacos*); (b) restoration of *Andalgalornis* attacking a small horse-like mammal. (Fig. (a) after Andrews 1901; (b) based on a restoration by Bonnie Dalzell.)

224

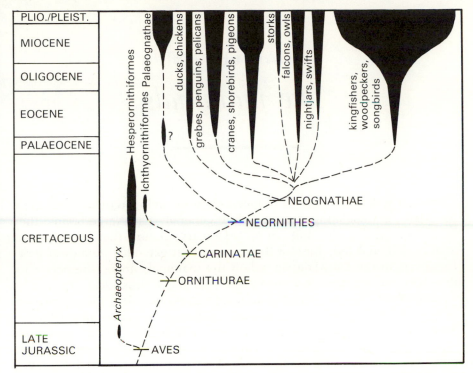

Figure 8.14 Phylogenetic tree of the birds, showing the relative importance of the different groups through time, their known fossil records, and postulated relationships. Abbreviations: PLEIST, Pleistocene; PLIO, Pliocene.

mammals of the day elsewhere. Some also entered North America in the Pliocene and Pleistocene. They were eventually replaced by predatory cats and dogs.

DIVERSIFICATION OF THE BIRDS

Birds have a patchy fossil record because their skeletons are generally rather delicate, and because many do not live near to lakes, rivers, and seashores, the sites of sedimentary rock deposition. Certain rare fossil formations, such as the Solnhofen limestone, contain beautifully preserved bird fossils, but many others are known only from scraps. Indeed most of the Tertiary record of birds consists of single bones, or small collections of fragments. However, certain elements, such as the humerus and the tarsometatarsus, are readily identifiable often to order or family level, so that incomplete remains can fill the gaps.

Overall bird diversity (Fig. 8.14) was low at first. Modest radiations in the Palaeocene and Early Eocene mark the origin of new neognath groups, while the later radiations in the Miocene and Pliocene reflect the burgeoning of passerine birds in particular.

CHAPTER NINE

The mammals

Mammals today are easy to distinguish from living reptiles. Mammals have hair, an insulating covering, they generally have large brains, they feed their young on milk from mammary glands (hence the name 'mammals'), and they care for their young over extended periods of time. However, in the Late Triassic, when the mammals arose, the boundary line between reptiles and mammals was much less clear. Indeed, a succession of Triassic carnivorous mammal-like reptiles, the cynodonts, progressively acquired 'mammalian' characters over a time-span of 30–40 Myr, and the exact point at which these reptiles became mammals can only be established by an arbitrary decision. The origin of the mammals is covered by Kemp (1982), Mesozoic mammals by Lillegraven *et al*. (1979), mammal evolution by Savage & Long (1986), and mammalian faunas by Savage & Russell (1983).

THE CYNODONTS AND THE ACQUISITION OF MAMMALIAN CHARACTERS

The cynodonts arose in the Late Permian, when forms such as *Procynosuchus* already showed mammalian characters in the cheek region and palate, and in the lower jaw. During the Triassic, a succession of cynodont families appeared, mostly weasel-sized to dog-sized carnivores, but including a herbivorous side branch. These may be treated as six major steps on the way to the origin of the mammals (Fig. 9.1), as argued by Kemp (1982, 1988a).

Thrinaxodon from the Early Triassic of South Africa and Antarctica shows several mammalian features: the zygomatic arch, formed from the jugal and the squamosal, forms a wide curve and bends up a little; the dentary makes up most of the lower jaw and it sends a high **coronoid process** up inside the zygomatic arch; the numbers of incisors are reduced to four above and three below; the cheek teeth are elaborated; and the secondary palate is nearly complete, formed by the fusion of the premaxillae, maxillae, and palatines.

226

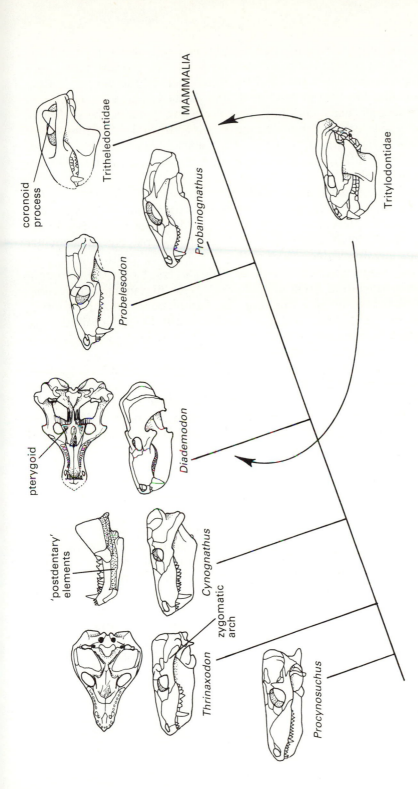

Figure 9.1 The evolution of the cynodont skull from the Late Permian *Procynosuchus* to the Late Triassic tritheledonts, tritylodonts, and mammals, showing postulated relationships. (After Kemp 1988a, and other sources.)

coronoid process

Tritheledontidae

MAMMALIA

Probelesodon

Probainognathus

Tritylodontidae

pterygoid

Diademodon

'postdentary' elements

Cynognathus

Thrinaxodon

zygomatic arch

Procynosuchus

The second stage is represented by *Cynognathus*, also from the Early Triassic of South Africa, but from younger beds. The dentary forms 90% of the length of the lower jaw, and the bones behind the dentary form a narrow rod that lies in a groove inside the dentary (Fig. 9.1). The cheek teeth have several cusps, and they show **occlusion** – that is, the surface of upper and lower teeth meet while food is being sheared – and they create **wear facets** which match exactly. This allows more effective food-processing by grinding and crushing during chewing, an impossible feat for a typical reptile.

The third stage was also achieved in the Early Triassic with the herbivorous *Diademodon* from the same beds as *Cynognathus*. In the palate (Fig. 9.1), the ectopterygoid and the lateral processes of the pterygoid are tiny. The cheek teeth of *Diademodon* occlude extensively, the smaller lower tooth forming a deep facet in the broad upper tooth.

The fourth stage was reached in the Middle Triassic in the chiniquo-dontid *Probelesodon* from Argentina (Fig. 9.1). New 'mammalian characters' seen in *Probelesodon* include the very thin bar of bone behind the orbit, made from the postorbital, and the long low **sagittal crest**, a raised ridge in the midline at the back of the skull roof. It provided enlarged sites of origin for the major jaw muscles.

Probainognathus from the same beds as *Probelesodon*, the fifth stage (Fig. 9.1), has a low zygomatic arch, additional cusps on the cheek teeth, and the beginnings of a second jaw joint. Incredible as it may seem, *Probainognathus* and some other advanced cynodonts have a double jaw joint. The sixth evolutionary stage was reached by the tritylodonts and tritheledonts.

JAW JOINTS, EARS, AND MAMMAL ORIGINS

In reptiles, the jaw joint is between the quadrate at the back of the skull, and the articular at the back of the lower jaw. In modern mammals, the jaw hinges on a new joint between the squamosal and the dentary. Intermediates are preserved in the fossil record to show how the transition happened. Basically, the four standard reptilian bones in the lower jaw behind the dentary were squeezed out of existence or into the middle ear. At the same time, the dentary sent its new coronoid process up inside the zygomatic arch where it formed a new contact with the squamosal which eventually became the sole jaw joint (see box).

JAW JOINT TO MIDDLE EAR

The jaw joint in *Thrinaxodon* (Fig. 9.2a & c) is the primitive quadrate-articular one, but the quadrate is much reduced. The surangular, just behind the coronoid process of the dentary, comes very close to the squamosal, and the stapes touches the quadrate.

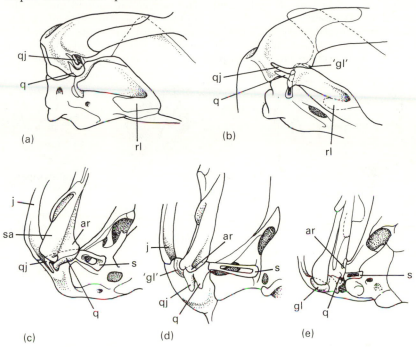

Figure 9.2 The evolution of the mammalian jaw joint: (a) & (b) posterolateral and (c)–(e) ventral views of the posterior right-hand corner of the skull and lower jaw of (a) & (c) the early cynodont *Thrinaxodon*, (d) the later cynodont *Probainognathus*, and (b) & (e) the early mammal *Morganucodon*, showing the move of the quadrate and articular towards the middle ear region. Abbreviations: ar, articular; gl, glenoid; j, jugal; q, quadrate; qj, quadratojugal; rl, reflected lamina; s, stapes; sa, surangular. (After Crompton & Hylander 1986.)

A few subtle changes in *Probainognathus* (Fig. 9.2d) mark the beginnings of the switch. The surangular now meets the squamosal in a special hollowed facet, the glenoid, which allowed rocking movements. The articular forms part of a narrow rod (including the reduced angular, prearticular, and surangular) which is loosely held in a groove on the inside of the dentary. By this stage the reptilian and the mammalian jaw joints are both present very close together, and apparently functioning in tandem.

The final stage, where the squamosal-dentary jaw came to dominate, is seen in *Morganucodon* (Fig. 9.2b & e). The surangular seems to have vanished, and its place is taken by a distinctive enlarged process of the dentary which fits into the glenoid on the squamosal.

Box – continued

In typical reptiles and birds, the eardrum, or tympanum, is a circular sheet of skin held taut in the curve behind the quadrate. Sound is transmitted to the inner ear within the braincase in the form of vibrations by the stapes, a rod of bone extending from the tympanum to the inner ear (Fig. 9.3a). In modern mammals, sound is transmitted via a set of three tiny bones within the middle ear, the **auditory ossicles**: the **malleus**, **incus**, and **stapes**, or hammer, anvil, and stirrup. The tympanum is held taut by the curved **ectotympanic** which sits just behind the squamosal-dentary jaw joint. The mammalian stapes is the same as the reptilian stapes, the malleus is the reptilian articular, the incus the quadrate, and the ectotympanic the angular. The reptilian jaw joint is present within our middle ear, and the

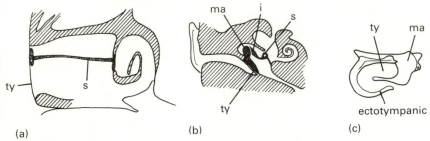

Figure 9.3 Structure of the ear, in vertical section, in (a) a typical reptile, and (b) a mammal; (c) the ear ossicles of a modern mammal in lateral view. Abbreviations: i, incus; ma, malleus; s, stapes; ty, tympanum. (Fig. (a) based on various sources; (b) & (c) after Hopson 1966.)

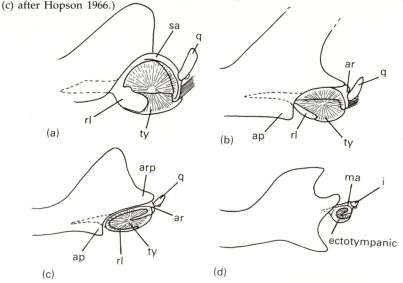

Figure 9.4 Allin's (1975) theory for the origin of the mammalian jaw joint and middle ear system; lateral views of the posterior portion of the lower jaw: (a) primitive cynodont; (b) advanced cynodont; (c) early mammal; (d) modern therian mammal. Abbreviations as in Figures 9.2 & 9.3; and: ap, angular process; arp, articular process. (After Allin 1975.)

close association explains why we can still hear our jaw movements when we chew.

The evolution of the cynodont and mammalian jaw joint and middle ear has been reconstructed as a four-stage process by Allin (1975). He assumes that the angular, articular, and quadrate bones were already involved in sound conduction in early cynodonts (Fig. 9.4a). He restores a large tympanic membrane beneath the dentary, held taut by the reflected lamina of the angular below and the surangular above. Vibrations of the tympanum passed through the articular and quadrate to the stapes. In other words, he regards the lower jaw as a key part of the hearing equipment of early cynodonts. The tympanum became smaller, and was pushed behind the new squamosal-dentary jaw joint in early mammals. At the same time, the articular-quadrate crank became reduced and separated from the rest of the skull and lower jaw, and moved fully into a separate auditory passage.

CYNODONT BIOLOGY

Cynodont jaw mechanics

The changes in jaw articulation and in the shape of the lower jaw and temporal region of the skull of cynodonts must have had profound effects on the action of the jaw muscles in feeding (Kemp 1982, Crompton & Hylander 1986). An early synapsid (Fig. 9.5a) had three main jaw closing muscles, the external adductor, the posterior adductor, and the internal adductor (which included the pterygoideus and other portions). In cynodonts and mammals, the internal and posterior adductors are much reduced, and two new muscles are derived from the external adductor (Fig. 9.5b), a deep **temporalis muscle**, and a superficial **masseter muscle**.

In more advanced cynodonts (Fig. 9.5c), the volume of adductor muscles is greater, as shown by the great outwards bowing of the zygomatic arch.

The shift in jaw articulation and the rearrangement of jaw muscles seen in the cynodonts and mammals allowed an important advance, namely chewing. We can move our jaws from side to side and back and forwards, and these actions are essential to the complex grinding activities of our cheek teeth (see box).

The strikingly mammalian jaw muscles and cheek teeth of cynodonts were associated with another major evolutionary step, the reduction of the number of cycles of tooth replacement. In reptiles and lower vertebrates in general, teeth are replaced more or less continuously as the old ones wear out. In placental mammals, on the other hand, there is only one replacement, when the milk teeth of the juvenile give way to the adult set. This step had seemingly already taken place in *Thrinaxodon* and *Diademodon*, and it was probably essential because of the complexity of cheek tooth occlusion.

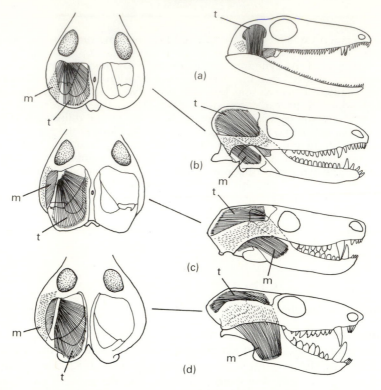

Figure 9.5 The evolution of cynodont jaw muscles: (a) lateral view of the skull of *Ophiacodon*, showing the small temporalis muscle; dorsal and lateral views of the skulls of (b) *Procynosuchus*, (c) *Thrinaxodon*, and (d) *Probelesodon*, showing progressive expansion of both muscles, and their invasion of larger and larger areas of the bones of the skull roof. Abbreviations: m, masseter muscle; t, temporalis muscle. (After Kemp 1982.)

The cynodont skeleton

The skeleton of cynodonts shows a number of major changes towards the mammalian condition (Jenkins 1971). *Thrinaxodon* has a double occipital condyle (Fig. 9.7a & c) as in mammals, while typical reptiles, including the pelycosaurs, have a single occipital condyle (Fig. 9.7b). The vertebrae of *Thrinaxodon* (Fig. 9.7a) also show mammal-like features. The dorsal vertebrae and ribs are divided clearly into two sets, the 13 **thoracic vertebrae** in front, and seven lumbar vertebrae behind which lack ribs, and the tail is very short, with only 10–15 vertebrae (Fig. 9.7a).

The main changes in the limbs and limb girdles of *Thrinaxodon* and later forms are connected with a major change in posture. Pelycosaurs, like most reptiles, had a sprawling posture, with the limbs held out sideways and the belly just above the ground (see p. 85), while *Thrinaxodon* shows an erect or upright posture with the limbs tucked under the body. This allowed improved efficiency in running since the limb muscles no longer

TOOTH OCCLUSION IN CYNODONTS

Occlusion is well developed in *Scalenodon*, a relative of *Diademodon*, from the Middle Triassic of Tanzania. The jaw cycle ends with a pronounced backwards pull of the lower jaw, and a powerful shearing and crushing movement is initiated in which all seven lower cheek teeth move tightly back into curved facets of the broad upper cheek teeth (Fig. 9.6a & b). Food items are sheared by a double cutting system, between the raised transverse ridges of lower and upper teeth, and between longitudinal ridges on the external side of both sets of teeth. Finally, as the backwards movement ends, the main faces of both teeth nearly meet, and any food particle caught between would effectively be crushed.

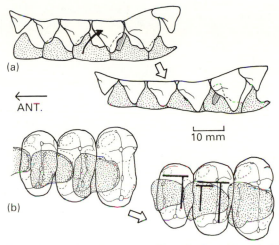

Figure 9.6 Tooth occlusion in cynodonts. The cheek teeth of the herbivore *Scalenodon*: (a) final stages of jaw closing and firm occlusion, in lateral view; (b) occlusal views of the same two jaw positions to show the backwards and sideways slide of the lower teeth (stippled) across the much broader upper teeth. In all cases, the front of the mouth is on the left (ANT, anterior). (After Crompton 1972.)

had to support the body as well as operating the arms and legs, and the effective stride length could also increase. The major joints changed their orientation and the shapes of the ends of the limb bones were much altered (Fig. 9.7d & e). The hip bones are also very different in shape because of major changes in the layout of the leg muscles. The pubis and ischium of *Thrinaxodon* (Fig. 9.7a & e) are reduced in size and they run backwards a little, while the blade of the ilium is relatively large, especially in front.

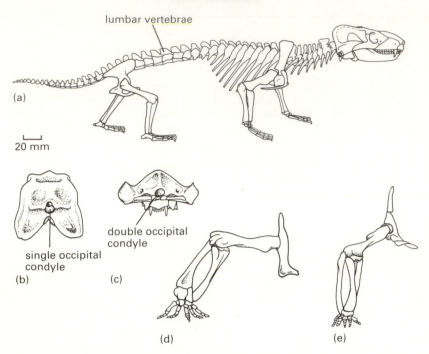

lumbar vertebrae

(a)

20 mm

single occipital
condyle

(b)

double occipital
condyle

(c)

(d)

(e)

Figure 9.7 The cynodont skeleton: (a) *Thrinaxodon*; (b) & (c) splitting of the single reptilian occipital condyle, seen in (b) pelycosaurs, into (c) two in later cynodonts; (d) & (e) postural evolution; (d) the sprawling hindlimb of a pelycosaur and (e) the semi-erect hindlimb of a cynodont. (Modified from Jenkins 1971.)

Tritheledonts and tritylodonts

Two cynodont groups, the tritheledonts and tritylodonts, appear to be very close to being mammals, but they do not show the full transition to a squamosal-dentary jaw joint. These represent the sixth stage in the march from *Procynosuchus* to the true mammals.

The tritheledonts, or ictidosaurs, are a rather poorly-known group of small animals that are mammal-like in many respects (Kemp 1982, Hopson & Barghusen 1986). Four or five genera are known from fragmentary skulls from the Late Triassic and Early Jurassic of South Africa and South America, which show a number of mammalian characters (Fig. 9.1) such as the loss of the postorbital bar between the orbit and the temporal fenestra, a slender zygomatic arch, and an external **cingulum**, or ridge, on the upper cheek teeth.

The tritylodonts, represented by about ten genera from the Early to Middle Jurassic of most parts of the world (Kühne 1956, Sues 1986), were highly successful herbivores that ranged in size from skull lengths of 80–220 mm. *Kayentatherium* from North America (Sues 1986) has the typical rodent-like tritylodont skull (Fig. 9.8a–c) with a deep lower jaw

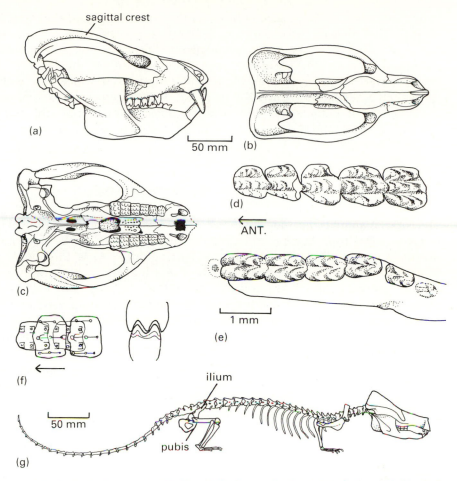

Figure 9.8 The tritylodonts (a)–(c) *Kayentatherium* and (d)–(g) *Oligokyphus*: (a)–(c) skull in lateral, dorsal, and ventral views; (d) cheek teeth of the upper jaw, in occlusal view; (e) cheek teeth of the lower jaw, in occlusal view; (f) occlusion of upper and lower cheek teeth, occlusal view showing the direction of movement (arrow), and vertical section; (g) skeleton (ANT, anterior). (Figs (a)–(c) altered from Sues 1986; (d), (e) & (g) after Kühne 1956; (f) after Crompton 1972.)

and high sagittal crest, indicating powerful jaw muscles, and a highly specialized dentition. There are elongate incisors and six to eight massive cheek teeth in straight rows. The upper cheek teeth all bear three longitudinal rows of crescent-shaped cusps, while the lower teeth bear two rows (Fig. 9.8d & e). When the jaws close, the lower teeth move back and the crescent-shaped cusps are drawn across the food, tearing it up along five parallel grating surfaces (Fig. 9.8f). *Oligokyphus* (Fig. 9.8g) has a long body and short limbs.

The tritylodonts have generally been allied with the other herbivorous cynodonts such as *Diademodon* because of the occluding broad cheek teeth with parallel rows of cusps (Hopson & Barghusen 1986). However, there are a number of remarkable parallelisms between tritylodonts and mammals such as the loss of the postorbital bar, loss of the postorbital and prefrontal bones, and cheek teeth with divided roots and well-developed wear facets, which Kemp (1982, 1983) interpreted as synapomorphies of tritylodonts and mammals. He noted many more in the limbs and limb girdles, and regards the the Tritylodontidae as a close outgroup of the Mammalia (Fig. 9.1).

MORGANUCODON – THE FIRST MAMMAL

The first mammals appeared in the Late Triassic, and *Morganucodon* (=*Eozostrodon*) from Europe and China is the best known. *Morganucodon* (Kermack *et al.* 1973, 1981, Jenkins & Parrington 1976) is tiny, with a 20–30 mm skull and a total body length of less than 150 mm. It probably looked generally like a shrew. The skull (Fig. 9.9a–e) shows all the mammalian characters of the tritheledonts and tritylodonts: loss of the postorbital bar, a slender low zygomatic arch, and a reduced ectopterygoid and lateral pterygoid processes in the palate. *Morganucodon* still has the reptilian quadrate-articular jaw joint, but it now functions largely as part of the middle ear system (Fig. 9.9d), and the mammalian squamosal-dentary joint is the main jaw hinge. This marks the transition from reptile to mammal, an arbitrary choice of character, but one that is generally accepted.

The braincase of *Morganucodon* (Fig. 9.9e) shows a number of mammalian features seen only in part in advanced cynodonts. The brain is enclosed almost completely in bone. Typical primitive reptiles have the brain enclosed at the side only by the prootic, opisthotic, exoccipital, and supraoccipital bones (see p. 74). In *Morganucodon*, the prootic sends a large sheet of bone forwards, the anterior lamina of the periotic (or petrosal), which meets the parietal and squamosal. In addition, the reptilian epipterygoid, typically a thin column of bone, has become a broad sheet, termed the **alisphenoid** in mammals.

The lower jaw of *Morganucodon* (Fig. 9.9a & d) is almost solely composed of the dentary bone, but the posterior bones are still present: a reduced splenial and coronoid, and a rod containing the surangular, prearticular, angular, and articular. *Morganucodon* has rather advanced teeth with several changes from those of the advanced cynodonts. The cheek teeth may be divided into **premolars** and **molars** (Fig. 9.9a), as in all mammals. The premolars are present in the milk dentition, and then replaced, while the molars are present only in the adult dentition. The teeth all occlude, and wear surfaces can be seen on the incisors as well as on the cheek teeth. The main chewing movement in *Morganucodon*

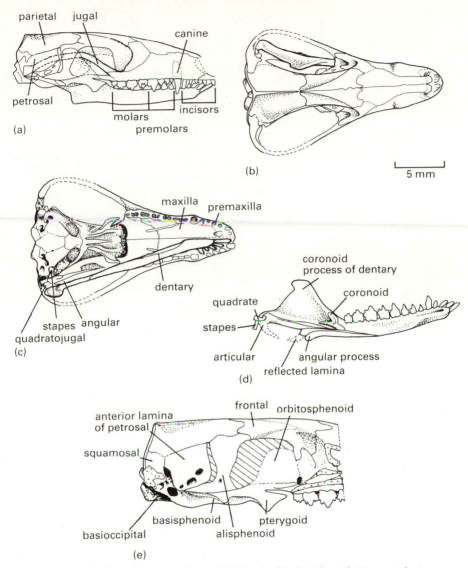

Figure 9.9 The skull of *Morganucodon* in (a) lateral, (b) dorsal, and (c) ventral views; (d) lower jaw in medial view; (e) lateral wall of the braincase (unossified areas shown with regular hatching). (After Kermack *et al.* 1981.)

followed a triangular route, rather than being simply back and forwards as in advanced cynodonts, and the food was sheared by the cross-cutting cusps on the teeth.

The skeleton of *Morganucodon* is poorly known, but its close relative *Megazostrodon* from South Africa (Jenkins & Parrington 1976) has a long low body, rather like that of *Oligokyphus*, but the limbs are rather longer (Fig. 9.10). The ribcage is restricted to the thoracic vertebrae, with no ribs

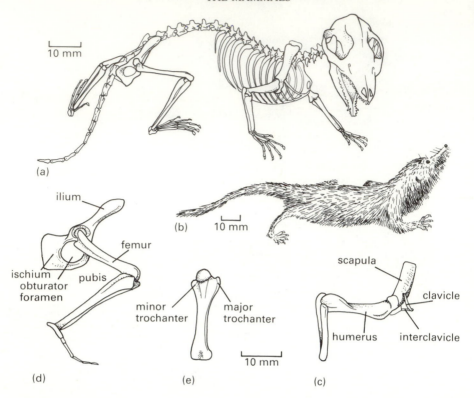

Figure 9.10 The skeleton of morganucodontids: (a) skeleton of *Megazostrodon*; (b) body restoration; (c) forelimb and pectoral girdle; (d) hindlimb and pelvic girdle of *Morganucodon*; (e) humerus. (Figs (a), (c), (d) & (e) after Jenkins & Parrington 1976; (b) after Crompton 1968.)

on the lumbars (cf. Fig. 9.7a). The forelimb and shoulder girdle (Fig. 9.10c) are rather cynodont-like, still sprawling, while the pelvis and hindlimb (Fig. 9.10d) are typically mammalian. The posture is erect, the ilium is a rod-like element pointing forwards, and fused to a reduced pubis and ischium, and there is a very large **obturator foramen**, a circular gap in the pelvis between the pubis and ischium. The femur (Fig. 9.10e) is also very mammalian, with a ball-like head which fits sideways into the acetabulum, a necessary feature in an erect animal, and seen also in dinosaurs and birds (see p. 149). In addition, there are two clear processes on either side, the minor and major trochanters, which provided sites for insertion of the important muscles that moved the leg back and forwards during walking.

Morganucodon and *Megazostrodon* were agile insectivores, as far as can be told. Their locomotion was fully mammalian, with the possibility of rapid and variable movements, even if sustained running might have been difficult. The well-developed pointed cutting teeth suggest a carnivorous diet, and the small size of these animals points to insects as the main food source. It is also most likely that *Morganucodon* and *Megazostrodon* were endothermic (fully warm-blooded) and nocturnal. Other mammalian characters include the large brain and probable possession of mammary glands. What is the evidence for all these features, some of which are not obviously fossilizable?

Endothermy, the generation of heat and control of body temperature by internal means (see p. 172) is indicated by several lines of evidence. Crompton *et al.* (1978) argued that endothermy is suggested in *Morganucodon* because of its close resemblance to modern insectivorous mammals in its body size and proportions, and because of the fully developed secondary palate, which allowed these animals to breathe rapidly while feeding since the air stream was separated from the mouth. In addition, certain cynodonts such as *Thrinaxodon* have tiny pits all over their snouts which are very like the passages for nerves and blood vessels to the sensory whiskers of modern mammals. If the advanced cynodonts had sensory whiskers, they must have had hair as well.

How could Crompton *et al.* (1978) say that *Morganucodon* was nocturnal in habits? *Morganucodon* has a greatly enlarged brain when compared to typical cynodonts, and the enlargement has mainly affected key elements in the senses of hearing and smell, both of which are useful for a nocturnal animal. Further, modern insectivorous mammals tend to be nocturnal, and hence avoid competing for food with birds and lizards.

Finally, did *Morganucodon* have mammary glands? If it had hair, it must have had sweat glands to dissipate excess heat. Mammary glands are thought to be modified sweat glands. A second line of argument relates to the precise tooth occlusion of mammals. Mesozoic mammals, like modern ones, probably delayed the appearance of their teeth until rather late when the head was near to its adult size, and hence needed only two sets of teeth, the milk and the adult, during their lives.

THE MESOZOIC MAMMALS

Morganucodon and *Megazostrodon* are typical morganucodontids, a family that existed in many parts of the world from Late Triassic to Late Jurassic times. Another 20 or so mammalian families have been recorded in the Mesozoic (Lillegraven *et al.* 1979, Kemp 1982), but many of these are based on incomplete material and their relationships are hard to assess. The main groups will be reviewed here in roughly chronological order.

Early forms

By Late Jurassic times, as many as eight mammalian lineages were in existence. The docodonts from the Middle and Late Jurassic of Europe and North America have been described only from isolated jaw bones

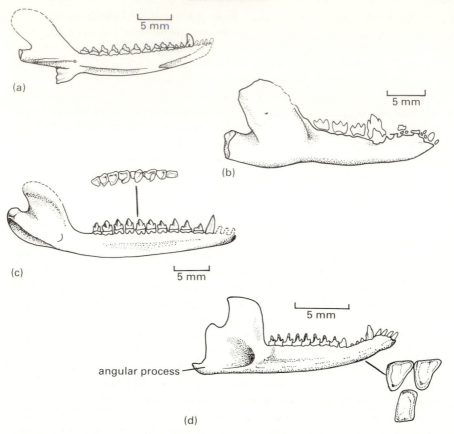

5 mm

5 mm

5 mm

angular process

5 mm

(a)

(b)

(c)

(d)

Figure 9.11 Mesozoic mammals: (a) lower jaw of the docodont *Docodon*, in lateral view; (b) lower jaw of the triconodont *Triconodon* in lateral view; (c) lower jaw of the symmetrodont *Spalacotherium* in lateral view, and occlusal view of the molars; (d) lower jaw of the dryolestid *Crusafontia* in lateral view, and occlusal view of two upper and one lower cheek tooth. (Fig. (a) after Kron 1979, (b) after Crompton & Jenkins 1979; (c) after Cassiliano & Clemens 1979; (d) after Krebs 1971.)

(Fig. 9.11a) and teeth. The triconodonts, from the Middle Jurassic to the Late Cretaceous of Europe, North America and Central Asia are known mainly from isolated teeth and jaw bones, although two partial skeletons have been found in the Early Cretaceous of North America (Jenkins & Schaff 1988). *Triconodon* (Fig. 9.11b) has pointed shearing molars with three main cusps in a line (hence 'triconodont').

Some Late Jurassic and Early Cretaceous mammals show hints of a new kind of tooth pattern in which the three main cusps on the lower and upper molars form a low triangular shape. Symmetrodonts like *Spalacotherium* (Fig. 9.11c) have the middle cusp set well-over from the other two. Four or five other families of mammals with similar molars, the 'eupantotheres', include the dryolestids from Europe and North America.

The lower jaw (Fig. 9.11d) has a larger coronoid process than in *Spalacotherium* and there is an angular process, a feature that relates to sideways rotations of the jaw during chewing.

The multituberculates

The largest group of Mesozoic mammals, and one which survived into the Oligocene, were the multituberculates, a group of rodent-like omnivores which arose in the Late Jurassic (Clemens & Kielan-Jaworowska 1979). *Kamptobaatar* has a broad flat skull (Fig. 9.12a–c) with large eyes that appear to have faced forwards over a short snout. There are large rodent-like incisors, generally no canines, and a long gap in front of the cheek teeth, as in rodents. The lower premolar 4 forms a large shearing blade that may have been used mainly for chopping tough vegetation, which was ground up by the batteries of molar cusps behind.

Ptilodus from the Palaeocene of Canada (Fig. 9.12d) may have been arboreal since it has a long prehensile tail for grasping branches, a

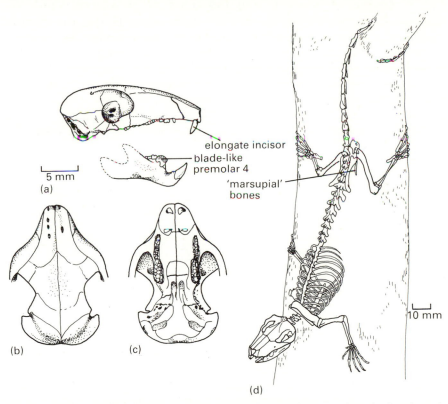

Figure 9.12 The multituberculates: (a)–(c) skull of *Kamptobaatar* in lateral, dorsal, and ventral views; (d) reconstructed skeleton of *Ptilodus* descending a tree trunk. (Figs (a)–(c) after Kielan-Jaworowska 1971; (d) after Krause & Jenkins 1983.)

reversible foot as in squirrels which allows it to descend a tree trunk head-first, and flexible elbow and knee joints (Krause & Jenkins 1983). Two curious little bones have been found attached to the front of the pelvis (Fig. 9.12d), and these are interpreted as **epipubic**, or marsupial, bones. These are present in modern marsupials and monotremes, as well as in the Late Cretaceous placental *Barunlestes* (see p. 246) and seem to be a primitive feature associated with egg-laying or the bearing of small live young, as in modern marsupials, which then move to a pouch.

The monotremes

Modern mammals fall into three groups, the monotremes, marsupials, and placentals. These three form a kind of reproductive continuum, since the monotremes lay eggs, and the marsupials give birth to tiny live young, which in both cases finish developing in a pouch, and placentals retain their young in the uterus to a more advanced stage. The monotremes (subclass Monotremata), represented today by the duck-billed platypus of Australia and the echidnas of Australia and New Guinea, share many primitive features, such as egg-laying, the retention of (tiny) prefrontal and postfrontal bones, an interclavicle, and two coracoids. Neither monotreme has teeth in the adult, although the juvenile platypus has molars (Fig. 9.13a) which are soon lost.

The fossil record of monotremes until recently extended back only to the Middle Miocene (c. 15 Myr), which was rather frustrating to palaeontologists since the monotremes were supposed to be the most primitive living mammals. However, Archer and colleagues (1985) have

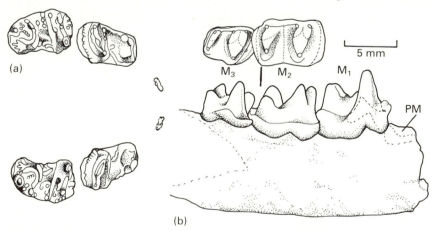

Figure 9.13   The monotremes: (a) the temporary upper molars of a juvenile *Ornithorhynchus*, in occlusal view of both sides; (b) jaw fragment of *Steropodon*, the possible Early Cretaceous monotreme, in lateral and occlusal views. Abbreviations: M_1–M_3, molars 1–3; PM, premolar. (Fig. (a) after Beer & Fell 1937; (b) after Archer *et al.* 1985; and Kielan-Jaworowska *et al.* 1987; copyright © 1985, 1987 Macmillan Magazines Ltd.)

reported a jaw fragment of a Lower Cretaceous monotreme from Australia, *Steropodon*. The molars (Fig. 9.13b) are like those of the modern platypus in the very short V-shaped array of cusps, the height of the transverse ridges, and other features.

The tribosphenic molar appears

The triangular array of cusps seen in symmetrodonts and dryolestids was once regarded as a direct prelude to the molar pattern of therians (i.e. marsupials + placentals). However, this tooth pattern is not seen in the monotremes or multituberculates. The true tribosphenic molar tooth pattern (see box) shows a triangle of three cusps with the point facing outwards in the lowers, and inwards in the uppers.

Cretaceous marsupials and placentals

Some Cretaceous therians are hard to classify as either marsupials or placentals. For example, *Deltatheridium* from the Late Cretaceous of Mongolia (Fig. 9.15a), a short-snouted animal with well-developed carnivorous cheek teeth, shares some characters with both the marsupials and the placentals. It has generally been classified as an intermediate third stock that died out at the end of the Cretaceous, but Kielan-Jaworowska & Nessov (1990) argue that it is a sister group of the marsupials. Other intermediate forms are known from the Early Cretaceous, but most are simply teeth or very fragmentary specimens.

The first marsupials (infraclass Metatheria) are known from the Late Cretaceous of North and South America in particular. The majority of these fossils are jaw fragments and teeth, assigned to as many as 30 species and three families (Clemens 1979). *Alphadon* (Fig. 9.15b) is a typical form, probably rather like the living opossum. The teeth give evidence of its marsupial nature since it has three premolars and four molars (placentals have four or five premolars and three molars). The upper molars (Fig. 9.15b) are not as wide as typical placental molars of the same length (cf. Fig. 9.16d) and they have several large cusps. Marsupials such as *Alphadon* form nearly half the species of many Late Cretaceous mammal faunas of North America, and it is interesting to speculate what might have happened had they not been virtually wiped out during the K–T event (see p. 203).

The dominant mammals today, the placentals (infraclass Eutheria), are represented by eight or nine families in the Late Cretaceous of North America, South America, and Mongolia. Indeed, the remains from Mongolia include some nearly complete specimens (Kielan-Jaworowska *et al.* 1979, Kielan-Jaworowska 1984). *Zalambdalestes* (Fig. 9.16a–e) is an agile hedgehog-sized animal with a long-snouted skull. The zygomatic arch is slender and there is no separation between the orbit and the

THE TRIBOSPHENIC MOLAR

A comparison of three molar teeth from different early mamals shows how the tribosphenic molar may have evolved. In *Kuehneotherium* from the Early Jurassic (Fig. 9.14a), there are three major shearing surfaces – surfaces which operate as cutting planes as the jaws close and the occluding teeth slide sideways across each other. In *Peramus*, a Late Jurassic relative of the dryolestids, there are four shearing surfaces (Fig. 9.14b), the new one being added by a medial expansion of the upper molar. The final stage (Fig. 9.14c) shows six shearing surfaces. The two new ones have arisen on the vastly expanded medial portion of the upper molar whose highest point, the protocone, fits into a depression on the lower, the **talonid basin**. The nomenclature of the cusps, ridges, and facets of typical mammalian molar teeth such as these is complex (Fig. 9.14d & e). The main terms to note are for the triangles (**trigons**) of three cusps: the **paracone** (anterior), **metacone** (posterior), and **protocone** (medial) in the uppers, and the **paraconid**, **metaconid**, and **protoconid** in the **trigonids** of the lowers. The talonid basin, occupying the posterior half of lower molars, is bounded by the **entoconid** (medial), **hypoconid** (lateral), and the **hypoconulid** (posterior).

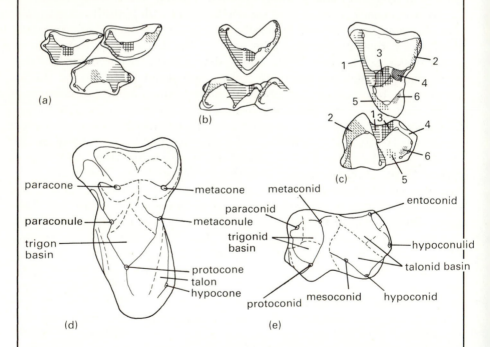

(a)

(b)

(c)

paracone

metacone

paraconule

metaconule

trigon basin

protocone
talon
hypocone

(d)

metaconid

paraconid

entoconid

trigonid basin

hypoconulid

talonid basin

protoconid   mesoconid   hypoconid

(e)

Figure 9.14 The tribosphenic molar: (a)–(c) sequence of stages in the evolution of the tribosphenic molar: (a) *Kuehneotherium*, (b) *Peramus*, and (c) *Didelphodus*, showing the addition of shearing surfaces on both upper (top) and lower (bottom) molars; (d) & (e) nomenclature of the main cusps, ridges, and basins, of *Gypsonictops* (d) upper and (e) lower molars. (After Bown & Krause 1979.)

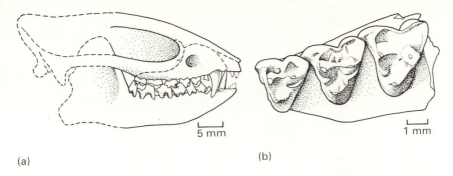

(a)

(b)

5 mm

1 mm

Figure 9.15 Late Cretaceous possible marsupials: (a) skull of *Deltatheridium* in lateral view; (b) maxillary fragment of *Alphadon* in occlusal view. (Fig. (a) after Kielan-Jaworowska *et al.* 1979; (b) after Clemens 1966.)

temporal region. The braincase (Fig. 9.16c) is primitively small. There are typical numbers of teeth for a placental (Fig. 9.16d & e), four premolars and three molars. The molars are broad, and they lack the specializations of marsupial molars. All teeth appear to be replaced once, apart from the molars, whereas in marsupials there is only one set of teeth, except for the third premolar which is replaced.

In the skeleton, the vertebrae of the neck are broad low-spined elements. The atlas forms a fused ring (Fig. 9.16f) with broad facets on either side for the two occipital condyles, while the axis has an unusual long spinal process.

The shoulder girdle is only incompletely known (Fig. 9.16a), but it shows evidence of the structure seen in modern marsupials and placentals, being held in an erect posture. Advances include the loss of the interclavicle and a modern scapula. The scapula also takes on an entirely new form, better seen in a modern mammal (Fig. 9.16g). The scapular blade is divided in two by a sharp ridge or spine which ends in the **acromion process** to which the clavicle is attached. The fields in front of the spine, and behind it, bear major new muscles that move the arm back and forwards with the elbows tucked well in. Further advances in the arm which relate to erect gait are that the humerus head fits into a glenoid that faces downwards instead of sideways as in early mammals, and the elbow joint is hinge-like. The hand (Fig. 9.16h) has long digits.

Although the zalambdalestid arm shows many new features, the hindlimb is rather like that of earlier mammals. It seems that erect gait was achieved in the hindlimb by advanced cynodonts in the Triassic, but in the forelimb only much later by Late Cretaceous marsupials and placentals. The foot of *Zalambdalestes* is long. In the ankle, the calcaneum has a long 'heel' (Fig. 9.16i), and the astragalus sits on top of it, out of contact with the ground, as in modern placentals. The fibula is reduced to a thin splint which is largely fused to the tibia. *Zalambdalestes* is

245

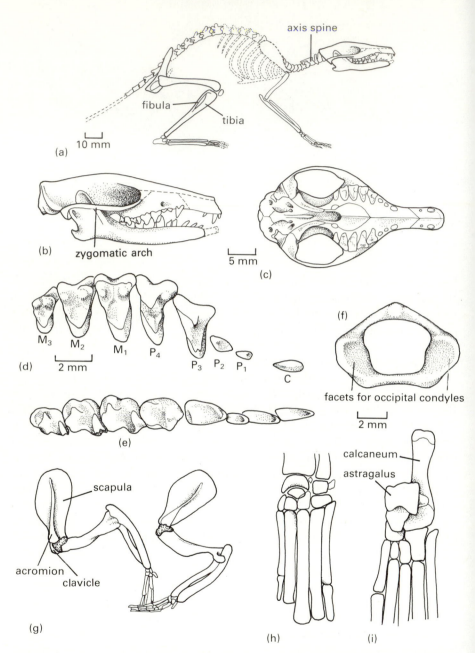

Figure 9.16 The Late Cretaceous placental mammals (a)–(e) & (i) *Zalambdalestes* and (f) & (h) *Barunlestes*: (a) restored skeleton; (b) & (c) skull in lateral and ventral views; (d) upper dentition in occlusal view; (e) lower dentition in occlusal view; (f) atlas vertebra in anterior view; (g) two positions of the forelimb of the living opossum *Didelphis* during a stride; (h) hand; (i) foot. Abbreviations as in Figure 9.15. (Figs (a), (f), (h) & (i) after Kielan-Jaworowska 1978; (b) after Kielan-Jaworowska 1975; (c) after Kielan-Jaworowska 1984; (d) & (e) after Kielan-Jaworowska 1968; (g) based on Jenkins 1971b.)

interpreted as a specialized jumping mammal which made great rabbit-like leaps using its powerful legs.

Other Late Cretaceous placental mammals include representatives of several groups that later rose to dominance, such as the insectivore-like leptictids, the ungulate-like acrtocyonids, the herbivorous pantodonts and notoungulates, the omnivorous 'condylarths', and a possible early primate (see pp. 260–5).

Relationships

The classical view on the relationships of Mesozoic mammals has been that they fall into two major lines, the 'therians' and 'prototherians'. The 'therians', those forms with a triangular array of cusps on the molars, included *Kuehneotherium*, the symmetrodonts, marsupials, and placentals. The 'prototherians' had the molar cusps arranged in rows, and included the morganucodontids, docodonts, triconodonts, multituberculates, and monotremes. A second line of evidence for this view concerned the side wall of the braincase. In monotremes, multituberculates, and morganuco-dontids, most of the side wall is made from an expansion of the periotic (prootic) bone (see pp. 236–7), whereas in marsupials and placentals the alisphenoid (epipterygoid) takes this role (Fig. 9.17).

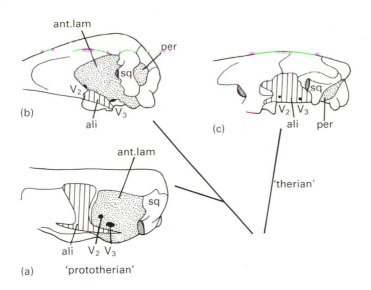

Figure 9.17 Two major patterns of side wall structure of the mammalian braincase, the 'prototherian', as seen in (a) morganucodontids and (b) monotremes, and the 'therian', as seen in (c) marsupials and placentals. Abbreviations: ali, alisphenoid; ant.lam, anterior lamina of the periotic; per, periotic; sq, squamosal; V_2, V_3, branches 2 and 3 of cranial nerve V. (After Kemp 1983.)

Kemp (1982, 1983) argued against this view and presented evidence that the 'prototherians' are part of a monophyletic Mammalia, and that the monotremes are the sister group of the marsupials and placentals. The teeth of multituberculates, docodonts, and embryonic monotremes are so different in detail that they cannot be compared closely. Further, the braincase evidence does not support a major split within the Mammalia. The side wall of the braincase in embryonic monotremes is essentially the same as in embryonic marsupials and placentals. A similar thin sheet of bone, the anterior lamina, covers most of the side in all embryos studied, and it simply fuses later with either the periotic in monotremes, or the alisphenoid in marsupials and placentals. Hence, the braincase of the Late Triassic morganucodontids could be ancestral to both modern patterns.

Within a monophyletic Mammalia, it is hard to arrange the various Mesozoic forms into a straightforward cladogram. Rowe's (1988) attempt (Fig. 9.18) is provisional. The Late Jurassic forms (Dryolestidae, Triconodonta, Docodonta, Symmetrodonta) seem to fall into two clusters, at the same level as the Monotremata. These are followed by multi-tuberculates, and then the two other modern groups, together the Theria. Kemp's (1982, 1983, 1988c) analysis of these groups gave slightly different results. For example, he linked the three living groups in a clade, with the Multituberculata as the outgroup of all three.

The level in the cladogram at which the name Mammalia is applied is now disputed. Rowe (1988) defines mammals as all of the descendants of the latest common ancestor of the monotremes and therians, the living forms. Others choose to include some of the more primitive groups within the Mammalia (Fig. 9.18), a larger clade called the Mammaliaformes by Rowe (1988).

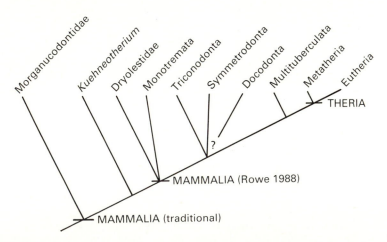

Figure 9.18 Cladogram showing the postulated relationships of the major groups of mammals, and of the Mesozoic groups in particular.

Marsupials arose in the Late Cretaceous of the Americas (see p. 243), and they spread from there to found two major living assemblages, one mainly in South America and the other in Australasia (Archer 1982). For years, this split distribution of marsupials was a mystery and numerous biogeographical theories have been proposed.

Geography and marsupial migrations

The Late Cretaceous and Tertiary marsupials of North and South America belonged to several families. In the Tertiary period, North American didelphid marsupials spread to Europe (Fig. 9.19b) and Africa (Fig. 9.19e), where they survived until the Miocene. In North America, the didelphids became extinct in the Miocene, but they have reinvaded from South America much more recently. The fossil record of marsupials in Australia is sparse, with only a few odd teeth and jaws known from the Late Oligocene onwards.

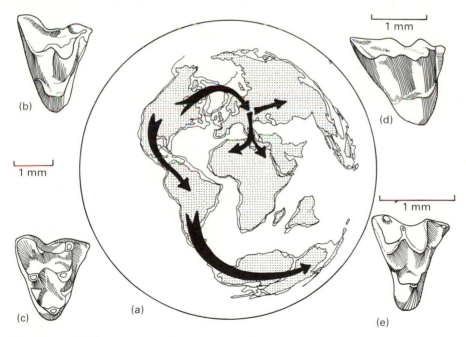

Figure 9.19 (a) The migration of the marsupials from an origin in the Late Cretacous of the Americas, into Antarctica and Australasia, and into Europe, North Africa, and Asia. (b)–(e) Typical opossum-like marsupial molars from all parts of the world: (b) *Amphiperatherium* from Europe; (c) *Alphadon* from North America; (d) cf. *Peratherium* from Kazakhstan, USSR; (e) *Garatherium* from Algeria, North Africa. (Fig. (a) much modified from Marshall 1980a; (b)–(e) modified from various sources.)

The present split distribution of marsupials has led to much debate among biogeographers. Formerly, a northern dispersal route was favoured, with the early marsupials travelling across Asia from the Americas to Australia, but no evidence of marsupials had ever been found in Asia. With the acceptance of continental drift (see p. 15), most people preferred a southern dispersal route from South America to Australia via Antarctica. A South American type of marsupial found in the Eocene of Antarctica seemed to tie in with the southern dispersal route theory. Recently, a small marsupial tooth was reported from the Oligocene of central Asia (Fig. 9.19d). It is a didelphid rather like the European *Peratherium*, with no particular Australian affinities. The new discoveries show evidence of an early Tertiary radiation of marsupials in Europe and their migration into Africa and Asia, but subsequent extinction in all three areas. Nevertheless, it is interesting to speculate that the Asian opossum might have been interpreted very differently by a northern dispersalist!

The Australian marsupials

Australian marsupials are more varied and diverse than those of South America since they have always been the main mammal group on the island continent. The fossil record of Australian marsupials extends back to the Oligocene, but it is best in the Pliocene and Pleistocene. The degree of convergence between Australian marsupials and placentals from other parts of the world is often striking. For example, the marsupial 'wolf' *Thylacinus* has a skull that seems at first sight to be identical to that of the dog or fox (Fig. 9.20a & b). It differs in details, however; the molars of *Thylacinus* have both shearing and grinding surfaces, whereas in *Canis* meat is cut and bones crushed by separate teeth. Similar convergences may be found in the marsupial moles, ant-eaters, climbing insectivores, leaf-eaters, and even grazing ungulates (even though a kangaroo looks very different from a deer or antelope, it lives in roughly the same way!).

The most spectacular faunas of the fossil marsupials are known from the Pleistocene when giant wombats, kangaroos, and others lived with giant echidnas and the heavily armoured turtle *Meiolania* (see p. 190). The scene (Fig. 9.20c) was dominated by great herds of the hippopotamus-sized wombat *Diprotodon* and its smaller relatives. This gentle giant has heavy limbs with broad plantigrade feet to bear its weight. The short-faced kangaroo *Procoptodon* (Fig. 9.21a) was a grazer which used its powerful jaws to chew tough leaves and grass. Its skull is much shorter and deeper than that of a modern kangaroo. Like them, it has four toes in its foot, but the fourth is the only functional one (Fig. 9.21b). Toes 2, 3 and 5 are reduced, and they are firmly bound together by connective tissue. *Procoptodon* no doubt moved fast by hopping, just as modern kangaroos do, an efficient mode of locomotion which allows them to

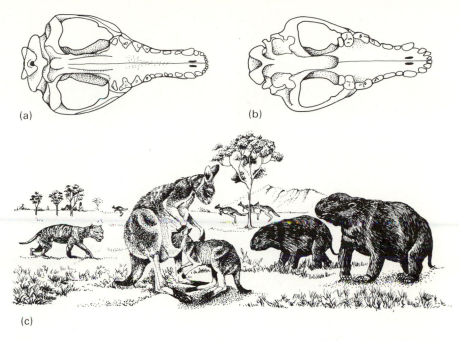

Figure 9.20 Convergent evolution of a dog-like form of the skull (ventral view) in (a) the marsupial 'wolf' *Thylacinus* and (b) the placental dog *Canis*; (c) scene in the Pleistocene of Australia, with the marsupial 'lion' *Thylacoleo* on the left, a mother and juvenile giant kangaroo *Procoptodon*, and two giant wombats *Diprotodon*. (Figs (a) & (b) redrawn from Keast 1972; (c) based on a painting in Benton 1986, copyright © Grisewood & Dempsey Ltd 1986, reproduced by permission of the publishers.)

achieve racehorse speeds of 45–55 km per hour over short distances. Most of these giant herbivores were probably safe from predation because of their size, but the young, and smaller species, were preyed on by the marsupial lion, *Thylacoleo*. Its heavy 250 mm long skull (Fig. 9.21c) has strong canine-like incisors and exceptionally long flesh-cutting blades extending across two teeth.

SOUTH AMERICAN MAMMALS – ANOTHER WORLD APART

For most of the Cenozoic (66–0 Myr), South America was an island, isolated from all other parts of the world. As in Australia, a spectacular **endemic** (geographically restricted) fauna of mammals evolved which shows little similarity to those of other parts of the world (Simpson 1948, 1967, 1980). South America had its own families of marsupials, some of which mimicked dogs, bears, sabre-toothed cats, and others in an uncanny way. The herbivores for most of the Cenozoic were rodents, some as large as deer (see p. 278), horse-mimics and hippo-mimics, and

251

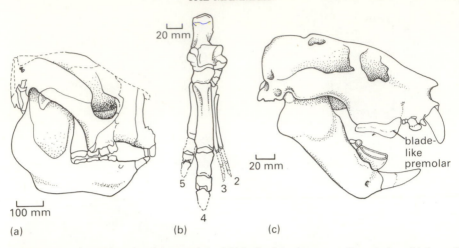

20 mm

100 mm

(a)

5   3 2

4

(b)

20 mm

blade-
like
premolar

(c)

Figure 9.21 Australian fossil marsupials: (a) skull of the kangaroo *Procoptodon*; (b) foot of
the kangaroo *Protemnodon*, showing the dominant fourth toe; (c) skull of the marsupial 'lion'
Thylacoleo, showing the blade-like cheek teeth. (Fig. (a) after Tedford 1966; (b) after Flannery
1982; (c) modified from Finch & Freedman 1982.)

the armadillos and sloths (edentates). Where did these remarkable
mammals come from, and what has happened to them now?

The Mesozoic mammals of South America

For much of the Mesozoic (245–66 Myr), South America was linked to
Africa (see p. 146), but this connection was lost during the Cretaceous
when the South Atlantic Ocean began to open up. There may have been a
geologically brief land bridge formed to Central and North America about
75 Myr ago when mammals were able to cross both ways, but the fossil
record gave almost no clues about this key episode in mammalian
evolution when the unique South American groups became established.
However, everything changed when three new Cretaceous beds were
discovered in Argentina and Bolivia (Bonaparte 1986, Marshall & Muizon
1988, Van Valen 1988).

The Argentinian faunas consist largely of relatives of northern Jurassic
groups such as triconodonts, symmetrodonts, and dryolestids, as well as
some unique South American taxa. The Bolivian specimens include two
insectivorous proteutherians, three omnivorous 'condylarths', and two
herbivores (a pantodont and a notoungulate). These fossils
demonstrate that North and South America shared similar mammals
before the isolation event, and they show the ancientness of at least two
groups, the opossums and 'notoungulates', which rose to prominence
later.

South American marsupials

Marsupials radiated in South America to a lesser extent than they did in Australia, but they dominated as insectivores, carnivores (partially), and included some small herbivores (Reig *et al*. 1987). The ten or so families of extinct insectivorous and carnivorous marsupials show remarkable convergences with placental shrews, cats, sabre-tooths and dogs (Marshall 1982). The borhyaenids, such as *Prothylacynus* (Fig. 9.22a), have short limbs and rather dog-like skulls. Their later relatives, the Late Miocene and Pliocene thylacosmilids, have skulls (Fig. 9.22b) which are almost indistinguishable from those of the sabre-toothed (placental) cats which lived in North America at the same time. The upper canine tooth is very long, and it grew continuously, unlike the canine of true cats. It was presumably used for puncturing the thickened hides of the large thick-skinned South American 'notoungulates' (see p. 256).

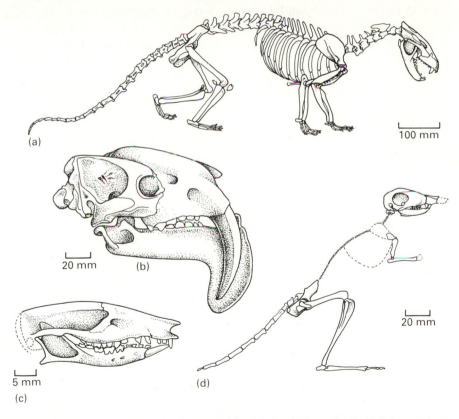

(a)

100 mm

20 mm   (b)

20 mm

5 mm   (d)

(c)

Figure 9.22 South American fossil marsupials: (a) the borhyaenid *Prothylacinus*; (b) the sabre-toothed thylacosmilid *Thylacosmilus*; (c) the caenolestid *Palaeothentes*; (d) the argyrolagid *Argyrolagus*. (Fig. (a) after Sinclair 1906; (b) after Riggs 1934; (c) after Marshall 1980b; (d) after Simpson 1970.)

Smaller South American marsupials include the caenolestids like *Palaeothentes*, a small insectivore or omnivore with an elongate lower incisor, and blade-like cheek teeth (Fig. 9.22c). The kangaroo rat lookalike *Argyrolagus* (Fig. 9.22d) has a narrow snout and broad cheek teeth for crushing tough plant food. The hindlimbs are long and powerful which suggests a jumping mode of locomotion.

Armadillos, sloths, and ant-eaters

Some of the most characteristic mammals of South America are the Xenarthra (or Edentata) which include the armadillos, tree sloths, and ant-eaters. This group has had a spectacular history which is not evident from the living forms (Simpson 1948, 1980). The name Xenarthra (literally 'strange joints') refers to supplementary articulations which are present in all forms between some of the vertebrae. In addition, they have a peculiar arrangement in the hip girdle in which the ischium, as well as the ilium, is fused to the anterior caudal vertebrae (Fig. 9.23c). The name Edentata (literally 'no teeth') refers to a third feature of the group, great reduction of the teeth. Xenarthrans have few or no incisors, and the ant-eaters have no teeth at all.

The armadillos (family Dasypodidae) first appear in the fossil record in the Late Palaeocene, but the remains are only armour scutes. They radiated in the Oligocene and Miocene when a variety of small and large forms appeared. Like the modern *Dasypus* (Fig. 9.23a), they all have a bony shield over their heads, a body armour that is partly fixed and partly formed of movable rings, and a bony tube over the tail.

The most spectacular armadillos were the glyptodonts (Fig. 9.23b & c) which reached very large size in the Pliocene and Pleistocene (Gillette & Ray 1981). The heavy armour, weighing as much as 400 kg in a 2 tonne animal, is clearly proof against voracious predators such as the sabre-toothed marsupials (see p. 253). The skull is short and deep (Fig. 9.23c) and the massive jaws accommodate long continuously-growing cheek teeth which were used to grind up abrasive grasses. The short tail is flexible and in some forms bears a spiked club (Fig. 9.23e) which was probably used to whack sabre-tooths.

The sloths date back to the Eocene, and they had a broad radiation, even though only a few species of tree sloths still survive (Webb 1986). A Miocene sloth, *Hapalops* (Fig. 9.23d) is a small semi-arboreal animal which has only four or five cheek teeth in the jaws. Sloth evolution followed two main ecological lines from the Miocene onwards. Some remained small and became adapted to life in the trees, like the modern tree sloths (family Bradypodidae), while the ground sloths (family Megatheridae), achieved giant size.

Megatherium, the largest ground sloth at 6 m in length, was a massive animal that may have fed on the leaves from tall trees (Fig. 9.23e). It

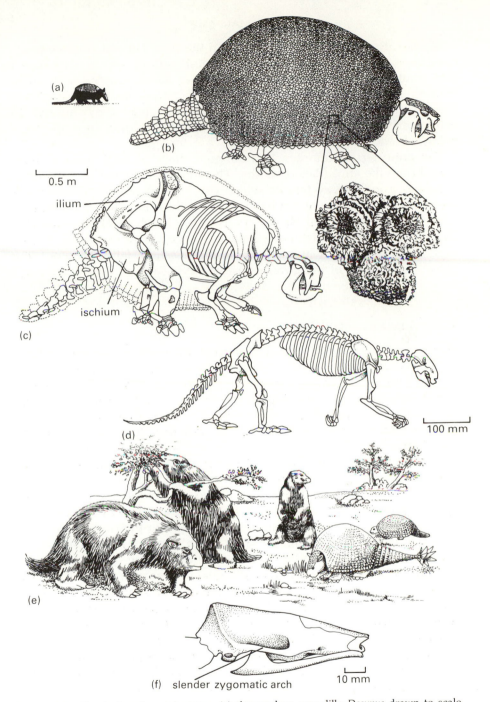

(a)

0.5 m

ilium

ischium

(b)

(c)

100 mm

(d)

(e)

(f) slender zygomatic arch

10 mm

Figure 9.23 South American edentates: (a) the modern armadillo *Dasypus* drawn to scale with (b) & (c) the glyptodont *Glyptodon*, showing the armour covering, a detail of the armour, and the skeleton; (d) the sloth, *Hapalops*; (e) a Pleistocene scene in South America showing the ground sloth *Megatherium* (left) and the glyptodont *Doedicurus*; (f) the oldest ant-eater *Eurotamandua*. (Fig. (a) altered from Patterson & Pascual 1968; (b) & (c) after Burmeister 1874; (d) after Matthew 1918; (e) based on a Charles Knight painting; (f) after Storch 1981.)

could rear up on its hind legs, resting on the short tail and massive bowl-like pelvis, and pulling branches to its mouth with its long hooked claws. The giant ground sloths ranged widely over South, Central and North America in the Pleistocene, and they died out only 11 000 years ago. They were no doubt encountered by early man: were they hunted to extinction? Specimens found in caves often have clumps of their yellowish and red hair still preserved.

The ant-eaters (infraorder Vermilingua), have a much poorer fossil record than the armadillos or sloths. The snout is long and toothless and it houses a long sticky tongue which can be shot out to capture small insects. The oldest fossil ant-eater remarkably comes from the famous Eocene lignite deposits of Messel in Germany (Storch 1981). *Eurotamandua* (Fig. 9.23f) is very like the living ant-eaters except that it retains a narrow zygomatic arch. What this unexpected new find tells us about the geographic history of the group is a mystery!

South American ungulates

There were four uniquely South American ungulate groups dating from the Palaeocene to the Pleistocene. The term ungulate (literally 'bearing hooves') refers to familiar moderate- to large-sized herbivores such as horses, cows, rhinos, pigs, elephants, and the like. The South American forms may be related to these ungulates, or they may form an independent radiation with no special affinities to large plant-eaters elsewhere.

The litopterns include a range of rabbit-, horse-, and camel-like forms (Cifelli 1983). *Diadiaphorus* from the Miocene and Pliocene (Fig. 9.24a) is a lightly built animal with many striking convergences with horses. The legs are long and only the middle toe (hoof) touches the ground (Fig. 9.24b). Litopterns have the nostrils set well back in the skull roof, which almost certainly indicates the presence of an elephant-like trunk.

The notoungulates are by far the largest of the South American herbivore groups, with well over 100 genera (Simpson 1948). Their ear region is greatly expanded, with additional chambers above and below the normal middle ear cavity. Some notoungulates were beaver-like or rabbit-like, while others were as large as bears. The largest, *Toxodon*, was first collected by Charles Darwin in the Pleistocene of Argentina (Fig. 9.24c). He described it as 'perhaps one of the strangest animals ever discovered'. As in many other South American forms, the roots of the teeth remained open throughout life so that they continued to grow to keep up with the wear produced by grazing.

The astrapotheres (Fig. 9.24d) are as large as rhinoceroses, but they had long bodies and short legs. The lower incisors stick out straight in front and may have been used in digging for water plants and roots. The

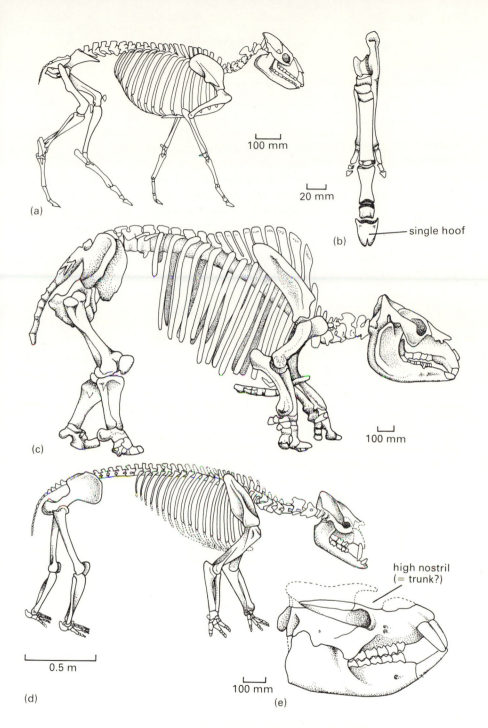

(a)

100 mm

20 mm

single hoof

(b)

(c)

100 mm

(d)

0.5 m

high nostril
(= trunk?)

100 mm

(e)

100 mm

Figure 9.24 The litoptern *Diadiaphorus*, (a) skeleton and (b) foot in anterior view showing the horse-like single hoof; (c) the notoungulate *Toxodon*; (d) *Astrapotherium*; (e) *Pyrotherium*. (Figs (a) & (b) after Scott 1910; (c) after Flower & Lydekker 1891; (d) after Riggs 1935; (e) after Loomis 1914.)

pyrotheres are also large long-bodied animals with trunks. The skull (Fig. 9.24e) is short and bears broad tusk-like incisors.

South American waifs

Several smaller groups of placental mammals invaded South America during the Tertiary. For example, bats arrived at least by Miocene times, and rodents appeared in the Early Oligocene. The South American rodents became important elements of the faunas, and some reached large size. Primates also reached South America in the Oligocene, and gave rise to a radiation of marmosets and monkeys (see p. 305). The invaders are termed 'waifs' since the first bats, rodents, and primates to reach South America were probably small populations that had wandered off-course in search of food.

The Great American interchange and extinction

All of the South American ungulates have gone, as have the larger carnivorous marsupials, glyptodonts, and ground sloths. These extinctions in the Pliocene and Pleistocene must relate to the opening up of the Central American land bridge about 3 Myr ago (Fig. 9.25). North American mammals such as raccoons, rabbits, dogs, horses, deer, camels, bears, pumas, and mastodonts headed south, while South American opossums, armadillos, glyptodonts, ground sloths, anteaters, monkeys, and porcupines headed north. The standard story has been that the 'superior' northern migrants wiped out the weaker southern mammals by intensive competition. However this view has been challenged, and a range of detailed studies show that the interchange was much more complex (Stehli & Webb 1985, Marshall 1988).

Marshall *et al.* (1982) have shown that, at the family level, the interchange was balanced. In terms of genera, the classic story at first seems to be confirmed: 50% of the present-day mammal genera in South America are derived from members of immigrant North American families, whereas only 21% of the present-day mammal genera in North America had their origins in South America. However, the total number

---

Figure 9.25 The biogeographic history of South America, and the land bridge. Maps showing the position of South America in the Early Cretaceous (135 Myr), Middle Eocene (50 Myr), and Early Miocene (20 Myr) across the top. Movements of major groups after the formation of the land bridge 3 Myr ago: sloths, ant-eaters, caviomorph rodents, armadillos, porcupines, opossums, ground sloths, and glyptodonts head north, and jaguars, squirrels, sabre-tooths, elephants, deer, wolves, rabbits, and horses head south. The graph (bottom left) shows how northern invaders to South America depressed the diversity of South American groups a little, but mainly added to the overall diversity by insinuation. Abbreviations: l, litopterns; n, notoungulates. (Based on various sources, including Marshall *et al.* 1982, Marshall 1988, and others.)

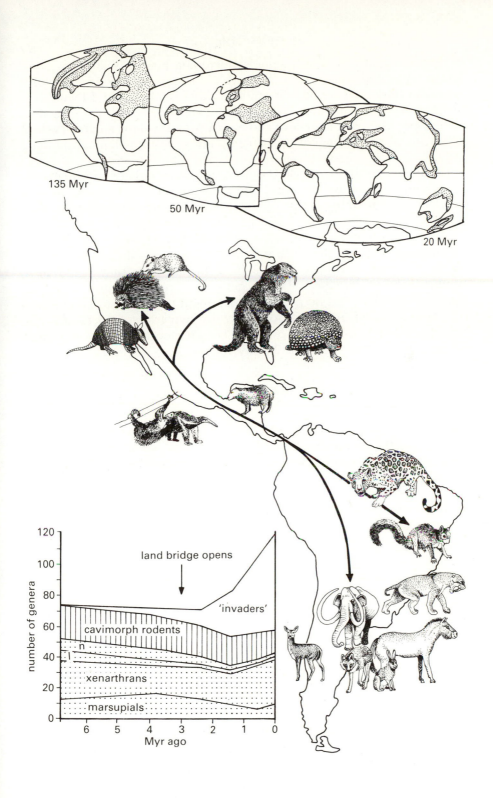

135 Myr

50 Myr

20 Myr

land bridge opens

'invaders'

cavimorph rodents

n

l

xenarthrans

marsupials

number of genera

120

100

80

60

40

20

0

6   5   4   3   2   1   0

Myr ago

of mammal genera in South America increased markedly after the land-bridge appeared (Fig. 9.25), and this increase consisted of North American immigrants which 'insinuated', that is, they found themselves niches without competing and causing extinctions among the genera already present. South America and North America show similar levels of extinction of native genera immediately after the formation of the land bridge.

The major extinctions affected South American ungulates and edentates. Were they inferior to the horses and deer from North America? The litopterns and notoungulates were already declining before the invaders arrived, and the surviving lines died out much later along with their supposed competitors, the invading mastodonts and horses. Further, the glyptodonts, ground sloths, and toxodonts were so different from the North American forms, that it is hard to see how they could have competed. Fourthly, when all of the genera of large herbivores are added together, it is clear that no gradual replacement took place – the numbers of genera of South American forms went down from 26 to 21 after the interchange, and then rose to 26 again. The Pleistocene extinctions cannot be explained by the invasions 2.5 Myr before.

THE BEGINNING OF THE AGE OF PLACENTAL MAMMALS

The Palaeocene Epoch (66–58 Myr) marks the first phases of the adaptive radiation of the mammals. While South America and Australia acquired their own largely unique fauna, the rest of the world (North America, Europe, Asia, Africa) shows a more uniform pattern (Krause 1984). The death of the dinosaurs must have left vast areas strangely empty of large land animals. A sample of life just after the world began to fill up with mammals may be seen by studying a well-known Palaeocene fauna from North America.

The Fort Union Formation of the Crazy Mountain Basin

Mammal fossils were first discovered in 1901 in the Fort Union Formation of the Crazy Mountain Basin, Montana, and large collections were made at a number of localities over the next few years. Simpson (1937) described a typical fauna of 79 species from the Gidley and Silberling Quarries (Fig. 9.26), dominated by multituberculates such as *Ptilodus*. Some of the mammals belong to familiar modern groups, such as the early lemur-like animal *Plesiadapis* (see p. 333), the shrew-like insectivore *Stilpnodon*, and the cat-like carnivore *Didymictis*. However, the other mammals belong to wholly extinct groups, the insect-eating leptictid *Prodiacodon*, the pig-like taeniodont *Conoryctes*, and the sheep-sized pantodont *Pantolambda*.

Figure 9.26 A typical mid-Palaeocene mammalian fauna, based on information from the Fort Union beds: two plesiadapiforms, *Plesiadapis*, crouch in the tree, top left, just above a multituberculate, *Ptilodus*, while two cat-like *Didymictis* feed on a carcass of the leptictid *Prodiacodon*. On the right, the pantodont *Pantolambda* looks over a low cliff at *Didymictis* below, the pig-like taeniodont *Conoryctes* feeding on wood, and the shrew-like insectivores *Stilpnodon* in the undergrowth at the front. (Based on various sources.)

The Fort Union fauna contains no large mammals, a feature typical of the Palaeocene. Very few exceeded sheep-size. Further, although some of the groups present are still with us today, most (about 75%) are wholly extinct. These major Palaeocene groups are described (see box).

PALAEOCENE MAMMALIAN DIVERSITY

Small Palaeocene mammals

The leptictids are small shrew-like insectivorous forms that existed from the Late Cretaceous to Oligocene in Asia and North America. *Leptictis*, a late form, has a long snout lined with small sharp teeth (Fig. 9.27a), evidently adapted for puncturing the skin of insects. The leptictids may be related to *Zalambdalestes* of the Late Cretaceous (see pp. 243–7) or to the living Insectivora (Novacek 1986).

The pantolestids are otter-sized animals that have broad thickly-enamelled molars which may have been used in crushing shellfish. The apatemyids are another small group of insect-eaters with no obvious descendants. *Sinclairella* (Fig. 9.27b) shows the strange dentition, part insectivore and part rodent, that characterizes the group. The cheek teeth are adapted for puncturing insect skins, while the incisors are extremely long and projecting, rather like the front gnawing teeth of a rat. The anagalids dominated the Asian Palaeocene faunas. Their broad molars indicate a diet of plant food, and the anagalids are reconstructed as being rather rabbit-like in habits and appearance.

Early rooters and browsers

The taeniodonts were a small group of North American herbivores that ranged up to pig-sized (Schoch 1986). *Stylinodon* (Fig. 9.27c) has short limbs and these are rather odd in that the forelimb and hand are larger than the hindlimb and foot. The claws are narrow and curved, and they were probably used for digging up succulent roots and tubers. The tillodonts are a second small group of herbivores whose relationships are as much a mystery as are those of the taeniodonts. They were up to bear-sized, and most evidently had a diet of tough plant material like the taeniodonts. The pantodonts may be related to the tillodonts. They were rooting and browsing forms that ranged in size and appearance from pig to hippo. *Titanoides*, one of the larger forms (Fig. 9.27d), has massive limbs, plantigrade feet (soles flat on the ground), and digging claws on its hands.

The arctocyonids include *Arctocyon* from the Palaeocene of Europe and North America (Fig. 9.28a), a sheep-sized animal that looks rather dog-like (D. E. Russell 1964). However, its molars are broad and adapted for crushing plant food, rather than slicing flesh.

The arctocyonids are usually associated with a larger group, the 'condylarths', an assemblage of five or six distinct lineages at the base of the radiation of later ungulates (Cifelli 1983, Prothero *et al*. 1988). *Phenacodus* (Fig. 9.28b), often interpreted as close to the ancestry of horses, is sheep-

---

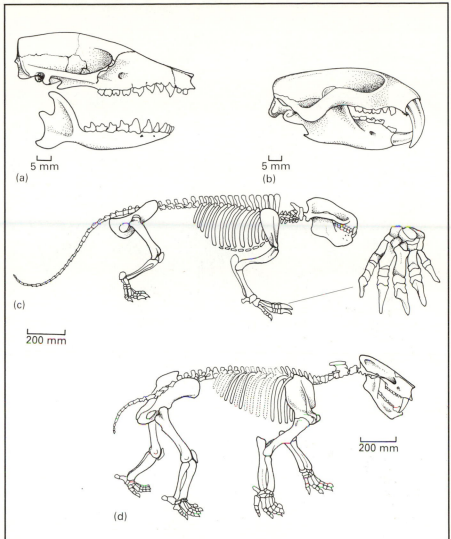

Figure 9.27 Palaeocene mammals: (a) the leptictid *Leptictis*; (b) the apatemyid *Sinclairella*; (c) the taeniodont *Stylinodon*; (d) the pantodont *Titanoides*. (Figs (a) & (b) after Scott & Jepsen 1936; (c) after Schoch 1982; (d) after Simons 1960.)

sized and the limbs are long and primitive. The outer toes are shorter than the middle three, and the cheek teeth have broad grinding surfaces as is seen in early horses (see p. 281).

The largest mammals in the Late Palaeocene and Eocene were the dinocerates, or uintatheres, of North America and Asia. *Uintatherium* (Fig. 9.28c & d) is as large as a rhinoceros, and has bony protuberances on its head. Males have 150 mm long canine teeth which may have been used in fighting, a possible explanation of the bony bumps too. Uintatheres have broad cheek teeth which were used to deal with plant food, and their brains

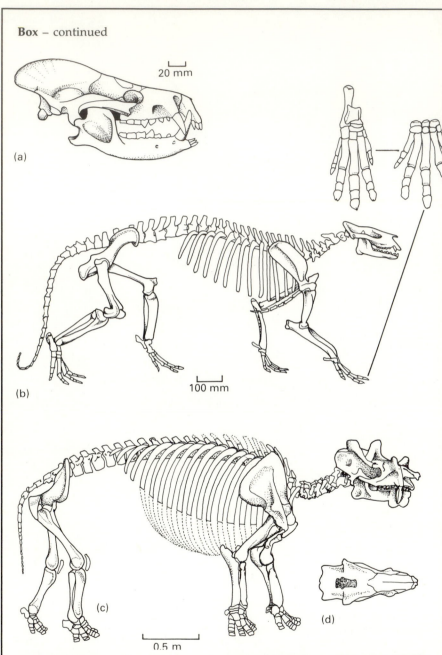

20 mm

100 mm

0.5 m

Figure 9.28 Palaeocene herbivores: (a) the arctocyonid *Arctocyon*; (b) the 'condylarth' *Phenacodus*, with anterior views of the foot and hand; (c) & (d) the dinocerate *Uintatherium*, skeleton and dorsal view of the skull, showing the area occupied by the brain shaded. (Fig. (a) after D. E. Russell 1964; (b) after Osborn 1910; (c) & (d) after Marsh 1885.)

are unusually small (Fig. 9.28d). Uintatheres appear to be unrelated to modern ungulate groups (Prothero *et al.* 1988), and their closer affinities have been sought with the arctocyonids, the anagalids, or the South American ungulates.

Palaeocene flesh-eaters

The largest mammalian meat-eaters in the Palaeocene were, strangely, ungulates called mesonychids. Early forms like *Mesonyx* (Fig. 9.29a) are about wolf-sized and have pointed molar teeth adapted for cutting flesh, just like those of a dog; they are still broad, and may also have been used for crushing bones. One of the later mesonychids, *Andrewsarchus* from the Late Eocene of Mongolia, has a vast skull, 830 mm long and 560 mm wide, larger than any other known terrestrial carnivore, and in life it must have been a terrifying 5–6 m or more long. The mesonychids are often classified as 'condylarths', and they show affinities with the arctocyonids and whales (Prothero *et al.* 1988).

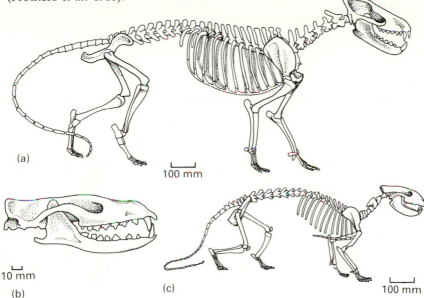

Figure 9.29 Palaeocene flesh-eaters: (a) the mesonychid *Mesonyx*; the creodonts (b) *Sinopa*, and (c) *Oxyaena*. (Fig. (a) after Scott 1888; (b) after Matthew 1909; (c) after Osborn 1895.)

The creodonts, the main meat-eaters in North America, Europe, and Asia in the early Tertiary, ranged from stoat- to bear-sized. *Sinopa*, an early fox-like creodont (Fig. 9.29b), has a low skull, and all of its cheek teeth are sharpened for cutting flesh. *Oxyaena* (Fig. 9.29c), a rather cat-like animal, has a long body and short limbs, while some later forms reached huge size, with skulls up to 650 mm long. The creodonts are generally regarded as including the ancestors of the modern carnivores (see p. 268).

The Palaeocene placental explosion

In North America, Europe, and Asia, the placental mammals underwent a rapid radiation during the 8 Myr of the Palaeocene (Rose 1981; Fig. 9.30). The 15 or so lines of extinct mammals just described diversified widely in most parts of the world. In addition, several living mammalian orders also arose in the Palaeocene, the insectivores, 'modern' carnivores, dermopterans ('flying lemurs'), bats, primates, perissodactyls, pholidotans (pangolins), and rodents. The other modern orders all arose during the subsequent 22 Myr of the Eocene.

The global diversity of mammalian families rose from 15 in the latest Cretaceous to 32 in the Early Palaeocene, 68 in the Late Palaeocene, and 78 in the Early Eocene (Benton 1987b, 1989a). Gingerich (1984) found a rise from 40 or so genera worldwide in the Late Cretaceous and Early Palaeocene, to about 120 in the Late Palaeocene, and 200 in the Early Eocene. He notes that well-preserved mammalian faunas in the Late Cretaceous of North America contain 20–30 mammalian species, while mid Palaeocene faunas have typically 50–60.

This phase of radiation of placental mammals during the Palaeocene and Early Eocene is usually treated as one of the best-known examples of an **adaptive radiation**. It is assumed that the placentals had some **key adaptation**, such as extended parental care, greater intelligence, or sharper jaws, that allowed, or even drove the radiation. As far as we

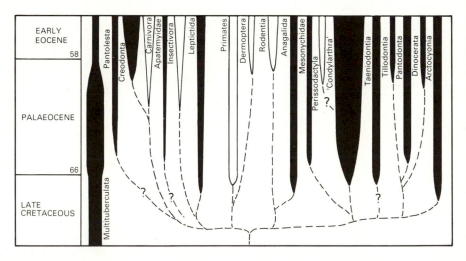

Figure 9.30 The radiation of the mammals in the Palaeocene of Europe and North America, showing two phases, one in the latest Cretaceous/earliest Palaeocene and one in the Late Palaeocene, as far as the fossil record indicates. Groups that are now extinct are shaded black, extant orders are left blank: this shows the extent of the early proliferation of diverse groups that became extinct soon after.

know, no competition took place between the dinosaurs and the mammals, and the radiation of the latter was purely opportunistic, just as was the radiation of the dinosaurs in the first place (see pp. 120–2). It is probable that the intelligence, adaptable dentitions, and extended parental care of the placentals allowed them to radiate more rapidly into a broad range of niches during the Palaeocene than say the frogs or the snakes which might equally well have taken over the world.

HEDGEHOGS, MOLES, AND SHREWS

The order Insectivora is often said to be the 'most primitive' living placental group. Certainly, living shrews and hedgehogs lack many specialized adaptations and they are ecologically close to some of the earliest mammals, but there is no reason at present to regard them as any more primitive than the primates or the carnivores.

The shrews (soricomorphs) arose in the Late Cretaceous, but they are known only from fragmentary jaws until the mid Palaeocene. The palate of the Oligocene shrew *Domnina* (Fig. 9.31a), shows the W-shaped pattern of ridges on the upper molar teeth which is typical of the group. The moles, closely related to the shrews, arose in the Eocene. The forelimbs, which are used in burrowing, are broad and paddle-like, and the mole humerus (Fig. 9.31b) is a very characteristic broad bone with large processes for the attachment of powerful muscles.

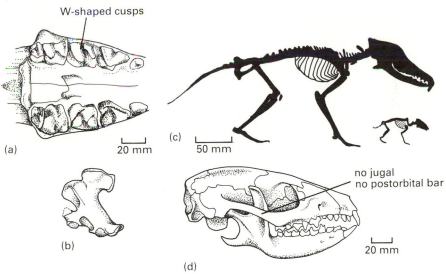

Figure 9.31 The insectivores: (a) palate of *Domnina*, an Oligocene shrew; (b) the broad humerus of the mole *Cryptoproctes*; (c) the giant Miocene hedgehog *Deinogalerix* drawn in proportion to the living *Erinaceus*; (d) skull of *Erinaceus*. (Fig. (a) modified from McDowell 1958; (b) & (d) redrawn from various sources; (c) after Butler 1981.)

The hedgehogs (erinaceomorphs) also arose in the Eocene. The most spectacular hedgehog was *Deinogalerix*, a long-limbed dog-sized animal (Fig. 9.31c) from the Late Miocene which was probably covered with stiff hair rather than spines (modified hairs). *Deinogalerix* was five times as long as the European hedgehog *Erinaceus*. The skull of *Erinaceus* (Fig. 9.31d) shows some derived characters of the Insectivora (Butler 1988), such as the loss of the jugal and the absence of a postorbital bar (reacquired in most placentals).

THE CARNIVORA

The living meat-eaters, cats, dogs, hyaenas, weasels, and seals are members of the order Carnivora. These animals are characterized by the possession of a pair of **carnassial** teeth on each side of the jaws: the upper premolar 4, and the lower molar 1 are enlarged as longitudinal blades that shear across each other like a powerful pair of scissor blades (Fig. 9.32a & b). Certain forms that crush bones, such as the hyaenas, have broad premolars with thick enamel and powerful jaw adductors. The canine teeth are generally long and used in puncturing the skin of prey animals, while carnivores use their incisors for grasping and tearing flesh, as well as for grooming, It has usually been assumed that the modern carnivores arose from the creodonts (see p. 265), but the exact origins of modern cats, dogs, bears, and seals are still controversial (Flynn *et al.* 1988).

The earliest true carnivores date from the Late Palaeocene and Early Eocene. The miacid *Vulpavus* has a long skull (Fig. 9.32c), and probably hunted small tree-living mammals.

The modern groups began to diverge in the Late Eocene and Early Oligocene, and they fall into two main groups. The first, the aeluroids or feliforms, includes the cats, hyaenas, and mongooses. The civets and mongooses (Viverridae) date back to the Palaeocene. They are abundant today in tropical Africa and Asia, and feed on a mixed diet of insects, small vertebrates, and fruit. Early viverrids gave rise to the hyaenas (Hyaenidae) in the Miocene, and the cats (Felidae) which are known from the Oligocene onwards.

During the evolution of the cats, dagger- and sabre-teeth arose independently several times (Martin 1980), and most extinct forms have larger canines than in modern lions and tigers (Fig. 9.32d & e). The sabre-toothed cats of North America and Europe are remarkably similar to the unrelated marsupial sabre-tooths of South America which share specific predatory adaptations: the lower jaw can be dropped very low; the sabre, up to 150 mm long, has a backwards curve; it is flattened like a knife blade, rather than round. *Smilodon* (Akersten 1985) used its sabres for cutting out chunks of flesh from its prey, rather than stabbing. It attacked a vulnerable young elephant, say, by sinking its teeth in superficially,

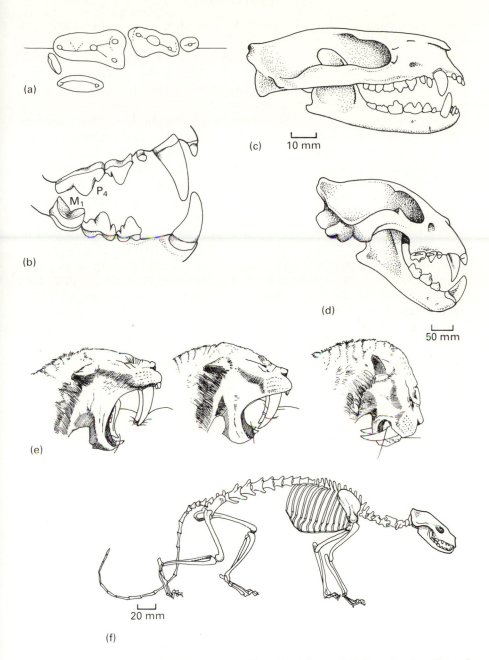

Figure 9.32 Carnivore teeth and jaws: (a) & (b) teeth of the cat *Felis* in occlusal and lateral views, showing the carnassials (upper premolar 4 (P$_4$) and lower molar 1 (M$_1$)); (c) skull of the miacid *Vulpavus*; (d) the modern tiger *Panthera*; (e) piercing and tearing flesh by the sabre-tooth *Smilodon*; (f) the early dog *Hesperocyon*. (Figs (a) & (b) after Savage & Long 1986; (c) & (f) after Matthew 1909; (d) redrawn from Thenius 1969; (e) based on Akersten 1985.)

closing the jaws, and levering a chunk of flesh off using its powerful neck muscles (Fig. 9.32e). The prey was left to bleed to death. When the abundant large elephants, rhinoceroses, toxodonts, pyrotheres, and the like died out in the Pleistocene, the sabre-tooths also disappeared.

The second carnivore group, the caniforms, includes the dogs, and the arctoids, the bears, raccoons, weasels, and seals. A typical early dog, *Hesperocyon* (Fig. 9.32f), has long limbs and digitigrade feet (the toes only touch the ground), which allowed it to run fast. The bears (Ursidae) arose in the Oligocene, and they were particularly successful in the northern hemisphere. The large extinct cave bear of Europe is known from extensive remains in the caves it used as a refuge from the icy plains over which it hunted. The raccoons (Procyonidae) and weasels (Mustelidae) are both known from the Oligocene onwards.

The seals, sealions, and walrus (Pinnipedia) appear to form a part of the arctoid group. Until recently, many zoologists split the pinnipeds into two independent groups, but a great deal of evidence now suggests (Wyss 1988) that the carnivores entered the seas once only. *Allodesmus* (Fig. 9.33a) from the Early Miocene of California is seal-like in many respects (Mitchell 1975). It is 2 m long, has broad paddle-like flippers, a very reduced tail, large eyes, and possibly some ability to detect the direction of sound underwater. The first seals, sealions, and walruses are known from the Middle and Late Miocene (15–10 Myr). *Thalassoleon*, an early sealion (Fig. 9.33b), has **homodont** teeth (undifferentiated single-cusped cheek teeth) and large orbits (Repenning & Tedford 1977).

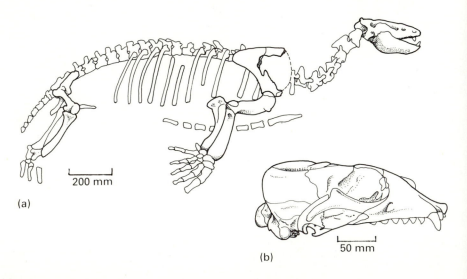

200 mm

(a)

50 mm

(b)

Figure 9.33 Fossil pinnipeds: (a) skeleton of *Allodesmus*; (b) skull of the sealion *Thalassoleon*. (Fig. (a) after Mitchell 1975; (b) after Repenning & Tedford 1977.)

THE ARCHONTA – BATS, TREE SHREWS, FLYING LEMURS AND PRIMATES

The primates (see Chapter 10) seem to have affinities with the bats and tree shrews (Novacek *et al.* 1988), based on that fact that these animals are graced with a 'pendulous penis suspended by a reduced sheath between the genital pouch and the abdomen'!

The tree shrews (order Scandentia) of south-east Asia look rather like emaciated squirrels, and yet they have often been associated with the primates. The skull (Fig. 9.34a) is primitive in many respects, but superficially primate-like in the enlarged brain and large eyes.

The flying lemurs (order Dermoptera) are a second group usually placed in the Archonta. The one living genus, the colugo *Cynocephalus* of south-east Asia, has a gliding membrane between its limbs, body, and tail, a broad flap of skin that allows it to leap for up to 100 m between trees.

The bats (order Chiroptera) include about 1000 species today, and the reason for their success is their advanced flying capabilities which make them effectively 'birds of the night' (Jepsen 1970). There are two groups of bats, the megachiropterans or fruit bats, and the more abundant microchiropterans, the small insect-eaters. Bat remains have been found in the Palaeocene, but the oldest well-known form is the Early Eocene *Icaronycteris* (Fig. 9.34b). Already all the key microchiropteran features are largely developed: the humerus, radius (and fused ulna), and digits are all elongated, and the flight membrane is supported by the spread fingers

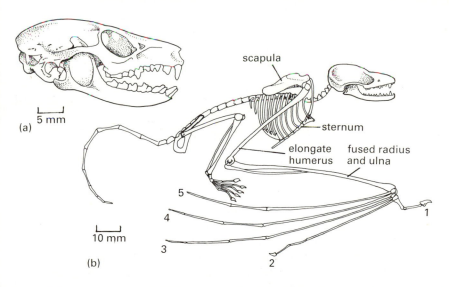

Figure 9.34 Tree shrews and bats: (a) the modern tree shrew *Ptilocercus*; (b) the Eocene bat *Icaronycteris*. (Fig. (a) modified from Evans 1942; (b) after Jepsen 1970.)

271

2–5 (digit 1, the thumb, is much shorter). The shoulder girdle is modified to take the large flight muscles on the expanded scapula on the back and the broad ribs and sternum on the front. The hindlimbs are strong, and the feet are turned backwards so that *Icaronycteris* could hang upside down as modern bats do. The eyes are large, and the ear region shows specializations for echolocation. The modern bat groups arose mainly in the Late Eocene and Oligocene, but remains are often scrappy. It is only rare conditions of preservation that give us the dramatic detail seen in *Icaronycteris*. It was found in the Green River Formation of Wyoming, a deposit better known for its remarkable fish faunas (see pp. 143–4). Other excellent specimens of bats, complete with skin impressions have been found in the renowned oil shale deposits of Messel in Germany.

THE MESSEL OIL SHALES – TOTAL PRESERVATION OF MAMMALIAN FOSSILS

The best-preserved fossils of mammals have been found in the Middle Eocene (c. 50 Myr) oil shales at Messel, near Frankfurt, West Germany (Schaal & Ziegler 1989, Franzen 1990). All details of their hair, stomach contents, and even internal organs are preserved in some cases! The Messel deposits contain abundant plant remains – laurel, oak, beech, citrus fruits, vines, and palms, with rare conifers, and ponds covered by water lilies, which indicate a humid tropical or subtropical climate. Invertebrate fossils include snails and insects, while fishes account for 90% of the vertebrate fossils. Rare frogs, toads, and salamanders have been found, as well as six genera of crocodilians, several tortoises and terrapins, and some large lizards and snakes. The birds include a wader as well as a large flightless form.

The mammal fossils, although constituting only 2–3% of vertebrates found, have attracted most attention. Thirty-five species belonging to 13 orders have been recorded so far. They include opossums, several primitive insect-eaters, and a few true insectivores.

A remarkable recent discovery is *Leptictidium*, a small leptictid (see p. 262), which was a biped, standing only 200 mm tall, that dashed about like a long-tailed leprachaun (Fig. 9.35). Three nearly complete skeletons (Storch & Lister 1985) show that it has a long tail, a strong but short trunk region, and relatively long hindlimbs and short forelimbs. The long tail suggests a balancing function as in kangaroos and in bipedal dinosaurs, and the short strong trunk also points to an ability to balance. *Leptictidium* was probably a facultative biped: it ran and walked on its hindlegs, but could have adopted a quadrupedal posture for slow locomotion and standing. The remarkable conditions of fossilization at Messel have allowed detailed studies of its diet. In one specimen, several dozen pieces of bone were found, some of which could be identified as limb bones and

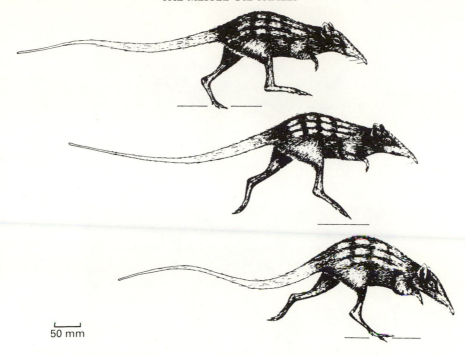

50 mm

Figure 9.35 The tiny bipedal insectivorous mammal *Leptictidium* from the Messel deposits, restoration of its running style. (After Storch & Lister 1985.)

vertebrae of a small reptile (possibly a lizard). The second skeleton contains bones of a small mammal, and the third contains fragments of chitin from the exoskeleton of large insects. The gut regions also show a variety of plant fragments, so that *Leptictidium* had a very varied diet.

Other small mammals from Messel include six species of bats, some of which have scales from butterfly wings and beetle exoskeletons preserved in their stomachs. There are two specimens of lemur-like primates and four of squirrel-like rodents. Carnivorous mammals include a creodont and two miacids (see p. 268), while ground-dwelling herbivores include a 'condylarth' (see p. 262), three perissodactyls (early horses and tapirs), and three artiodactyls (relatives of modern cattle and deer). Two of the most remarkable finds are the ant-eater *Eurotamandua* (see p. 256) and the pangolin *Eomanis* (see p. 294). The former belongs to a South American group and the latter to a south-east Asian, so that central Europe must have been a migratory cross-roads for mammals in the Eocene.

The Messel site seems to represent an Eocene lake which filled with organic matter periodically. Cadavers of land animals were washed in and birds and bats fell into the lake, and sank to the bottom. The anoxic bottom waters prevented putrefaction and scavenging and the corpses were slowly covered by organic clays and preserved as near-perfect fossils (Fig. 9.36).

273

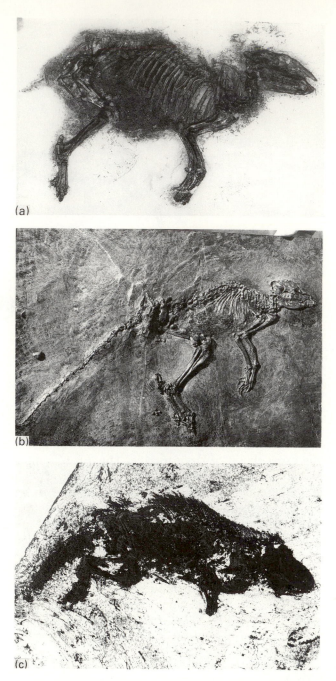

Figure 9.36 Exceptional preservation of mammalian fossils in the Messel deposits, West Germany: (a) the early horse-like animal *Propalaeotherium parvalum*, shoulder height 350 mm; (b) the dichobunid artiodactyl *Messelobunodon schaefferi*, shoulder height 220 mm; (c) the insectivore *Pholidocercus hassiacus*, length of head and trunk 190 mm, showing a clear silhouette of the fur. (Photographs courtesy of Dr J. L. Franzen, with permission of the Natur-Museum Senckenberg.)

The success of the rodents is legendary. They are a diverse and widespread order of mammals with over 1700 living species (40% of all living mammals). Their adaptability seems to know no bounds, as can be seen from the way in which mice, rats, and squirrels have modified their behaviour in order to coexist in a human landscape. Rodents are characterized by their remarkable teeth and jaws (see box) which have allowed a rapid evolutionary radiation, but their origins are still controversial (Luckett & Hartenberger 1985).

RODENT TEETH AND JAWS

Rodents have remarkable deep-rooted incisor teeth, one pair in the upper jaw and one in the lower, which grow continuously throughout life, an unusual feature among mammals. In cross-section a typical rodent skull (Fig. 9.37a) seems to be largely occupied by the deep open roots of the incisors which curve back round the snout region, and fill up most of the

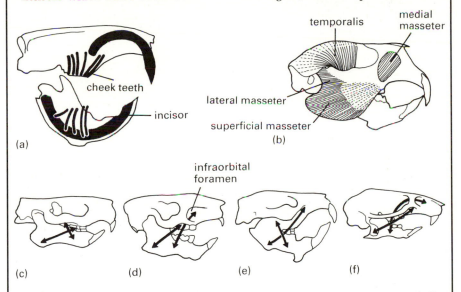

Figure 9.37 Rodent teeth and jaw muscles: (a) cross-section of a beaver skull showing the deeply rooted cheek teeth and ever-growing incisors in black; (b) main jaw muscles of the living porcupine *Erethizon*, showing the temporalis muscle, and the masseter muscle, which falls into three main portions; (c)–(f) the main lines of action of the segments of the masseter muscle in rodents with the (c) protrogomorph, (d) hystricomorph, (e) sciuromorph, and (f) myomorph patterns; in the last three, the medial masseter invades further and further forwards on the side of the snout. (Fig. (a) & (c)–(f) after Savage & Long 1986; (b) modified from Vaughan 1972.)

Box – continued

lower jaw. The incisors are used to gnaw wood, nuts, husks of fruit, and so on. They are triangular in cross-section and bear enamel only on the front face, so that the dentine behind wears faster and gives a sharp enamel cutting edge. Behind the incisors is a long diastema, a gap representing the missing second and third incisors, canine, and two or three premolars, followed by a single premolar, and three molars.

The main jaw actions of rodents are **propalinal**, that is, back and forwards. These are produced by the pterygoideus muscle which runs from the palate to the inside of the jaw and the masseter muscle, whose main portions originate generally in the snout area and run back to the outside of the lower jaw (Fig. 9.37b). The strength and effectiveness of the propalinal movements depends on the size and angle of the masseter muscle in particular. Four patterns occur in rodents (Fig. 9.37c–f):

(a) **protrogomorph**, seen in primitive forms, in which the middle and deep layers of the masseter attach to the zygomatic arch;
(b) **hystricomorph**, seen in porcupines, in which the deep masseter passes through the infraorbital foramen to attach to the side of the snout in front of the eye;
(c) **sciuromorph**, seen in squirrels and others, in which the middle masseter attaches in front of the eye; and
(d) **myomorph**, seen in rats and mice, in which the middle masseter is attached in front of the eye (as in sciuromorph), and the deep masseter passes up into the orbital area and through the infraorbital foramen.

The four muscle patterns appear to have arisen independently several times, and they do not characterize unique monophyletic groups.

Rodent evolution

Equipped with their ever-growing incisors and powerful low-angle masseters, the rodents have chewed their way through wood, tough plant fibres, and nuts for the past 40 Myr. The first rodents, the ischyromids of the Late Palaeocene and Eocene of North America and Eurasia, (Fig. 9.38a) show primitive characters in the protrogomorph jaw muscle pattern and in the teeth (Wood 1962). The cheek teeth (Fig. 9.38b) still have mound-like cusps instead of the ridges of later rodents (Fig. 9.38c), and the last molar is not fully part of the grinding dental battery. *Paramys*, and most other Eocene rodents have a primitive jaw arrangement in which the area of attachment of the masseter muscle on the dentary is a vertical surface in the same plane as the incisor tooth. This is the **sciurognathous** jaw pattern (Fig. 9.38d). A second pattern is seen in porcupines and the South American rodents in which the masseter insertion is deflected outwards, the **hystricognathous** (Fig. 9.38e) condition, that seemingly arose once only.

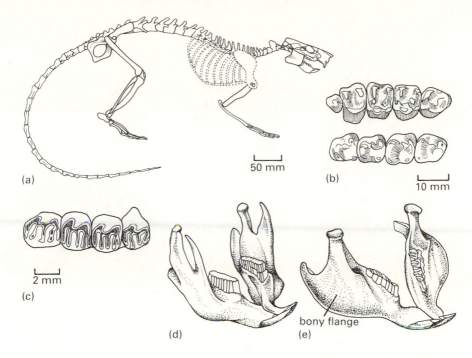

Figure 9.38 Early rodents: (a) & (b) the ischyromid *Paramys*, skeleton and cheek teeth from the upper (top) and lower (bottom) jaws, seen in occlusal view; (c) upper cheek teeth of the mouse *Theridomys* in occlusal view; (d) the sciurognathous lower jaw with vertical sides; (e) the hystricognathous jaw, with a deflected horizontal bony flange. (Figs (a) & (b) after Wood 1962; (c) after Stehlin & Schaub 1951; (d) & (e) after Savage & Long 1986.)

The sciurognathous rodents are the largest group, and include squirrels, dormice, beavers, hamsters, mice, rats, voles, and some extinct lineages. While modern beavers are known for their dam-building and tree-felling activities, some fossil forms excavated remarkable burrows. Large helical burrows named *Daimonelix* have been known for some time from the Oligocene and Miocene of Nebraska, USA. They extend to 2.5 m deep and have an upper entrance pit, a middle vertical spiral, and a lower living chamber (Fig. 9.39a). The burrow diameter is constant, and the helix may be dextral or sinistral in the same locality. These burrows have been ascribed to *Palaeocastor* (Fig. 9.39b), a primitive beaver, on the basis of complete and incomplete skeletons found in the living chamber (Martin & Bennett 1977).

The hystricognaths include several lines that radiated in Africa and South America in the Oligocene and Miocene. The early porcupine, *Sivacanthion* (Fig. 9.39c) from the Middle Miocene, is unusual in that it occurs outside Africa, in Pakistan. The largest hystricognath group is the Caviomorpha, the South American guinea pigs, capybaras, chinchillas,

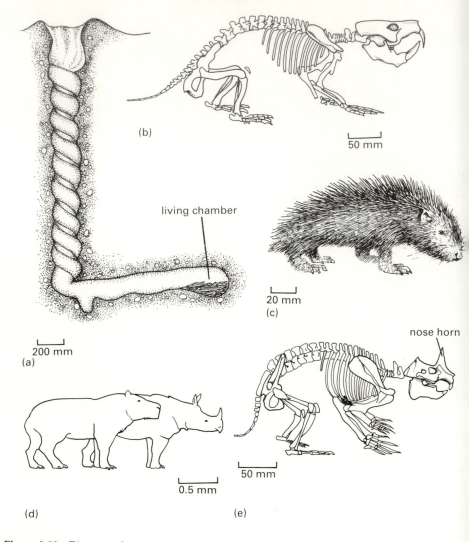

Figure 9.39 Diverse rodents: (a) spiral burrows, termed *Daimonelix*, made by (b) the Miocene beaver *Palaeocastor*; (c) restoration of the Miocene porcupine *Sivacanthion*; (d) relative size of the giant caviomorph *Telicomys* and a small rhinoceros; (e) the horned marmot *Epigaulus*. (Fig. (a) altered from Martin & Bennett 1977; (b) after Peterson 1934; (c) & (d) after Savage & Long 1986; (e) after Gidley 1907.)

and New World porcupines. The largest living caviomorph, the capybara, weighs 50 kg, and fills an ecological niche more akin to a deer than a rat or squirrel. However, it is a midget compared to some of the giant caviomorphs of the past. *Telicomys* from the Late Miocene and Pliocene (Fig. 9.39d) reached the size of a rhinoceros.

The oddest rodents were the primitive mylagaulids of the Miocene of the Great Basin, USA. *Epigaulus* (Fig. 9.39e) has broad paddle-like hands with long claws, used in digging, and small eyes, so it probably lived underground in burrows. It has a pair of small horns on the snout just in front of the eyes, whose function is a mystery, unless they were used in premating fights; not all specimens have the horns, so they may have been restricted to males only.

Rodent relatives – rabbits and elephant shrews

The closest relatives of the rodents may be the order Lagomorpha (rabbits and relatives), which share numerous derived characters of the skull and dentition (Novacek *et al.* 1988), such as the large open-rooted incisor teeth. Rabbits have a second small pair of incisors in the upper jaw while rodents have only one, but the similarities in the skull otherwise are very striking. *Palaeolagus* from the Oligocene of North America (Fig. 9.40a) is very like a modern rabbit. The tail is short, the hindlimb is long (for the characteristic jumping mode of locomotion), and the limb girdles are strong (to take up the impact of landing). The long incisors are used for nipping grass and leaves from bushes, and the broad cheek teeth are adapted for side to side grinding, rather than the propalinal movements seen in rodents.

The rare elephant shrews also seem to be related to rodents and rabbits. These small African insectivorous mammals (order Macroscelidea) date back to the Oligocene. The skull (Fig. 9.40b) is superficially shrew-like, but it shares some characters with rodents and rabbits.

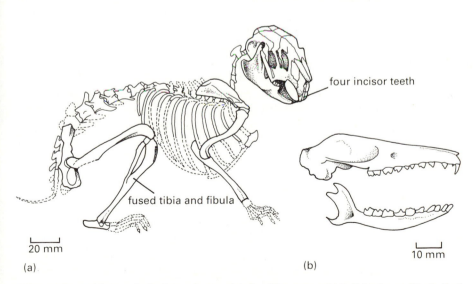

four incisor teeth

fused tibia and fibula

20 mm

10 mm

(a)

(b)

Figure 9.40 Rabbits and elephant shrews: (a) the Oligocene rabbit *Palaeolagus*; (b) skull of the elephant shrew *Elephantulus*. (Fig. (a) after Wood 1957; (b) redrawn from Lawlor 1979.)

BROWSERS AND GRAZERS – HORSES AND COWS

Palaeocene landscapes contained very few large plant-eaters. New hoofed herbivores arose in the Early Eocene, possibly from among the 'condylarths' (see p. 262) and two main groups radiated, the perissodactyls or odd-toed ungulates (1, 3, or 5 toes), such as horses, tapirs and rhinoceroses, and the artiodactyls, or even-toed ungulates (2 or 4 toes), such as pigs, cattle, deer, sheep, and camels.

The evolution of horses

Some of the first perissodactyls were horses, no larger than a terrier admittedly, but the first in what has come to be regarded as an evolutionary classic (Simpson 1961). Major changes can be observed during the history of the horses (Fig. 9.41): a reduction in the number of toes from four (front) and three (back) in the first horse *Hyracotherium*, to three in *Mesohippus*, and one in *Pliohippus* and the modern *Equus*; and a deepening of the cheek teeth from small leaf-crushing molars, to the deep-rooted grass-grinders of modern horses.

The changes in limb structure and teeth are linked to the overall increase in body size that occurred during horse evolution. The changes have been explained by a major environmental change which took place during the Late Oligocene and Early Miocene: the spread of grasslands in North America. Early horses, such as *Hyracotherium*, *Mesohippus*, and *Parahippus*, were browsers which fed on leaves from bushes and low trees. As the humid forests retreated and grasslands spread, new horse lineages, such as *Merychippus* and *Pliohippus*, stepped out on to the plains and put their reinforced molars to work. *Hyracotherium* was a cryptic animal that escaped predators by being small and blending into the background. On the open grasslands, the horses had to become large so that they could see predators some distance off, and they had to improve their running ability in order to escape. Long legs and single hooves allowed the later horses to achieve greater speeds. The dental changes were brought on by the major change in diet from leaves to grasses. Grass contains a high proportion of silica and it is very abrasive, so that grazers need high-crowned teeth which last for a long time, and they usually have complex infoldings of enamel and dentine to provide a better grinding surface.

The story of the horses has become a textbook example of 'progressive evolution' or a 'trend' since there seems to be a clearcut one-way line of change from the small leaf-eating *Hyracotherium* to the large grazing *Equus*. However, there is no evidence for uniform change, and the pattern of evolution is rather more complex than it might at first seem. There was no single line of evolution from *Hyracotherium* to *Equus*, and many sidelines branched off in the Oligocene and Miocene.

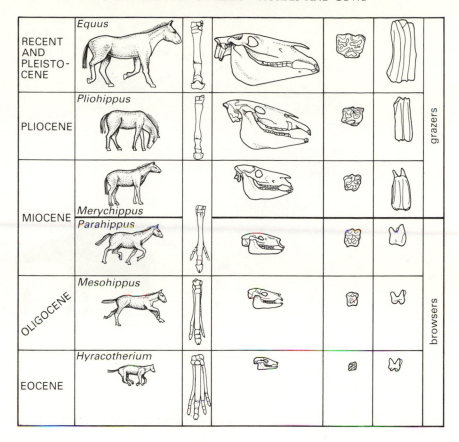

Figure 9.41 Horse evolution: sketches of body form, front limb, skull, and upper molar in occlusal and lateral views. The whole body restorations, skulls, and teeth are drawn to scale, while the legs are drawn to a standard length. Note the major changes in the skull and teeth when dietary habits changed from browsing to grazing. (Based on Savage & Long 1986 and other sources.)

Tapirs and rhinoceroses

The other living perissodactyls, the tapirs of Central and South America and south-east Asia, and the rhinoceroses of Africa and India, are probably related since they share some tooth characters. Early tapirs, such as *Heptodon* from the Eocene of North America (Fig. 9.42a) probably looked rather like the contemporary horses. The tapirs radiated in Eocene times, but settled down to a single lineage after that (Radinsky 1965). The main evolutionary change was the development of a proboscis or short trunk (Fig. 9.42b).

The rhinoceroses had a much more varied history, with a variety of spectacular families, now extinct, in the Oligocene and Miocene of North America and Asia in particular (Prothero *et al.* 1986). The Eocene and

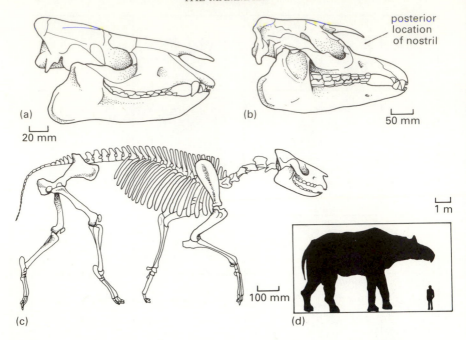

posterior
location
of nostril

(a)

(b)

50 mm

20 mm

1 m

(c)

100 mm

(d)

Figure 9.42 Tapirs and rhinoceroses: (a) the Eocene tapir *Heptodon*; (b) the modern *Tapirus*; (c) the Oligocene rhinoceros *Hyracodon*; (d) the giant Oligocene rhinoceros *Indricotherium* in silhouette, and to scale with a human. (Figs (a) & (b) altered from Radinsky 1965; (c) after Scott 1941; (d) after Savage & Long 1986.)

Oligocene rhinoceroses, such as *Hyracodon* (Fig. 9.42c), were moderate-sized hornless running animals, not unlike the early horses and tapirs. *Indricotherium* (=*Baluchitherium*), the largest land mammal of all time (Fig. 9.42d), was 5.4 m tall at the shoulder and probably weighed 30 tonnes (the largest elephants today weight 6.6 tonnes). The horned rhinoceroses radiated widely in the Miocene, but died out in North America in the Pliocene. A variety of rhinos lived in the Old World during the Pleistocene, including the extinct woolly rhino *Coelodonta* of Europe and Russia.

Brontotheres and chalicotheres

Two other lines of unusual perissodactyls, the brontotheres and the chalicotheres, which share some dental characters (Prothero *et al.* 1988), arose in the Eocene, but are now extinct. The brontotheres, or titanotheres, look very like the uintatheres (see p. 264). *Brontops* from the Early Oligocene of the USA (Fig. 9.43a) is a heavily built animal, 2.5 m high at the shoulder, and with a horn on its snout like a thickened catapult. The horn was probably covered with skin in life, and it may have been a sexual display structure. The brontotheres died out in the

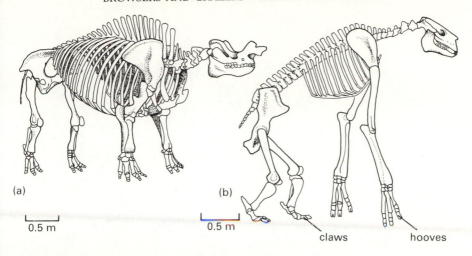

(a)

(b)

0.5 m

0.5 m

claws   hooves

Figure 9.43 Brontotheres and chalicotheres: (a) *Brontops*; (b) *Chalicotherium*. (Fig. (a) after Scott & Osborn 1887; (b) after Zapfe 1979.)

Middle Oligocene, and they were replaced ecologically by the giant rhinoceroses.

The chalicotheres, which died out in the Pleistocene, are even odder-looking than the brontotheres (Zapfe 1979). *Chalicotherium* (Fig. 9.43b), looks rather like a cross between a horse and a gorilla! The head is horse-like, but the forelimbs are very long and hindlimbs short. The pelvis is low and broad, and it is likely that *Chalicotherium* could stand bipedally and pull down leaves from high branches. The fingers bear small 'hooves', and the toes small claws which may have been useful in digging for roots. It seems that *Chalicotherium* walked with its hands curled up, a kind of knuckle-walking seen elsewhere only in chimps and gorillas.

Cattle, deer, and pigs

The even-toed ungulates, the artiodactyls, fall into two main groups, the cattle, deer, giraffes, camels, and antelopes, and the pigs and hippos (Gentry & Hooker 1988). The oldest artiodactyls were small rabbit-sized animals that fed on leaves, and had toes 3 and 4 enlarged to bear most of the weight of the body, a characteristic feature of all even-toed ungulates. The first major artiodactyl radiation, the Selenodontia, occurred in the Oligocene and Miocene with the oreodonts of North America. These low pig-sized animals (Fig. 9.44a) have four toes on each foot, and were probably not very fast-moving. Their cheek teeth (Fig. 9.44b) show the **selenodont** pattern of cusps, a feature of all cattle, deer, sheep, and camels: the cusps form pairs of crescent-shaped ridges (selenodont means 'moon tooth') which were long-lasting grinders, effective for side to side

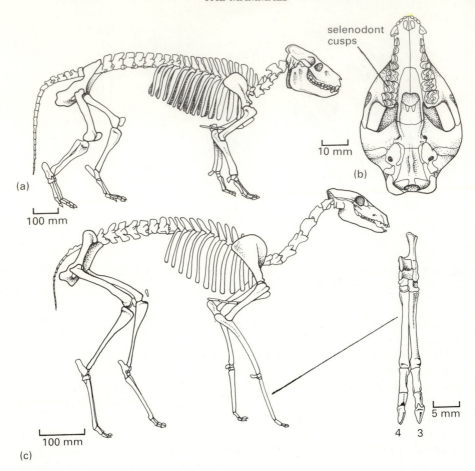

selenodont
cusps

10 mm

(b)

(a)

100 mm

5 mm

4   3

100 mm

(c)

Figure 9.44 Early artiodactyls: (a) the oreodont *Merycoidodon*; (b) ventral view of the skull of the oreodont *Bathygenys*; (c) the early camel *Poebrotherium*, skeleton and hindfoot in anterior view, showing the divergent toes 3 and 4. (Figs (a) & (c) after Scott 1940; (b) after Wilson 1971.)

chewing. Large numbers of oreodonts have been collected in the Badlands of South Dakota, and they evidently wandered the early North American wooded savannas in huge herds, browsing on low bushes.

Relatives of the oreodonts include the camels and llamas. An early camel, *Poebrotherium* from the Early Oligocene of North America (Fig. 9.44c), is a slender goat-sized animal. Like all camels, it has a long neck, long limbs, and two toes (3 and 4) which diverge slightly. It still has hooves on these toes, but by Miocene times, camels had broad pads as in modern forms. It is a strange fact that most of the evolution of camels took place in North America (Harrison 1985), and it was only in the Pliocene that they entered their present areas of North Africa, the Middle East, and South America (llamas).

The main selenodont group is the Ruminantia, cattle, sheep, antelope, and deer, so-called because they all **ruminate**, or regurgitate their food. The cow has a four-chambered stomach. A mouthful of grass enters the rumen where it is partially broken down by bacteria. The food is returned to the mouth for rumination or 'chewing the cud' and it then passes through the other three stomachs, which allows a cow to extract the maximum nutritive value from its food. Camels also have a ruminating system, but other plant-eaters, such as pigs, rhinos, and horses, lack it. Ruminants have also reduced or lost their upper incisors and have only a horny pad against which the lower incisors nip off food items. *Hypertragulus*, an early form from the Oligocene of North America, is a small rabbit-sized animal that shows the ruminant horny pad (Fig. 9.45a). Its lower canine tooth looks like an incisor, and the first premolar has taken on the canine role.

The ruminants remained small and rather rare until the Middle and Late Miocene when the modern groups radiated (Janis & Scott 1987). These, the tragulids ('mouse deer'), deer, giraffes, cattle, and antelopes, nearly all have horns of one kind or another (Fig. 9.45b–g): a bony horn core that is surrounded by a permanent horny sheath (cattle), or a bony structure that is shed annually (deer antlers), or permanent bony horns covered with skin (giraffes). These types of horns probably evolved independently in the three main groups of ruminants as fighting structures. Males of the ruminant groups use their horns in head-butting (sheep), or 'antler-wrestling' (deer), which may follow displays establishing social dominance rank, winning females, and patrolling feeding territories. Plant-eaters such as horses or camels do not have horns or antlers since they live in open grasslands and eat less clumped food resources, so that territories are unnecessary (Janis 1986).

Pigs and hippos

The pig and hippo line of artiodactyls, called the Bunodontia, radiated from Late Eocene times onwards, but never achieved the diversity of the selenodonts. During the Oligocene, North America was populated by giant pig-like animals called entelodonts. These 2–3 m long animals had long heavy skulls (Fig. 9.46a), and they may have fed on a broad range of plants near bodies of water. The deep lappets on the zygomatic arch and the knobs beneath the lower jaw may have been associated with specialized muscles for chewing, but their function is uncertain. *Perchoerus*, an early peccary (Fig. 9.46b) from the Oligocene of North America, has long canines, used in feeding and in fighting.

The replacement of perissodactyls by artiodactyls

The history of hoofed terrestrial plant-eaters seems to show a replacement

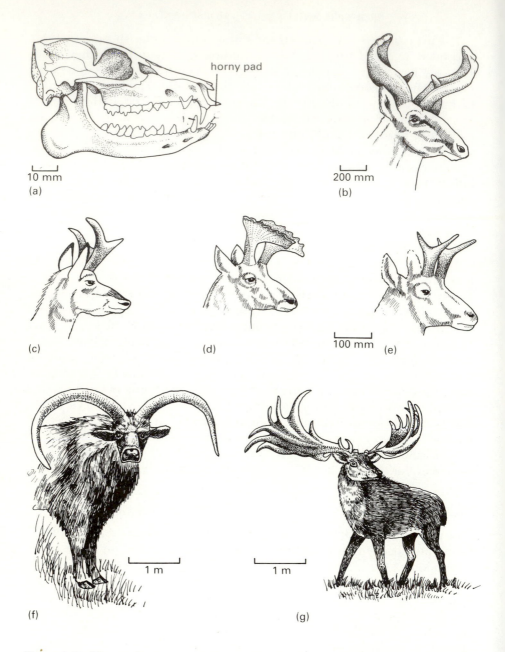

horny pad

10 mm

(a)

200 mm

(b)

(c)

(d)

100 mm

(e)

1 m

(f)

1 m

(g)

Figure 9.45 The ruminant artiodactyls: (a) the Oligocene *Hypertragulus*; (b)–(e) restored heads and horns of (b) the Pliocene giraffe *Sivatherium*, (c) the modern pronghorn *Antilocapra*, (d) & (e) the Miocene pronghorns *Ramoceros* and *Meryceros*; (f) the giant Pleistocene sheep *Pelorovis*; (g) the giant Pleistocene Irish deer *Megaloceros*. (Fig. (a) after Scott 1940; (b)–(e) after Halstead 1973b; (f) & (g) after Savage & Long 1986.)

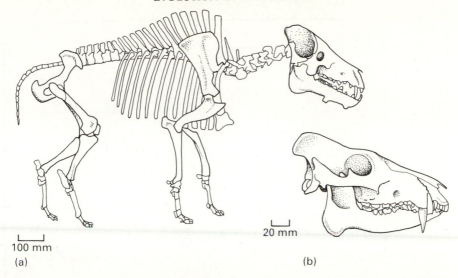

100 mm

20 mm

(a)   (b)

Figure 9.46   Fossil pigs: (a) the Oligocene entelodont *Dinohyus*; (b) the Oligocene peccary *Perchoerus*. (Fig. (a) after Peterson 1909; (b) after Scott 1940.)

of the perissodactyls by the artiodactyls. The Oligocene savannas and forests of North America and Asia were dominated by early horses, rhinos, brontotheres, and chalicotheres, while from mid Miocene times onwards, the camels, pigs, and ruminants rose to prominence. Today, there are 79 genera of artiodactyls, and only six of perissodactyls. The story of how this happened is often taken as a classic example of competitive ecological replacement on a large scale. The omnivorous suiforms and the ruminating selenodonts were able to sweep away all other plant-eaters in their path (Fig. 9.47a).

However, the statistics do not support this view. Cifelli (1981) found no evidence of a matching decline of one group and a rise in the other. In fact, the patterns of radiation and extinction of both perissodactyls and artiodactyls run more in parallel with each other than in opposition (Fig. 9.47b), and it is likely that each group was evolving independently and responding similarly to a variety of environmental stimuli. Artiodactyl success is said to have resulted from their superiority to perissodactyls, but such scenarios turn out to be seriously flawed (Janis 1976). The digestion of perissodactyls is not inferior to the ruminating digestion of the selenodont artiodactyls in all situations, since it is better adapted for coping with highly fibrous fodder.

EVOLUTION OF THE WHALES

The whales (order Cetacea) are some of the most spectacular living mammals. Looking at a great blue whale, 30 m long, or a fast-swimming

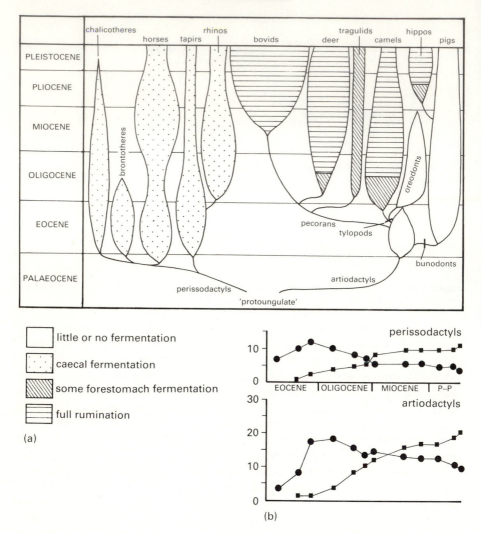

Figure 9.47 The supposed replacement of perissodactyls by artiodactyls: (a) phylogeny of the major perissodactyl and artiodactyl groups, showing their relative importance through time and their digestive mode: ruminants dominate today; (b) origination (circles) and extinction (squares) rates for perissodactyls and artiodactyls throughout the Tertiary show that both groups diversified and declined in tandem. Abbreviation: P-P, Pliocene, Pleistocene. (Fig. (a) based on Janis 1976; (b) after Cifelli 1981.)

dolphin, it is hard to imagine how they evolved from terrestrial mammal ancestors, and yet that is what happened in the Palaeocene. The oldest-known whale, *Pakicetus* from the Early Eocene of Pakistan (Gingerich & Russell 1981), has a long-snouted skull with primitive carnivorous teeth lining its jaws (Fig. 9.48a). These teeth are long and have three cusps, just as in the mesonychids (see p. 265) which appear to be the closest relatives

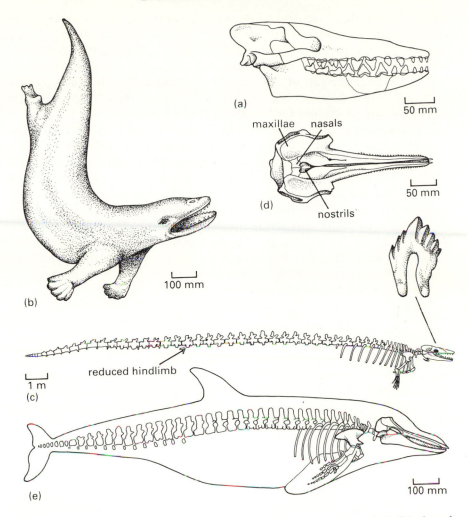

(a)

50 mm

maxillae   nasals

50 mm

(d)

nostrils

reduced hindlimb

1 m

(c)

100 mm

(b)

100 mm

(e)

Figure 9.48 The whales: (a) & (b) the first whale *Pakicetus*, reconstructed skull in lateral view, and tentative life restoration; (c) the first giant whale, *Basilosaurus*, skeleton and typical triangular-crested tooth; (d) telescoping of the skull elements in a dorsal view of the skull of *Kentriodon*; (e) skeleton of *Kentriodon*, a Miocene dolphin. (Fig. (a) altered from Gingerich & Russell 1981; (b) after Savage & Long 1986; (c)–(e) after Kellogg 1936.)

of the whales. The skeleton of *Pakicetus* is unknown, but a very tentative reconstruction (Fig. 9.48b) shows a semi-aquatic coast-dwelling carnivore which could still move on land.

By Late Eocene times, whales had become fully aquatic and very large. *Basilosaurus* (Fig. 9.48c) is over 20 m long, and has virtually lost its hindlimbs. The head is relatively small, and the teeth have a comb-like pattern of small pointed cusps.

After the Eocene, the whales radiated into two main groups (Barnes

1984), the toothed whales, such as dolphins, porpoises, and sperm whales (suborder Odontoceti), and the baleen whales such as the blue whale and humpback (suborder Mysticeti). In all modern whales, the bones of the top of the snout (premaxilla, maxilla, nasal) have moved right back over the top of the skull (Fig. 9.48d & e). This is associated with a backwards move of the nostrils to lie above the eyes (the blowhole), an adaptation for breathing at the surface, which has had the effect of telescoping the rest of the skull elements backwards. The toothed whales radiated in the Miocene, and dozens of fossil dolphin-like forms are known (Fig. 9.48d & e), with up to 300 simple pointed peg-like teeth. The toothed whales show a second advance in developing an echolocation system. The splayed bowl-like nasal region over the snout houses a fatty cushion-like mass which focuses whistles, clicks, and squeaks produced in the nasal passages and sends them out as a directed beam of sound. The echoes are picked up in the narrow lower jaw and transmitted through bone to the ear. The mysticetes have lost their teeth and have instead baleen, or whalebone, a modified protein akin to horn, which is used for filtering planktonic organisms out of the seawater.

ELEPHANTS AND THEIR RELATIVES

The two living species of elephant, the Indian and the African, are a sorry remnant of the former diversity of the group (order Proboscidea). Its early evolution took place mainly in Africa. *Moeritherium*, from the Late Eocene and Oligocene of North Africa (Fig. 9.49a) has a deep skull with the upper and lower second incisors enlarged as short projecting tusks. The skeleton indicates a long animal that was about 1 m tall and probably lived in freshwaters rather like a small hippo.

Several lines of mastodont evolution, which culminated in the modern elephants, continued through the Oligocene and Miocene (Coppens *et al.* 1978), with trends to larger size, few functional teeth, tusks, and a trunk. These changes appear to be linked. As the mastodonts became taller (modern elephants are up to 3.5 m at the shoulder), the head became heavier not least because of the large tusks. The vast head is supported on a very short neck, and so the modern elephant cannot reach the ground with its mouth. Hence, the short trunk of the early proboscideans became much longer. Modern elephants have long lives, up to 75 years, and this leads to problems of tooth wear by abrasive plant material. Whereas *Moeritherium* had all six cheek teeth operating together, as in other mammals, the modern elephant has only one in place at a time. They still have six cheek teeth, but these come into action in sequence, numbers 1–3 in the young animal, number 4 at age 4–5, number 5 at age 12–13, and number 6 at age 25 or so. This last tooth has to last for up to 50 years, and it is true to say that old elephants die when their last tooth is worn to the bone.

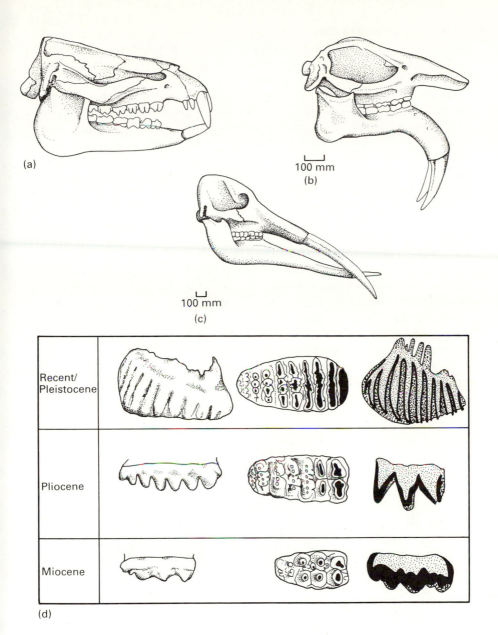

Figure 9.49 Proboscidean evolution: (a) *Moeritherium*; (b) *Deinotherium*; (c) *Gomphotherium*; (d) evolution of elephant molars from the low mounded teeth of the Miocene *Gomphotherium* (bottom), through the more incised teeth of the Pliocene *Stegodon* (middle), to the deeply ridged teeth of the living *Elephas*; teeth are shown in lateral, occlusal, and section views; enamel is black, cementum heavy stipple, and dentine light stipple. (Figs (a) & (c) after Andrews 1906; (b) after Harris 1978; (d) after Savage & Long 1986.)

One extinct proboscidean line, the deinotheres, lived until the end of the Pliocene in the Old World (Harris 1978). They have a pair of tusks curling under the chin from the lower jaw (Fig. 9.49b), which may have been used in scraping the bark from trees.

All other proboscideans fall into a major group that radiated widely in the Miocene to Pleistocene. The Miocene gomphotheres (Fig. 9.49c) have four short tusks, but they still have several cheek teeth in place at any time. They spread from Africa to Europe, Asia, North America, and even South America. Some Plio-Pleistocene lineages of mastodonts retained the primitive pattern of rounded mound-like cusps on the cheek teeth, while the mammoths and elephants further elaborated the teeth to enhance their efficiency in grinding tough plant food (Fig. 9.49d). The mounds and valleys, covered by hard crystalline enamel, become extremely deep, and they increase in number to 10–25 transverse lines of low cusps. The valleys between the cusp rows are filled with cement, so that a worn tooth is made from an alternating series of transverse lines of enamel, dentine, enamel, cement, enamel, dentine, and so on. The hard enamel forms ridges, and the whole tooth is like a large rasping file.

Mammoths, the most potent images of the Pleistocene Ice Ages (Fig. 9.50), spread from Africa over much of Europe and Asia, and later, North America. The woolly mammoth is known from many bones, as well as near-complete carcasses preserved for thousands of years in the frozen tundra of Siberia and Alaska. These show a 2.8 m tall elephant, covered with an 80 mm thick fat layer and shaggy red hair. The broad sweeping

Figure 9.50 Three woolly mammoths in a typical Ice Age scene. (Based on a painting by John Long in Savage & Long 1986.)

tusks may have been used to clear snow from the grasses and low plants which they ate. The flesh of mammoths can still be eaten, and the preservation is often good enough to yield the remnants of their last meal in the stomach or even in the mouth. Mammoths lived side by side with early humans, and died out only 12 000 years ago in Europe, and 10 000 years ago in North America.

The closest relatives of the elephants are a small group of aquatic mammals, the seacows (order Sirenia), large fat animals that live in coastal seas or freshwaters of tropical regions, and feed on water plants. They arose in the Early Eocene, and radiated during the Eocene to Miocene (Domning 1978). The Miocene dugong *Dusisiren* (Fig. 9.51a) shows the strange down-turned snout, the reduced dentition (only four cheek teeth on each side), the broad thickened ribs, front paddle, reduced hindlimb, and whale-like tail. The proboscideans, seacows, and extinct relatives share a number of derived characters of the skull, and they are known as the Tethytheria.

The hyraxes are even less likely-looking relatives of elephants than are the sirenians. These rabbit-sized animals (Fig. 9.51b) live in Africa and the Middle East, feeding on a mixed diet. They have short limbs, five-fingered hands and three-toed feet. The fossil record of hyraxes dates back to the Eocene and the group radiated in the Oligocene and Miocene, before declining to its present diversity of two genera. Hyraxes share derived characters with the tethytheres (Novacek *et al.* 1988, Tassy & Shoshani 1988), but some workers find close relationships with certain extinct perissodactyls (Prothero *et al.* 1988).

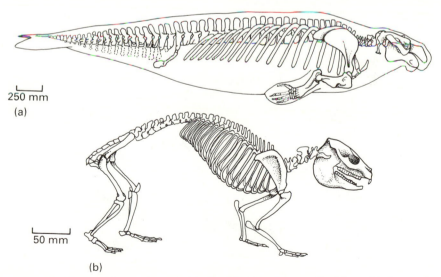

250 mm

(a)

50 mm

(b)

Figure 9.51 Proboscidean relatives: (a) the Miocene dugong *Dusisiren*; (b) the modern hyrax *Heterohyrax*. (Fig. (a) after Domning 1978; (b) after Blainville 1839.)

ODD ANT-EATERS

Two unrelated mammalian groups remain to be described. The aardvark from Africa is the sole living representative of the order Tubulidentata. It is a pig-like animal with reduced teeth, which lives in burrows and digs for termites. Fossil aardvarks date back to the Miocene (Fig. 9.52a).

The pangolins, order Pholidata, have similar long-snouted skulls for the purpose of ant-eating. They live now in Africa and south-east Asia. The skull is reduced to a tubular structure with a narrow lower jaw and no teeth. Fossil pangolins are known from North America and the oldest is *Eomanis* from the Eocene Messel pit in Germany (Storch 1978; see pp. 272–4), a surprisingly modern-looking form (Fig. 9.52b).

EXTINCTIONS IN THE ICE AGE

Many fossil mammals of the Pleistocene are regarded as typical of the Ice Ages that affected large parts of the world – animals such as the mammoth, woolly rhinoceros, giant Irish deer, giant cattle, and cave bear. However, these have all disappeared in relatively recent times, and there is considerable research interest now in trying to establish just what happened and why (Martin & Klein 1984).

The Pleistocene epoch (2–0 Myr) is marked by five or more major Ice Ages, during which the ice covering the North Pole advanced southwards, and blanketed parts of Europe as far south as Germany and England, northern Asia, and Canada. Ice also advanced outwards from the

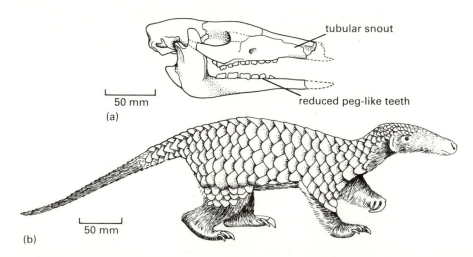

Figure 9.52 Odd ant-eaters: (a) the Miocene aardvark *Orycteropus gaudryi*; (b) the Eocene pangolin *Eomanis*. (Fig. (a) after Andrews 1896; (b) after Storch 1978.)

Himalayas and the Alps, and there were major climatic changes throughout the rest of the world. Between the Ice Ages, there were intervals of warmer weather, the main ones being interglacials, during which elephants and hippos roamed around England. The last Ice Age ended about 11 000 years ago.

At this time, somewhere between 12 000 and 10 000 years ago, the mammalian faunas of all continents underwent major changes. In North America, for example, 73% of the large mammals (33 genera) died out, including all of the proboscideans (mammoths, mastodonts), the horses, tapirs, peccaries, camels, ground sloths, and glyptodonts, as well as various predators and deer (Fig. 9.53). In South America, 46 genera died out (80%), including edentates, rodents, carnivores, peccaries, camels, deer, litopterns, notoungulates, horses, and mastodonts. In Australia, 55

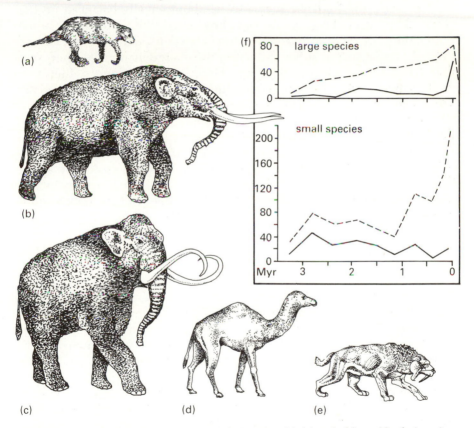

Figure 9.53 Pleistocene extinctions in North America: (a)–(e) typical large North American mammals before the extinctions: (a) the Shasta ground sloth, *Nothrotheriops*, (b) the American mastodon, *Mammut*, (c) the Columbian mammoth, *Mammuthus*, (d) the camel, *Camelops*, and (e) the sabre-toothed cat, *Smilodon*; (f) patterns of diversity (dashed line) and extinction (solid line) of mammals in North America during the past 3 Myr: large species show more dramatic extinctions in the Late Pleistocene than do small species. (Figs (a)–(e) after drawings by Mr J. Fuller, by permission; (f) redrawn from Martin & Klein 1984.)

species vanished, including echidnas, marsupial carnivores, wombats, diprotodonts, kangaroos, and wallabies. In Europe, on the other hand, the losses were less severe. True, the woolly rhino, mammoth, and giant deer died out, but others, such as the horse, hippo, musk ox, hyaena, and saiga antelope, simply contracted their ranges to other parts of the world. Extinctions in Africa and Asia at this time were seemingly modest.

Palaeontologists blame these extinctions on a number of causes, and these have polarized into two main camps. One explanation is that climates and environments changed rapidly as the ice sheet retreated, and that the large mammals in particular were vulnerable to such disturbances. The second view is that spreading human populations exerted pressure on the larger mammals in particular, and they were wiped out by hunting, the so-called 'overkill hypothesis'.

At present, the evidence seems to be about equally balanced on both sides. The 'overkillers' point out how well the spread of human populations seems to correlate with the extinctions, and also that virtually the only organisms to suffer extinction were large mammals. They argue that if there were major climatic and environmental changes, then there ought to have been extinctions among the smaller animals and plants. They also ask why the climatic changes in Europe had little effect, and why earlier glacial retreats did not cause extinction. On the other hand, the 'climatists' point to the lack of archaeological evidence of kill sites, and the fact that humans entered North America and Australia long before the bulk of the extinctions took place. They also ask why species that were probably not hunted also died out. Of course, the extinctions might be the result of a combination of both models; climatic deterioration, followed by human slaughter as the final straw.

PHYLOGENY OF THE PLACENTALS

The phylogeny of the placental mammals has been studied in more detail than that of any other group of organisms, and yet there are few points of agreement. The evolutionary tree is still a fairly unresolved 'bush' with few precise theories of relationship, possibly because many of the branching events occurred in close succession during the Palaeocene and Eocene radiation of the mammals.

A number of cladistic analyses of morphological characters of the Eutheria (placental mammals) have been published. McKenna (1975) found synapomorphies in support of a pairing of macroscelideans (elephant shrews) and lagomorphs (rabbits), and (tentatively) those two with rodents (Fig. 9.54). He also grouped primates, chiropterans, dermopterans, and scandentians in an unresolved tetratomy. He established an 'ungulate' group consisting of artiodactyls + tubulidentates, cetaceans, hyracoids + perissodactyls, and sirenians + proboscideans. He

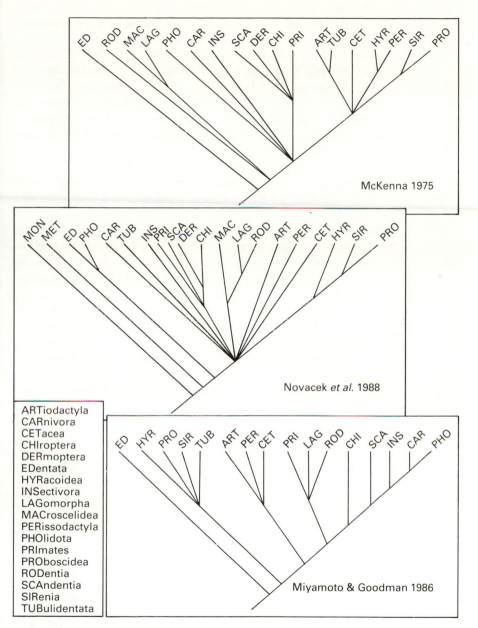

Figure 9.54 Three cladograms showing postulated relationships of the placental mammals, according to McKenna (1975), Novacek *et al.* (1988), and Miyamoto & Goodman (1986). The first two cladograms are based on morphological characters, the third on molecular data.

was unable to place the pholidotans, carnivores, or insectivores with any degree of confidence, but argued that the xenarthans were the most primitive order of placentals. Rather different cladograms were presented by other authors. Novacek *et al.* (1988) reviewed all morphologically-based cladograms of Eutheria in some detail, and their final cladogram (Fig. 9.54) is less resolved than that of McKenna! They pair edentates and pholidotans, but no longer divide the rest of the eutherians into two major groups: the 'ungulates' and the insectivores, primates, carnivores, and so on are now all united in a single 'bush'. The only resolved groups within this clade are the Archonta (scandentians, primates, dermopterans, chiropterans), an unnamed group (macroscelideans, lagomorphs, rodents), and the Paenungulata (hyracoids, sirenians, proboscideans). Novacek *et al.* (1988) could not find any evidence to support an 'ungulate' clade, while Prothero *et al.* (1988) prefer to regard it as valid on the basis of a small number of characters.

An independent line of enquiry has focused on the elucidation of eutherian relationships: molecular sequencing. Specific proteins, such as haemoglobin from the blood or myoglobin from the muscles, vary among the species of vertebrates. Closely related species have near-identical haemoglobins while distantly related species have rather different chemical patterns. The large molecules of haemoglobin can be broken up and analyzed to establish the order, or sequence, of their constituent amino acids and it is assumed that the number of differences between the haemoglobins of two species is directly proportional to the time since they shared their nearest common ancestor. Thus human and chimp haemoglobin is nearly identical (date since split 5–7 Myr), while human haemoglobin is about as different from that of a cow as it is from that of a mouse (date since split = 70–60 Myr).

Recent molecular phylogenies of Mammalia incorporate species from nearly all extant orders (excluding Dermoptera and Macroscelidea) for a cross-section of seven polypeptides (α- and β-haemoglobin, myoglobin, lens α-crystallin A, fibrinopeptides A and B, cytochrome c, and ribonuclease. Miyamoto & Goodman (1986) gave a cladogram (Fig. 9.54) which picks out five major mammalian groups. The edentates are separated from all other eutherians, as in most of the cladistic analyses. The Paenungulata are paired with tubulidentates, the artiodactyls and perissodactyls with the cetaceans, the primates with Glires (rodents + lagomorphs), and a large assemblage consisting of (chiropterans + scandentians + insectivores + carnivores + pholidotans) was established. The degree of resolution of this molecular phylogeny is similar to the morphologically-based cladograms, and many unresolved multitomies remain. Hence neither the morphological nor the molecular techniques have shed much light on the pattern of eutherian phylogeny. It is somewhat disheartening to find, indeed, that as more data become available, the less resolved is the phylogenetic tree!

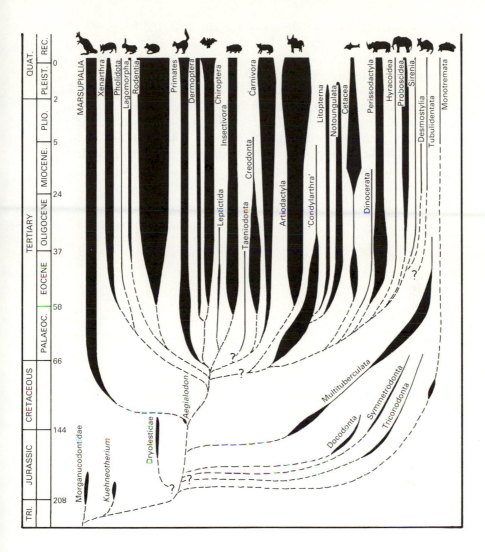

Figure 9.55 Phylogeny of the mammals, showing relative abundance, known fossil record (solid lines) and postulated relationships (dashed lines). The time scale (left) is not to scale. Abbreviations: PALAEOC, Palaeocene; PLEIST, Pleistocene; PLIO, Pliocene; QUAT, Quaternary; REC, Recent; TRI, Triassic. (Much modified from Gingerich 1984.)

THE PATTERN OF MAMMALIAN EVOLUTION

Mammals remained at low diversity for the first 160 Myr or so of their existence, during the Mesozoic, but they underwent a major radiation in the Palaeocene and Early Eocene (see pp. 260–7). Indeed, those 10 Myr or so are often regarded as the best example of a rapid evolutionary radiation, during which 20 or more new lineages arose. Since the Eocene, the rate of mammalian diversification has slowed; no major new body plans have arisen in the last 50 Myr (Fig. 9.55). However, the relative fates of the mammalian orders show changes, with great diversification of the rodents, insectivores, bats, and artiodactyls, and apparent declines of the xenarthrans, whales, perissodactyls, and proboscideans.

CHAPTER TEN

Human evolution

───────

The fossil evidence for human evolution is patchy and the early stages are poorly known. There has been a great deal of controversy over primate and human relationships, partly because of the limited number of good fossils, but probably mainly because of the intense research activity associated with them. There are more palaeoanthropologists than there are good fossils, and each of course has his own theories!

In this chapter, the fossil evidence on primate evolution is presented, with critical assessments of some of the major controversies over relationships. Fuller accounts of primate evolution may be found in Szalay & Delson (1979), Gingerich (1984), and Fleagle (1988). Primate phylogeny is surveyed by papers in Wood *et al.* (1986) and by P. Andrews (1988), and readable accounts of human evolution are given by Leakey (1981), Reader (1988), and Lewin (1989).

WHAT ARE THE PRIMATES?

There are about 200 species of living primates, of which modern humans are but one, *Homo sapiens*. All primates, from bush babies and tarsiers to gorillas and humans (Fig. 10.1), may be characterized by 30 or so characters that relate to three major sets of adaptations: (a) agility in the trees; (b) acute brain and eyesight; and (c) parental care.

Primates are essentially tree-dwellers, although many lack the remarkable agility of certain South American monkeys and the gibbons. Anatomical changes to permit this kind of activity include a very mobile shoulder joint in which the arm can be rotated in a complete circle, grasping hands and feet in which the thumb or big toe may be opposable, flat nails instead of claws, and sensitive tactile pads on all digits.

Primates have larger brains, in proportion to body size, than most mammals. In addition, the eyes are generally large and close together on the front of the face, and the snout is reduced. The flattened face of most primates allows them to look forwards and to have a large amount of overlap between the fields of vision of both eyes, which makes

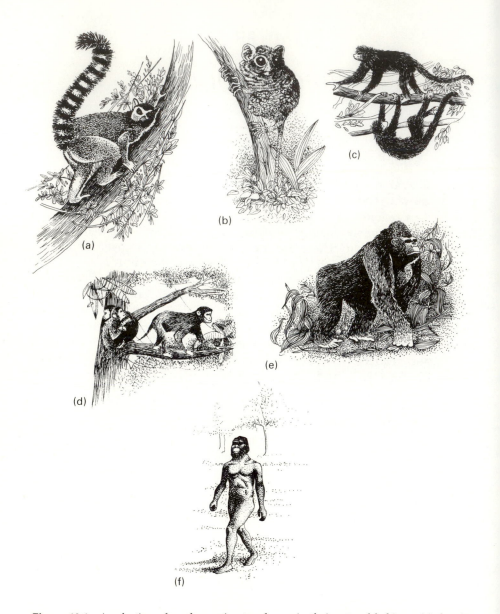

Figure 10.1 A selection of modern primates shown in their natural habitats: (a) the ring-tailed lemur, *Lemur catta*; (b) the spectral tarsier, *Tarsius*; (c) the spider monkey, *Ateles*; (d) the rhesus monkey, *Macaca*; (e) the gorilla, *Gorilla*; (f) the early human *Australopithecus*. (Based on various sources.)

stereoscopic, or three-dimensional sight possible. Primates need stereoscopic vision in order to judge distances when they leap from branch to branch, and the enlarged brain is necessary to allow them to cope with the complexities of forest life.

The third set of derived characters of the primates are to do with improved parental care of their offspring. Primates usually have only one baby at a time, the foetus is retained longer in the womb than in other mammals, and there is an extended period of parental care of the offspring. In addition, primates usually have only two mammary glands. Sexual maturity comes late, and the total life span is long. Primates have opted for a high parental investment approach which may have been essential so that the young could learn the complexities of forest life.

THE EARLY FOSSIL RECORD OF PRIMATES

The oldest reputed record of primates is a single tooth (Fig. 10.2a) from the Late Cretaceous of Montana, a lower molar with a square outline and blunt cusps. It is regarded by some as an early representative of *Purgatorius*, a member of the Plesiadapiformes, a group of five or six families that radiated in the Palaeocene and Eocene of North America and western Europe. The best known example is *Plesiadapis* itself from the Early Eocene of France (Fig. 10.2b), a squirrel-like animal with strong claws on its digits, and possible adaptations for tree-climbing (Simons 1964). The eyes are large, but face sideways, a primitive character. The long snout bears large rodent-like incisors, with large gaps behind, and broad cheek teeth for grinding plant food. Many palaeontologists doubt whether the plesiadapiforms are primates (see p. 305).

The primates radiated extensively during the Eocene, and most were adapids, which probably resembled the lemurs in life. Their long hindlimbs, grasping hands and feet, and long tail were presumably used for balancing during climbing. The snout is shorter than in *Plesiadapis* and the large eyes face forwards and were probably capable of true stereoscopic vision.

The omomyids, also largely from the Eocene of North America and Europe (Szalay 1976), were smaller than the other contemporaneous primates. *Tetonius* (Fig. 10.2c & d) has a short snout, huge stereoscopic eyes, and a bulbous braincase.

The plesiadapiforms, adapiforms, and omomyids are generally reckoned to be most closely related to the living 'prosimians' (literally 'pre-monkeys'), animals such as the lemurs of Madagascar, and the lorises of Africa and southern Asia (infraorder Lemuriformes), which are rather squirrel-like in appearance (Fig. 10.1). Their fossil record is very poor, consisting of a few Miocene specimens, and a number of recently extinct

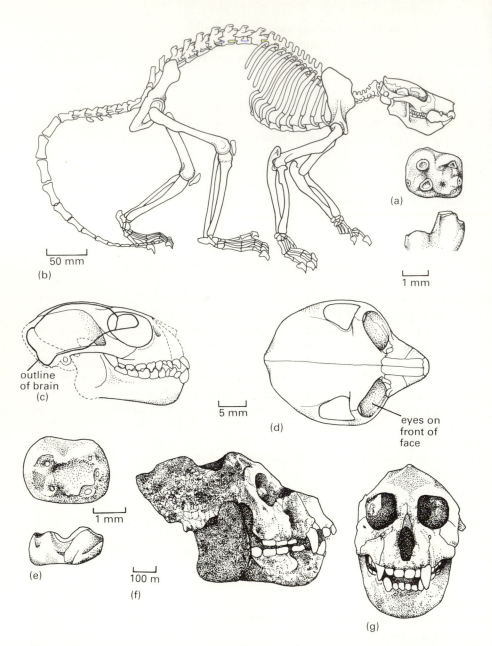

Figure 10.2 The plesiadapiforms and early primates: (a) tooth of the plesiadapiform *Purgatorius* from the Late Cretaceous; (b) skeleton of the plesiadapiform *Plesiadapis*; (c) & (d) skull of the omomyid *Tetonius* in lateral and dorsal views; the enlarged brain and forwardly-facing orbits are shown; (e) lower first molar of *Biretia* from the Eocene of Egypt in occlusal and medial views; (f) & (g) the skull of *Propliopithecus* from the Oligocene of Egypt in lateral and anterior views. (Fig. (a) after Szalay & Delson 1979; (b) modified from Tattersall 1970; (c) & (d) modified from Szalay 1976; (e) after Bonis *et al.* 1988; (f) & (g) extensively redrawn from Simons 1967.)

forms from Madagascar. The tarsier, another 'prosimian' from south-west Asia (Fig. 10.1), has a very poor fossil record.

Monkeys

The monkeys are divided into two groups which evolved separately in the New World (mainly South America) and the Old World (Africa, Asia, Europe). The New World monkeys, the platyrrhines (literally 'broad nose') have broadly-spaced nostrils that face forwards, and some have a prehensile tail. The fossil record of platyrrhines extends back to the Oligocene, but remains are sparse (Ciochon & Chiarelli 1980).

The catarrhines, or Old World monkeys and apes, with narrow snouts and non-prehensile tails, are known from fossils dating back to the Eocene, such as *Biretia* from Egypt, a single tooth (Fig. 10.2e) which appears to have some catarrhine characteristics (Bonis *et al.* 1988). Fossil catarrhines from the Oligocene and Miocene of Africa seem ancestral to both the Old World monkeys (cercopithecoids) and the apes (hominoids). For example, *Propliopithecus* (=*Aegyptopithecus*) from the Oligocene of Egypt (Fig. 10.2f & g) has a short snout, large forward-facing eyes, and an enlarged braincase. The heavy jaw and broad cheek teeth suggest a diet of fruit, while the limb bones show that *Propliopithecus* probably climbed trees and ran along stout branches (Simons 1984).

Modern cercopithecoids include the macaques of Africa, Asia, and Europe (the barbary 'ape' of Gibraltar), as well as various tree-climbing leaf-eaters, and the terrestrial baboons and mandrills. In the Pleistocene, such monkeys reached as far north as England.

Relationships of early primates

The Palaeocene and Eocene primate groups have been allied with the modern lemurs, lorises, and tarsiers in the suborder Prosimii, a group that is regarded as more primitive than the monkeys and apes (collect-ively, the suborder Anthropoidea). This division of the order Primates (Fig. 10.3a) has been challenged by some (e.g. Szalay & Delson 1979) who regard the tarsiers as more closely related to the anthropoids than to the lemuriforms, the extinct adapiforms, and the plesiadapiforms (Fig. 10.3b). The analysis by P. Andrews (1988) retains the anthropoid clade (Fig. 10.3c), virtually the only area of phylogenetic stability. The tarsiers and extinct omomyids are linked and placed as outgroups of the Anthropoidea; more distant outgroups are the lemuriforms and adapiforms. The sister group of the primates is the Scandentia (tree shrews, see p. 271), which have tactile pads on their fingers and some primate-like characters of the skull. The Plesiadapiformes are placed outside the primates here, since they seem to lack all the diagnostic characters. The main unresolved problems concern the status of the plesiadapiforms, the position of the

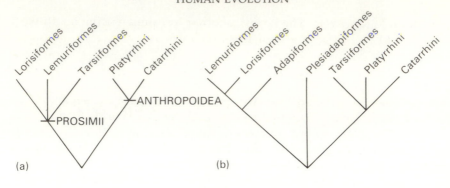

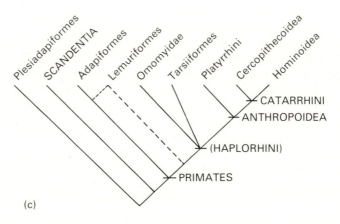

Figure 10.3 Three cladograms showing different patterns of postulated relationships of the major groups of primates: (a) the classical split into prosimians and anthropoids; (b) an alternative view in which the prosimians are divided up (Szalay & Delson 1979); (c) a recent analysis in which the prosimians are split, and the plesiadapiforms are excluded from the Order Primates (P. Andrews 1988).

tarsiers, seen by some as prosimians, and the position of the extinct adapiforms, seen by others as the sister group of the lemuriforms (Fig. 10.3c, dashed line).

APE EVOLUTION IN THE MIOCENE

The apes, superfamily Hominoidea, today include the gibbons and orang-utan of southern and eastern Asia, the gorilla and the chimpanzee from Africa, and humans. In the Miocene of East Africa, apes formed a major part of the mammalian fauna. A typical form is *Proconsul* (Fig. 10.4), now known from much of its skull and skeleton (P. Andrews 1978, Walker & Teaford 1989). *Proconsul* has a long monkey-like trunk, and the arm and hand bones share the characters of modern monkeys and apes. Many different modes of locomotion have been proposed, ranging from nearly-fully bipedal walking (when it was thought to be an early human), through knuckle walking, as seen in modern chimps and gorillas, to full **brachiation**, swinging hand over hand through the trees as in modern gibbons. The present view is that *Proconsul* could move on the ground on all fours, and run quadrupedally along heavy branches. The elbow and foot anatomy of *Proconsul* is fully ape-like, but the head is primitive, with small molar teeth, and long projecting canines (Fig. 10.4b). Its diet was probably soft fruit.

Proconsul is regarded as a true ape since it shows a number of derived characters shared with the modern forms, such as the absence of a tail, the relatively large brain size (150 cm^3), and a specialized pattern of cusps on the molar teeth (Fig. 10.4c): there are five cusps, separated by deep grooves in a Y-shape, the so-called 'Y-5 molar'.

The story of ape evolution continued in Africa during the Middle and Late Miocene (16–5 Myr), but some lines branched off and followed separate lines in Asia. These apes are generally referred to as ramamorphs, because of confusions in their nomenclature. The oldest, *Kenyapithecus* from Kenya (c. 15 Myr) was a 1 m tall animal that climbed trees and lived on the ground. *Sivapithecus* (now including *Ramapithecus*)

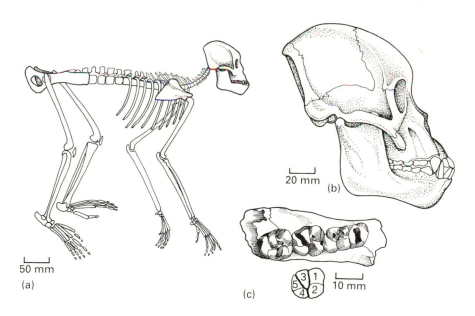

50 mm

(a)

20 mm   (b)

10 mm

(c)

Figure 10.4 The Miocene ape *Proconsul*: (a) skeleton; (b) skull; (c) jaw fragment with molar teeth, and diagrammatic representation of the Y-5 pattern (Fig. (a) after A. Walker *in* Lewin 1989, courtesy of Blackwell Scientific Publications; (b) modified from Walker *et al.* 1983; (c) after Gregory & Hellman 1929.)

is known from Turkey, northern India, Pakistan, and China (13–7 Myr). *Sivapithecus* (Fig. 10.5a) was rather like the modern orang-utan (Pilbeam 1984) with heavy jaws and broad cheek teeth covered with thick enamel, all of which suggest a diet of tough vegetation. *Sivapithecus* had a generalized quadrupedal locomotion which may have been used both in trees and on the ground.

Until about 1980, *Sivapithecus* was generally regarded as being on the line to humans, a view confirmed by a superficial comparison of palates (Fig. 10.5b–d). Apes have a rectangular dental arcade, humans have a rounded tooth row, and the palate of *Kenyapithecus* seems to form a perfect intermediate. However, the anatomical evidence suggests that the later species of *Sivapithecus* are a sideline of hominoid evolution, allied to the orang-utans.

The ramamorphs seem to have survived into the Pleistocene of China, with one of the strangest fossil apes, *Gigantopithecus*, a monstrous animal, with massive heavily worn teeth (Fig. 10.5e). It was ten times the size of *Ramapithecus*, and adult males might have reached heights of 2.5 m and weights of 270 kg. This huge animal stalked the forests of south-east Asia

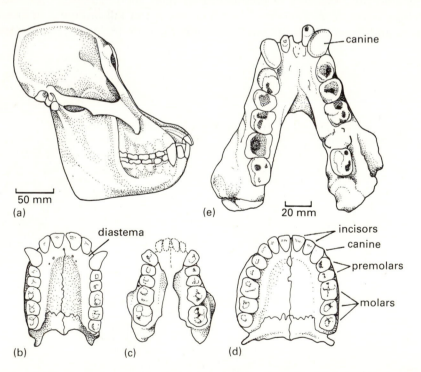

Figure 10.5 The ramamorphs: (a) skull of *Sivapithecus*; palates of (b) the chimpanzee; (c) *Ramapithecus*; and (d) modern human; (e) lower jaw of *Gigantopithecus* in occlusal view. (Fig. (a) modified from Ward & Pilbeam 1983; (b)–(d) after Lewin 1989, courtesy of Blackwell Scientific Publications; (e) based on Simons & Chopra 1969.)

from 5–1 Myr ago, and many regard it as the source of stories of yetis in Central Asia, and the big foot of North America.

Relationships of apes and humans

Until recently, most anthropologists assumed that humans formed a distinct lineage from the great apes, with forms such as *Proconsul* and *Sivapithecus* being placed on the direct line to humans. The split between apes and humans was dated to about 14 Myr, with some putting it even earlier at 15–25 Myr, thus in the Late Oligocene or Early Miocene.

However, new evidence from several lines has forced a major reconsideration of hominoid phylogeny since 1980. The main stimulus came from the findings of molecular biologists who sought to establish primate relationships. Early attempts at molecular sequencing (see pp. 297–8) in the 1960s and 1970s showed that humans were much more similar to chimps and gorillas than had been expected, and the branching point was dated at about 5 Myr (range of estimates, 7–4 Myr; Sibley & Ahlquist 1987, Miyamato *et al.* 1988). At first, these dates were regarded as gross underestimates by most anthropologists, but tests of the phylogenies using a dozen different proteins came up with the same answer. Further, a different molecular technique, comparing the similarity of the DNA of various primates, produced comparable results (Fig. 10.6a, reviewed by P. Andrews 1985). The DNA molecular clock suggested a slightly earlier divergence of the human and ape lines of 9–7 Myr ago.

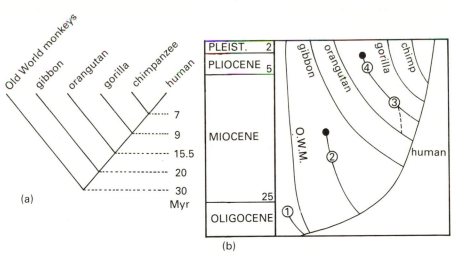

Figure 10.6 Relationships of the livng apes and humans: (a) cladogram showing postulated relationships and divergence times, based on studies of mitochondrial DNA; (b) phylogeny of the living and fossil apes and humans, based on recent morphological and molecular studies: 1, *Propliopithecus*; 2, *Proconsul*; 3, *Ramapithecus, Sivapithecus*; 4, *Gigantopithecus*. Abbreviations: O.W.M., Old World monkeys; PLEIST, Pleistocene.

A final line of evidence came from the discovery of the better specimens of *Sivapithecus*, which confirmed its relationship with orang-utans rather than humans (see p. 308). The fossil evidence is therefore consistent with a more recent ape–human split at 9–7 Myr, compatible with anatomical and molecular evidence (P. Andrews & Martin 1987). There is still some controversy over the relationships of the four living apes and humans, but the chimp–human pattern seems to be most generally accepted (Fig. 10.6a).

When this cladogram is expanded into a phylogenetic tree by the addition of fossil evidence (Fig. 10.6b), it becomes clear that *Proconsul* is a generalized hominoid, that the ramamorphs evolved side by side with the orang-utan in Asia, and that humans are part of an African ape group that has had an independent history for only 9–7 Myr.

EVOLUTION OF HUMAN CHARACTERISTICS

Two main sets of characters seem to set humans apart from the apes – bipedalism and large brain size. Bipedalism, walking upright on the hindlimbs, has led to anatomical changes in all parts of the body (Fig. 10.7). The foot became a flat platform structure with a non-opposable big toe, and straight phalanges in the toes. Apes and monkeys have a grasping foot with curved phalanges and an opposable big toe. The angle of the human knee joint shifts from being slightly splayed to being a straight hinge, and all the leg bones are longer. The hip joint faces downwards and sideways and the femur has a ball-like head that fits into it. The pelvis as a whole is short and bowl-like since it has to support the guts, and the backbone adopts an S-shaped curve. In apes, the pelvis is long and the backbone has a C-shaped curve to brace the weight of the trunk between the arms and legs. Remarkably, bipedalism introduced changes in the skull, since it now sat on top of the vertebral column, instead of at the front. The occipital condyles and the foramen magnum are placed beneath, rather than behind, the skull roof.

The evidence for the evolution of bipedalism includes the oldest hominid skeletons, dated as up to 3.6 Myr old (see p. 313), and a trackway of footprints in volcanic ash dated as 3.75 Myr old. However, bipedalism probably arose in the hominid line 10–5 Myr ago, when it split from the African apes. The origin of bipedalism appears to be tied to major environmental changes occurring in Africa from 15–10 Myr, when the tropical and subtropical forests retreated to the western part of central Africa, the Congo Basin. A rift valley opened up down the middle of Africa, running from Ethiopia to southern Africa, and habitats to the east of this line changed to more open woodlands, in places with only arid scrubby vegetation. The forest-dwelling Miocene apes became restricted to the west of Africa, where they gave rise to the gorillas and chimps,

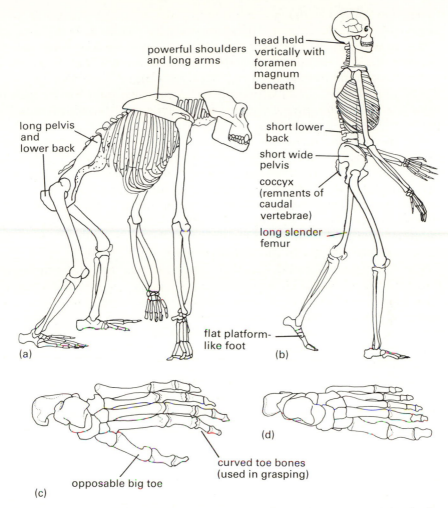

powerful shoulders
and long arms

head held
vertically with
foramen
magnum
beneath

long pelvis
and
lower back

short lower
back

short wide
pelvis

coccyx
(remnants of
caudal
vertebrae)

long slender
femur

flat platform-
like foot

(a)

(b)

curved toe bones
(used in grasping)

opposable big toe

(d)

(c)

Figure 10.7 Comparison of (a) the skeleton and (c) foot of a gorilla with those (b) & (d) of a modern human, to show major changes in posture, and the anatomical changes associated with bipedalism. (Based on Lewin 1989, courtesy of Blackwell Scientific Publications Ltd.)

while the apes that remained in the east had to adapt to life in the open.

A key adaptation to life in the open habitats was to stand upright in order to spot dangerous predators. Bipedal movement became an efficient way of covering large distances in the open, and it allowed these apes to carry food and other objects with them. This East African ape line is the family Hominidae, and all early human fossils come from the strip of land running from southern Ethiopia, through Kenya and Tanzania, to Zimbabwe and South Africa.

The second key human character was the increase in brain size which occurred much later, only about 2 Myr ago with the origin of the genus

311

Homo. The early bipedal hominids still had rather ape-like heads with brain sizes of 400–550 cm³. Modern humans have a brain size of 1000–2000 cm³ (mean, 1330 cm³) and this was approached by early species of *Homo*. Various anatomical characters changed as a result of the increase in brain size. The back of the head became enlarged to accommodate it, the face became flat, and placed largely beneath the brain, rather than in front of it. Thus, the projecting face of the apes was lost with increasing brain size in the human line, and this led to a shortening of the tooth rows. The rounded tooth row with a continous arc of teeth and no gap (diastema) between the incisors and canines (Fig. 10.5d) is a human character.

Present fossil evidence then suggests that human evolution followed a 'locomotion-first' pattern, with bipedalism arising 10–5 Myr ago, and the enlarged brain only 3–2 Myr ago. However, during most of this century, many experts held to the more comforting 'brain-first' theory, and the fossil evidence seemed to confirm their view.

The first fossil hominid skeleton was found in the mid-nineteenth century in Germany; a slouched and rather deformed specimen named Neanderthal man after the Neander Valley where it was found. This poor individual became the type 'cave man', our brutish forebear, coarse of limb, hairy of body, and small of brain. He grunted at his fellows, tore raw meat from the bones of prey animals, and huddled miserably in caves to keep warm.

An older human skeleton, found in the 1890s in Java, was hailed as the 'missing link' and named *Pithecanthropus erectus* (now *Homo erectus*), a seemingly brainy but primitive form. However, the key support for the 'brain-first' theory came in 1912 when a remarkable skull was found by an amateur, Charles Dawson, in southern England, at the village of Piltdown. The skull (Fig. 10.8a) showed a large brain size of modern proportions, but the jaw was primitive, having large coarse teeth, and suggesting a rather ape-like face. This specimen was a godsend to the leading anthropologists of the day, the true 'missing link', clearly ancient, and yet a brain forebear. Not only that, he was English!

In 1924, Raymond Dart announced an even more ancient skull from South Africa, which he named *Australopithecus africanus*. It was a child's skull (Fig. 10.8b), with rather ape-like features, and a rather small braincase. Dart's new fossil was greeted widely with scepticism. Surely it was only a fossil ape, with nothing to do with our ancestry? Piltdown man proved the 'brain-first' model.

During the 1950s, three important chains of events overthrew the whole structure of received wisdom on our ancestry. First of all, Piltdown man was shown to be a forgery – a subfossil human braincase with a modern orang-utan's jaw. The second set of events took place in southern Africa, where many specimens of *Australopithecus* had been coming to light, and the weight of new material was proving harder to discount by

312

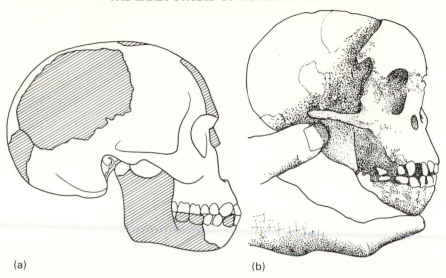

(a)   (b)

Figure 10.8 Two controversial hominid skulls of the early twentieth century: (a) Piltdown man, found in 1912, and subsequently shown to be a hoax; (b) the first skull of *Australopithecus africanus*, the Taungs baby, reported in 1924. (Modified from photographs.)

the supporters of Piltdown. Thirdly, the great champions of Piltdown man, the anatomists Professors Elliott Smith and Arthur Keith, and the palaeontologists Professor Arthur Smith Woodward, and Mr W. P. Pycraft, had died, and the unmasking of Piltdown in 1953 passed without any major public dispute.

THE EARLY STAGES OF HUMAN EVOLUTION

The line to modern humans includes at least six hominid species, three species of *Australopithecus*, and three of *Homo*. Over the last 50 years, ever-increasing numbers of hominid fossils have been found, and the story of their nomenclature and supposed place in the pattern of human evolution has been intricate.

Australopithecus afarensis – *the first hominid*

The oldest human fossils were found in the early 1970s in the Hadar region of Ethiopia by Donald Johanson and his team. Their most celebrated discovery was the skeleton of a young female, nicknamed Lucy, which consisted of 40% of the bones, remarkably complete by usual standards (Fig. 10.9a). She is 1–1.2 m tall, with a brain size of only 400 cm^3 and a generally ape-like face. Other primitive characters include the presence of a small diastema (Fig. 10.9b), long arms and rather short

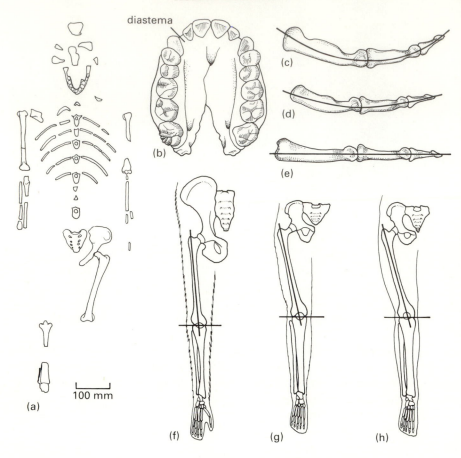

diastema

(b)

(c)

(d)

(e)

(a)

100 mm

(f)

(g)

(h)

Figure 10.9 The australopithecines: (a) skeleton of 'Lucy', the oldest reasonably complete hominid, *Australopithecus afarensis*; (b) palate of 'Lucy'; fingers of (c) an ape, (d) *Australopithecus*, and (e) a modern human, showing the loss of curvature, used for grasping branches; the hindlimbs of (f) an ape, (g) *A. afarensis*, and (h) a modern human, showing changes in pelvic shape, limb bone length and angle. (Fig. (a) modified from photographs; (b) & (f)–(h) after Lewin 1989, courtesy of Blackwell Scientific Publications Ltd.; (c)–(e) adapted from Napier 1962, copyright © 1962 by Scientific American, Inc. All rights reserved.)

legs, and curved finger and toe bones (Fig. 10.9c–e). These curved bones imply that Lucy still used her hands and feet in grasping branches, as apes do. However, *Australopithecus afarensis* is fully hominid in some significant ways: the tooth row is rounded (Fig. 10.9b), and hindlimbs and pelvis are fully adapted for a type of bipedal locomotion (Fig. 10.9f–h). Lucy is dated as slightly over 3.0 Myr old, and further specimens from the Laetoli Gorge in Tanzania are dated as 3.75 Myr old. These include some bones and the famous trackway of bipedal footprints.

The later australopithecines

The australopithecines lived on in Africa from about 3–1 Myr, and there were at least three species; *Australopithecus africanus* and *A. robustus* (=*Paranthropus robustus*) from southern Africa, and *A. boisei* from eastern Africa, a robust form rather like *A. robustus* (Walker *et al.* 1986). There were two size classes of australopithecines living in Africa at the same time (Fig. 10.10), the lightly built, or gracile, *A. africanus*, which was typically 1.3 m tall, 40 kg in body weight, and had a brain capacity of 480 cm³, and the heavier *A. robustus* (Fig. 10.11) and *A. boisei*, which were 1.75 m tall, 70 kg in body weight, and had a brain capacity of 550 cm³.

These australopithecines show advances over *A. afarensis* in the flattening of the face, the loss of the diastema, the small canine teeth, and further improvements in bipedality. They show some specializations that place them off the line to modern humans. For example, the molars and premolars and more massive than in *A. afarensis* or *Homo*, and they are covered with layers of thick enamel, adaptations in this lineage to a diet of tough plant food.

Homo habilis – *the first of our line?*

A lower jaw and other skull and skeletal remains found in 1960 and 1963 in the Olduvai Gorge, Kenya by Louis Leakey, is probably the oldest species of our own genus, *Homo*. This hominid had a large brain, in the

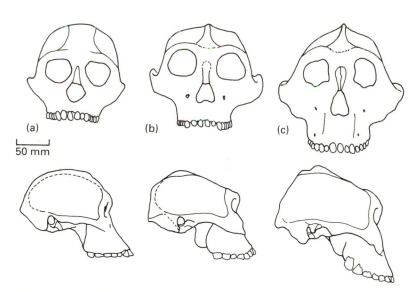

(a)

50 mm

(b)

(c)

Figure 10.10 Skull proportions of the australopithecines: skulls of (a) *A. africanus*, (b) *A. robustus*, and (c) *A. boisei* in anterior (top) and lateral (bottom) views. (Modified from Tobias 1967.)

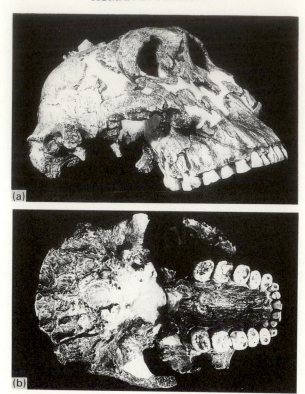

Figure 10.11 Photographs of a skull of *Australopithecus robustus* in (a) lateral and (b) ventral views. (Photographs courtesy of Mr R. E. H. Reid.)

range of 650–800 cm^3, and its hands had the manipulative ability to make tools, hence its name *Homo habilis* (literally 'handy man'). A more complete skull (Fig. 10.12) found ten years later near Lake Turkana (=Lake Rudolph) in Kenya, by Richard Leakey, showed a brain size of nearly 800 cm^3. For a hominid who was only 1.3 m tall, this is very close to the modern human range. The remains of *H. habilis* are dated as 2.5–1.5 Myr ago, and they have been found in association with the remains of various species of australopithecine. This conjures up the remarkable notion of two or three different human species living side by side, and presumably interacting in various ways.

The pattern of early hominid evolution

Most palaeoanthropologists accept that there are two separate lines of hominid evolution, *Australopithecus* and *Homo*. The key disputes concern the interpretation of *A. afarensis*. Firstly, are all the hominid species dated between 3.75 and 3.0 Myr ago one species or more than one? Secondly, are they all on the *Australopithecus* line, or both lines, or midway between

316

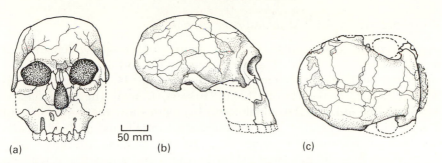

50 mm

(a) (b) (c)

Figure 10.12 The skull of *Homo habilis* in (a) anterior, (b) lateral, and (c) dorsal views. (Modified from Day *et al.* 1974.)

the two? Johanson & White (1979) argue that one species of hominid, *A. afarensis*, existed 4–3 Myr ago, and that it is ancestral to the *Australopithecus* and the *Homo* lines (Fig. 10.13a). The Leakeys and their collaborators argue that both lines were already separate 3.75 Myr ago, and that the common ancestor is much more ancient (Fig. 10.13b). Other palaeoanthropologists support a variety of other schemes with, for example, *A. africanus*, as the common ancestor of the two lines, or with the robust australopithecines forming a line quite separate from *A. africanus* which is seen as a true species of *Homo*.

THE LAST MILLION YEARS OF HUMAN EVOLUTION

Human beings spread out of eastern and southern Africa about 1 Myr ago, seemingly for the first time. Until then, all phases of evolution of *Australopithecus* and *Homo* seem to have taken place in the part of Africa between Ethiopia and South Africa.

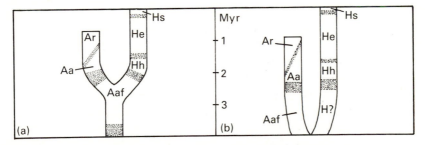

Figure 10.13 Two views of the position of the early australopithecines and the *Homo* line: (a) the view of Johanson & White (1979); (b) the view of the Leakeys. Abbreviations: Aa, *Australopithecus afarensis*; Aaf, *A. africanus*; Ar, *A. robustus*; H?, *Homo?*; He, *Homo erectus*; Hh, *H. habilis*; Hs, *H. sapiens*.

Homo erectus – *the first widespread human*

A new hominid species arose in Africa about 1.6 Myr ago, which showed advances over *H. habilis*. The best specimen, and one of the most complete fossil hominid skeletons yet found (Fig. 10.14a), was collected in 1984 by Richard Leakey beside Lake Turkana, Kenya. The pelvic shape shows that the individual is a male, and his teeth show that he was about 12 years old when he died. He stood about 1.6 m tall, and probably had a brain size of $850 \, cm^3$. The skull (Fig. 10.14b) is more primitive than *H. sapiens* since it still has large eyebrow ridges and a heavy jaw with no clear chin. However, the skeleton seems largely modern and fully bipedal in adaptations (hence the name *erectus*).

Other remains of *H. erectus* in eastern and southern Africa date from 1.6–0.2 Myr. After about 1 Myr ago, *H. erectus* seems to have started wandering worldwide, reaching north Africa by 1 Myr, China and Java by 0.8 Myr, and Europe by 0.5 Myr ago (Fig. 10.14c).

One of the richest sites for *H. erectus* is the Zoukoutien Cave near Beijing in China, the source of over 40 individuals of 'Peking Man'. They were found in cave deposits dating from 0.6–0.2 Myr, and seem to show an increase in mean brain size from $900–1100 \, cm^3$ during that time. The cave also provided evidence for a number of major cultural advances displayed by *H. erectus*: the use of fire, the use of a semipermanent home base, and tribal life of some sort. Elsewhere, *H. erectus* sites show that he manufactured advanced tools and weapons, and Acheulian hand axes of France, and that he hunted in a cooperative way. Acheulian tools date from 1.5–0.1 Myr in East Africa, and 0.8–0.1 Myr in Europe, and they are known otherwise as Old Stone Age or Palaeolithic. They show significant control in their execution with continuous cutting edges all round (Fig. 10.14d). The older Oldowan tools of East Africa, dated from 2.5–1.0 Myr, and generally ascribed to *H. habilis* are simple and rough, with usually only one cutting edge.

Neanderthal man

Neanderthal man, first found in Germany last century (see p. 312), and originally regarded as a dim-witted slouching brute, actually had a larger brain capacity (mean $1400 \, cm^3$) than modern humans (mean $1360 \, cm^3$). The heavy eyebrow ridges, massive jaws, and large teeth compared to modern *H. sapiens* (Fig. 10.15a & b) are now regarded as variations of a subspecific nature. In other words, Neanderthals were at one end of the spectrum of modern human variation and, it has been remarked that if a Neanderthal man were shaved and dressed in modern clothes, he would pass unnoticed on a busy city street (Fig. 10.15c)!

The Neanderthals have been found only in Europe and the Middle East in sites dated as 100 000 to 35 000 years old. The most abundant remains

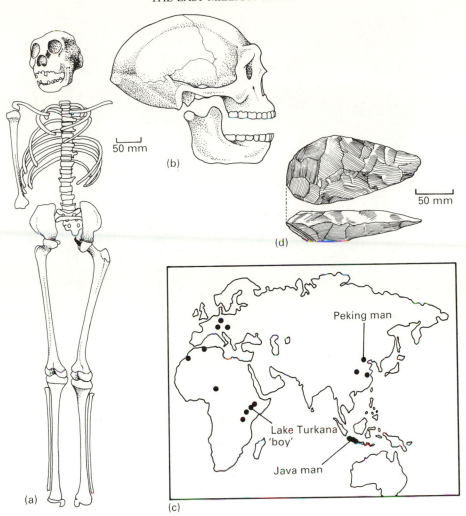

Figure 10.14 Finds of *Homo erectus*: (a) the skeleton of a youth from Lake Turkana, Kenya; (b) skull of Peking man; (c) map showing the distribution of finds of *H. erectus*; (d) Acheulian hand axe, found associated with European *H. erectus*. (Fig. (a) based on photograph by R. Leakey *in* Carroll 1987; (b) after Black 1934; (c) modified from Delson 1985; (d) after Savage & Long 1986.)

come from northern and central Europe and, in their most extreme form, they are associated with phases of the later Ice Ages that covered much of the area. A robust compact body is better able to resist the cold than our generally more slender form.

Neanderthals were culturally advanced in many ways (Trinkhaus 1986). For example, they made a variety of tools and weapons from wood, bone, and stone, the Mousterian (Middle Stone Age, Middle Palaeolithic) culture of Europe. These include delicate arrowheads, hand axes,

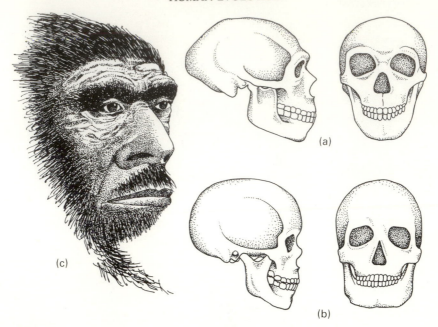

Figure 10.15 Neanderthal man: skulls of (a) Neanderthal man and (b) modern man, in lateral and anterior view; (c) restoration of the head of a Neanderthal man. (Figs (a) & (b) after Lewin 1989, courtesy of Blackwell Scientific Publications Ltd.; (c) after Savage & Long 1986.)

scrapers for removing fat from animal skins, and pointed tools for making holes in skins and for engraving designs on bone and stone, a total of 60 or so tool types. Neanderthals also made clothes from animal skins, used fire extensively, lived in caves or bone and skin shelters, and had religious beliefs. At Le Moustier in France, a teenage boy was buried with a pile of flints for a pillow and a well-made axe beside his hand. Ox bones were nearby, which suggests that he was buried with joints of meat as food for his journey to another world.

This human race seems to have disappeared about 35 000 years ago. It is not clear whether the Neanderthals were seen off by the loss of cold-weather habitat as the ice sheets retreated, whether they were killed off by more modern *H. sapiens* of our own type, or whether they interbred with the interlopers and their genes were spread amongst us all.

Modern Homo sapiens

The oldest true *H. sapiens* may be anything from 400 000 to 150 000 years old, when our species evolved from *H. erectus*. The dispute over timing depends on certain European fossils that could belong to either form, such as the Petralona skull from Greece (Fig. 10.16a), dated at 300–400 000 years old. Undisputed modern *H. sapiens* fossils are known from a variety

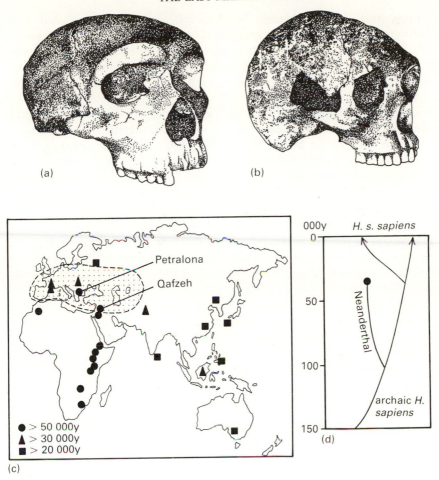

Figure 10.16 Early *Homo sapiens*: (a) the Petralona skull from Greece; (b) the Qafzeh skull from Israel; (c) distribution of early finds of *Homo sapiens*; the area in which 'classic' Neanderthals have been found is stippled; (d) postulated phylogeny of *H. sapiens*, with Neanderthal man as a short-lived subspecies. (Based on various sources.)

of sites in Africa that are 100–120 000 years old, while a specimen, dated at 92 000 years old from Qafeh in Israel (Fig. 10.16b), demonstrates that true *H. sapiens* preceded Neanderthals in the Middle East, and not the other way round.

The early evolution of modern *H. sapiens* took place in Africa and the Middle East before 110–190 000 years ago. The Neanderthals branched off and became established in Europe and western Asia before 100 000 years ago (Fig. 10.16d). After the extinction of the Neanderthals, modern *H. sapiens* spread into Europe from 40–30 000 years ago. The early European forms, often known as Cro Magnon Man, brought their advanced Upper Palaeolithic tools and filled the caves of France with paintings and carved objects. Modern *H. sapiens* then spread truly worldwide from about 40 000 years ago, reaching Russia, and travelling across Asia to the south-east Asian islands and Australia by 32 750 years ago, to North America (via Siberia and Alaska), possibly as early as 40 000 years ago, but this date is much disputed since artefacts become common only about 11 500 years ago.

The palaeontological and archaeological evidence then suggests that modern *H. sapiens* has populated the world, from a birthplace in Africa or the Middle East, in the last 40 000 years or so. This would imply that the modern human races have differentiated in this very short time. Confirming evidence has come from molecular studies, which find minute interracial genetic differences. Several studies of human DNA have also suggested an African origin for all human races 200–100 000 years ago (reviewed, Stringer & Andrews 1988).

The record of human evolution seems to show an ever-quickening pace of change. Major innovations have occurred ever more rapidly: bipedalism (10–5 Myr), enlarged brain (3–2 Myr), stone tools (2 Myr), wide geographic distribution (1 Myr), fire (0.5 Myr), art (35 000 yr), agriculture and the beginning of global population increase (10 000 yr). The rate of population increase was about 0.1% per annum at that time, rising to 0.3% per annum in the eighteenth century, and about 2.0% per annum today. In other words, the total global human population will more than double during the lifetime of any individual born today. In numerical terms at least, *Homo sapiens* have been spectacularly successful!

Appendix

GEOLOGICAL TIME

Several geological time-scales are currently in use, but dates in this book are taken from the summary scheme of A. R. Palmer 1983 (*Geology* **11**, 503–4).

Cenozoic Era

Quaternary Period	1.6–0	Myr
Pleistocene Epoch		
Tertiary Period	66.4–1.6	Myr
Pliocene Epoch	5.3–1.6	Myr
Miocene Epoch	23.7–5.3	Myr
Oligocene Epoch	38.6–23.7	Myr
Eocene Epoch	57.8–38.6	Myr
Palaeocene Epoch	66.4–57.8	Myr

Mesozoic Era

Cretaceous Period	144–66.4	Myr
Jurassic Period	208–144	Myr
Triassic Period	245–208	Myr

Palaeozoic Era

Permian Period	286–245	Myr
Carboniferous Period	360–286	Myr
Devonian Period	408–360	Myr
Silurian Period	438–408	Myr
Ordovician Period	505–438	Myr
Cambrian Period	570–505	Myr

CLASSIFICATION OF THE VERTEBRATES

The classification given below is a 'conservative cladistic' scheme based upon the cladograms described in this book. The hierarchical ranking (indenting) of the group names gives an indication of the ranking of taxa

323

in the cladogram. However, groups retain their traditional ranks as far as possible (based on Romer 1966, Carroll 1987). Fishes, amphibians, reptiles, birds, and mammals are tabulated separately in that order. As far as possible, all groups named below are monophyletic, but a small number of commonly used paraphyletic group names are given (marked*). All groups have living members, unless they are marked †.

Classification of the fishes

Phylum Chordata
  Subphylum Cephalochordata (Acraniata)
  Subphylum Tunicata (Urochordata)
  Subphylum Vertebrata (Craniata)
    *Class Agnatha
      Subclass Myxinoidea
      †Subclass Heterostraci
      Subclass unnamed
        Order Petromyzontiformes
        *†Order Thelodonti
        †Order Anaspida
      Subclass unnamed
        †Order Galeaspida
        †Order Osteostraci
        Infraphylum Gnathostomata
          Class Chondrichthyes
            †Family Cladoselachidae
            †Order Eugeneodontida
            †Order Symmoriida
            Subclass Holocephali
            Subclass Elasmobranchii
              †Order Xenacanthida
              †Order Ctenacanthiformes
              †Order Hybodontiformes
            Superorder Neoselachii
              Order Squalomorpha
              Order Galeomorpha
              Order Batoidea
        †Class Placodermi
          Order Ptyctodontida
          Order Rhenanida
          *Order Acanthothoraci
          Order Petalichthyida
          Order Phyllolepida
          Order Antiarchi
          Order Arthrodira

Superclass unnamed
 †Class Acanthodii
 Class Osteichthyes
  Subclass Actinopterygii
   *Infraclass Chondrostei
    *†Order Palaeonisciformes
    Family Polypteridae
    Family Acipenseridae
    Family Polyodontidae
    †Order Perleidiformes
    †Order Redfieldiformes
   Infraclass Neopterygii
    *Division Holostei
     †Family Lepisosteidae
     *†Family Semionotidae
     †Family Pycnodontidae
     †Family Macrosemiidae
     Subdivision Halecomorphi
    Division Teleostei
     †Family Pachycormidae
     †Family Aspidorhynchidae
     †Family Pholidophoridae
     †Family Leptolepidae
     †Order Ichthyodectiformes
     Subdivision Osteoglossomorpha
     Subdivision Elopomorpha
     Subdivision Clupeomorpha
     Subdivision Euteleostei
      *Order Salmoniformes
      Superorder Ostariophysi
      Order Stomiiformes
      †Order Aulopiformes
      Order Myctophiiformes
      Infradivision Acanthomorpha
       Superorder Paracanthopterygii
       Superorder Acanthopterygii
       Order Atherinomorpha
       Order Percomorpha
  Subclass Sarcopterygii
   Order Actinistia
   Order unnamed
    Superfamily Dipnoi
    †Superfamily Porolepiformes
   Order unnamed
    †Superfamily Osteolepiformes
    Superclass Tetrapoda

325

Classification of the amphibians

Superclass Tetrapoda
  *Class Amphibia
    Subclass Batrachomorpha
      ?†Family Ichthyostegidae
      ?†Order Aïstopoda
      †Order Nectridea
      †Family Colosteidae
      †Order Microsauria
      *Order Temnospondyli
        †Family Dendrerpetontidae
        †Family Brachyopidae
        †Suborder unnamed
          Family Capitosauridae
          Superfamily Rhinesuchoidea
          Family Rhytidosteidae
          Family Metoposauridae
          Family Trematosauridae
        †Family Eryopidae
        †Family Plagiosauridae
        †Family Dissorophidae
        †Family Branchiosauridae
        †Family Doleserpetontidae
        Infraclass Lissamphibia
          Order Gymnophiona
          Order Urodela
          Order Anura
    Subclass Reptiliomorpha
      †Family Loxommatidae
      †Family Crassigyrinidae
      †Suborder Anthracosauroidea
      †Suborder Seymouriamorpha
      †Suborder Diadectomorpha
      Series Amniota

Classification of the reptiles

Series Amniota
  *Class Reptilia
    Subclass Synapsida
      *†Order Pelycosauria
        Family Eothyrididae
        Family Caseidae
        Family Varanopseidae
        Family Ophiacodontidae

Family Edaphosauridae
Family Sphenacodontidae
Order Therapsida
  †Suborder Biarmosuchia
  †Suborder Dinocephalia
  †Suborder Dicynodontia
  †Suborder Gorgonopsia
  Suborder Cynodontia
    †Family Procynosuchidae
    †Family Galesauridae
    †Family Cynognathidae
    †Family Diademodontidae
    †Family Chiniquodontidae
    †Family Tritylodontidae
    †Family Tritheledontidae
  Class Mammalia (see below)
?†Subclass Parareptilia
  Family Mesosauridae
  Family Procolophonidae
  Family Millerettidae
  Family Pareiasauridae
Subclass Anapsida (*sensu stricto*)
  †Family Captorhinidae
  Order Testudines (Chelonia)
    †Family Proganochelyidae
    Suborder Pleurodira
    Suborder Cryptodira
      Superfamily Baenoidea
      †Family Meiolaniidae
      Superfamily Chelonioidea
      Superfamily Trionychoidea
      Superfamily Testudinoidea
  †Family Protorothyrididae
  Subclass Diapsida
    †Family Petrolacosauridae
    †Family Weigeltisauridae
    Infraclass Lepidosauromorpha
      †Order Younginiformes
      Superorder Lepidosauria
        Order Sphenodontida
          Family Sphenodontidae
          †Family Pleurosauridae
        Order Squamata
          *Suborder Lacertilia (Sauria)
            Infraorder Gekkota

Infraorder Iguania
Infraorder Scincomorpha
Infraorder Anguimorpha
Infraorder Amphisbaenia
Suborder Serpentes (Ophidia)
Infraclass Archosauromorpha
†Family Trilophosauridae
†Family Rhynchosauridae
†Order Prolacertiformes
Division Archosauria
†Family Proterosuchidae
†Family Erythrosuchidae
†Family Euparkeriidae
Subdivision Crocodylotarsi
†Family Phytosauridae
†Family Stagonolepididae
†Family Rauisuchidae
†Family Poposauridae
Superorder Crocodylomorpha
†Family Saltoposuchidae
†Family Sphenosuchidae
Order Crocodylia
†Family Protosuchidae
*†Suborder Mesosuchia
Family Teleosauridae
Family Metriorhynchidae
Family Sebecidae, etc.
Suborder Eusuchia
Family Gavialidae
Family Crocodylidae
Family Alligatoridae
Subdivision Ornithosuchia
†Family Ornithosuchidae
†Family Lagosuchidae
†Order Pterosauria
*Suborder Rhamphorhynchoidea
Suborder Pterodactyloidea
*†Superorder Dinosauria
Family Herrerasauridae
Order Saurischia
Suborder Theropoda
Infraorder Ceratosauria
Infraorder Carnosauria
Family Ornithomimidae
Infraorder Maniraptora

Family Compsognathidae
Family Coeluridae
Family Oviraptoridae
Family Dromaeosauridae
Family Troodontidae
Class Aves (see below)
Suborder Sauropodomorpha
*Infraorder Prosauropoda
Family Thecodontosauridae
Family Plateosauridae
Family Melanorosauridae
Infraorder Sauropoda
*Family Cetiosauridae
Family Camarasauridae
Family Brachiosauridae
Family Diplodocidae
Family Titanosauridae
Order Ornithischia
Family Pisanosauridae
Family Fabrosauridae
Suborder Cerapoda
Infraorder Ornithopoda
Family Heterodontosauridae
Family Hypsilophodontidae
*Family Iguanodontidae
Family Hadrosauridae
Infraorder Pachycephalosauria
Infraorder Ceratopsia
Family Psittacosauridae
Family Protoceratopsidae
Family Ceratopsidae
Suborder Thyreophora
Family Scelidosauridae
Infraorder Stegosauria
Infraorder Ankylosauria
Family Nodosauridae
Family Ankylosauridae
Diapsida *incertae sedis*
†Superorder Sauropterygia
Order Placodontia
Order Nothosauria
Order Plesiosauria
Family Plesiosauridae
Family Cryptoclididae
Family Elasmosauridae

329

Family Pliosauridae
†Order Ichthyosauria

Classification of the birds

Class Aves
  †Family Archaeopterygidae
  Subclass Ornithurae
    †Order Hesperornithiformes
    †Order Ichthyornithiformes
    †Order Enantiornithiformes
    Infraclass Neornithes
      Division Palaeognathae
      Division Neognathae
        Superorder unnamed
          Order Anseriformes
          Order Galliformes
        Superorder unnamed
          Order Podicepiformes
          Order Gaviiformes
          Order Sphenisciformes
          Order Pelecaniformes
          Order Procellariformes
        Superorder unnamed
          Order Gruiformes
          Order Charadriiformes
          Family Ardeidae
          Order Columbiformes
        Superorder unnamed
          Order Ciconiiformes
        Superorder unnamed
          Order Falconiformes
          Order Strigiformes
        Superorder unnamed
          Order Caprimulgiformes
          Order Apodiformes
        Superorder unnamed
          Order Coraciiformes
          Order Piciformes
          Order Passeriformes

Classification of the mammals

Class Mammalia
  †Family Sinoconodontidae

330

†Family Morganucodontidae
†Family Kuehneotheriidae
†Family Dryolestidae
Subclass unnamed (=Mammalia *sensu stricto*)
  Order Monotremata
  †Order Triconodonta
  †Order Symmetrodonta
  †Order Docodonta
  †Order Multituberculata
  Infraclass Theria (Tribosphenida)
    Division Metatheria
      ?†Family Deltatheridiidae
      Order Marsupialia
        1. Australian groups
          Suborder Dasyuroidea
          Suborder Perameloidea
          Suborder Phalangeroidea
          Suborder Phascolarctoidea
          Suborder Vombatoidea
        2. South American groups
          Suborder Didelphoidea
            Family Didelphidae
            Family Microbiotheriidae
            †Family Borhyaenidae
            †Family Thylacosmilidae
            †Family Argyrolagidae
          Suborder Caenolestoidea
    Division Eutheria (Placentalia)
      Cohort Edentata
        Order Xenarthra
          Infraorder Loricata
            Family Dasypodidae
            †Family Glyptodontidae
          Infraorder Pilosa
            †Family Megatheriidae
            Family Bradypodidae
          Infraorder Vermilingua
        Order Pholidota
      Cohort Epitheria
        Superorder Insectivora
          †Order Leptictida
          Order Lipotyphla
            Suborder Erinaceomorpha
            Suborder Soricomorpha
        Superorder unnamed

†Order Creodonta
Order Carnivora
  †Family Miacidae
  Suborder Feliformia (Aeluroidea)
    Family Viverridae
    Family Hyaenidae
    Family Felidae
  *Suborder Caniformia
    Family Canidae
    †Family Amphicyonidae
    Family Ursidae
    Family Procyonidae
    Family Mustelidae
  Suborder Pinnipedia
Order Tubulidentata
Superorder Archonta
  Order Scandentia
  Order Dermoptera
  Order Chiroptera
    Suborder Megachiroptera
    Suborder Microchiroptera
  †Order Plesiadapiformes
  Order Primates
    *Suborder Prosimii
      †Infraorder Adapiformes
      Infraorder Lemuriformes
      †Family Omomyidae
      Infraorder Tarsiiformes
    Suborder Anthropoidea
      Infraorder Platyrrhini
      Infraorder Catarrhini
        Superfamily Cercopithecoidea
        Superfamily Hominoidea
          †Family Pliopithecidae
          Family Hylobatidae
          Family Pongidae
          Family Hominidae
Superorder 'Anagalida'
  †Order Anagalida
  Order Macroscelidea
  Order Rodentia
    Suborder Sciurognathi
      †Superfamily Ischyromyoidea
      Infraorder Sciuromorpha
      Infraorder Myomorpha

Suborder Hystricognathi
  Infraorder Hystricomorpha
  Infraorder Phiomorpha
  Infraorder Caviomorpha
Order Lagomorpha
?Superorder Ungulata
  Order Artiodactyla
    †Family Dichobunidae
    Suborder Bunodontia (Suina)
      †Family Entelodontidae
      Family Suidae
      †Family Anthracotheriidae
      Family Hippopotamidae
    Suborder Selenodontia
      Infraorder Tylopoda
        Family Camelidae
        †Family Merycoidodontidae
      Infraorder Ruminantia
        †Family Hypertragulidae
        Family Tragulidae
        Family Cervidae
        Family Giraffidae
        Family Antilocapridae
        Family Bovidae
  †Order Arctocyonia
  *†Order Condylarthra
  Order Perissodactyla
    Superfamily Equoidea
    Suborder unnamed
      †Superfamily Brontotherioidea
      †Superfamily Chalicotherioidea
    Suborder Ceratomorpha
      Superfamily Tapiroidea
      Superfamily Rhinoceratoidea
  Grandorder Cete
    †Family Mesonychidae
    Order Cetacea
  ?†Grandorder Meridiungulata
    Order Litopterna
    Order Notoungulata
    Order Astrapotheria
    Order Pyrotheria
  Grandorder Paenungulata
    †Order Dinocerata
    Order Hyracoidea

Mirorder Tethytheria
   Order Sirenia
   Order Proboscidea
      †Family Moeritheriidae
      †Family Deinotheriidae
      †Family Mammutidae
      †Family Gomphotheriidae
      †Family Stegodontidae
      Family Elephantidae
Superorder unknown
  †Order Pantolesta
  †Order Apatemyida
  †Order Taeniodonta
  †Order Tillodontia
  †Order Pantodonta

Glossary

acrodont Teeth fused to the jaw bones.

adductor muscles Jaw-closing muscles that run from the skull roof or braincase region to the back of the lower jaw.

amniote A tetrapod that lays cleidoic eggs (i.e. a reptile, bird or mammal).

amphistylic Jaw suspension in which the upper jaw is attached at two points to the cranium.

analogy Comparable biological structures or functions that arose independently.

antorbital fenestra A skull opening between the nostril and the orbit; characteristic of archosaurian reptiles.

aorta Major blood vessel carrying oxygenated blood from the heart to the body.

apatite The crystalline component of bone; calcium phosphate.

auditory ossicles The small bones in the middle ear that transmit sound from the tympanum to the inner ear.

bicuspid Two-pointed tooth crowns, as seen in living amphibians.

brachiation Locomotion by swinging with the arms.

buccal Of the mouth cavity.

calcite Calcium carbonate.

cartilage Non-mineralized skeletal material, often developmentally a precursor of bone.

caudal Of the tail region.

centrum The cotton-reel-shaped lower portion of a vertebra.

cervical Of the neck.

character A describable feature of an organism that may be used in phylogenetic analysis.

choana An opening through bone, usually with a depressed periphery.

cilium (pl. **cilia**) Hair-like projection from a cell.

clade A monophyletic group.

cladistics Phylogenetic analysis involving the search for monophyletic groups by means of character analysis.

cladogram A dendrogram (tree-like diagram) produced by cladistic analysis, showing the relationships of groups.

claspers Pelvic elements in sharks and some other fishes found in males, and used during mating.

cleidoic Enclosed, as in the cleidoic (amniotic) egg.

community An assemblage of organisms that live in close contact and interact with each other.

coprolite Fossilized excrement.

cranial nerves The nerves of the head that run directly from the brain to particular sensory structures.

dentine The main constituent of teeth, lying within the enamel crown and root regions.

depressor muscle The muscle that opens the jaws, running from the back of the skull to the retroarticular process of the lower jaw.

dermal bone Bone formed embryologically in the outer portions of the body, within the skin.

deuterostomes Animals in which, embryologically, the opening at the cup-shaped (gastrula) stage becomes the anus: chordates and echinoderms are the main deuterostome groups.

digitigrade stance Posture in which the animal stands only on the tips of its toes.

diphycercal tail Narrow symmetrical tail of an aquatic vertebrate, in which there are only modest fins above and below the middle line.

dorsal Of the back.

ectotherm An animal that uses external means to control its body temperature.

embryology The study of embryos; development from the egg to hatching/birth.

enamel The crystalline material covering the crown of a tooth.

endemic Restricted in distribution to a single area.

endochondral bone Bone formed from cartilage, usually deep within the body.

endotherm An animal that uses internal means to control its body temperature.

faunal province A geographic area that is typified by one or more characteristic species.

fenestra An opening through bone.

gastrolith A stomach stone, swallowed by reptiles and birds to aid digestion.

gill arches The rods of cartilage or bone that support and surround the gills.

glenoid The mobile articulation between the shoulder girdle and the arm, and between the lower jaw and the skull.

heterocercal An asymmetrical aquatic tail, in which the upper portion is larger than the lower.
histology The study of biological tissues.
homeotherm An animal with a uniform body temperature.
homocercal A symmetrical aquatic tail, in which both upper and lower portions are equal in size and mirror images of each other.
homology Resemblance in biological structure or function that is the result of shared common ancestry.
hyostylic The jaw suspension of modern fishes, in which the upper jaw bone (palatoquadrate) contacts the cranium in only one place at the front, and moves against the hyomandibular behind.

interpterygoid vacuity Gap in the palate in the midline, between the pterygoids.

kinetic Mobile, in which separate bones may move relative to each other (usually of the skull).

larva A juvenile stage that differs from the adult (e.g. a tadpole).
lateral line A sensory line around the skull and along the side of the body, found in fishes and some aquatic amphibians.
lepidotrichia Small jointed bones in the fins of bony fishes.
lingual Of the tongue; the inside face of the jaw bones.
living fossil An animal with a long history and which has apparently not changed much over millions of years.
lumbar Of the lower back region.

mandible The lower jaw.
metamorphosis. Change from a juvenile larva to a rather different adult form.
monophyletic A group that contains all the descendants of a single common ancestor.
muscle scar A roughened area on the surface of a bone that indicates the site of a muscle attachment.
myotomes The muscle blocks along the length of the body of a chordate.

neural arch The upper portion of a vertebra, above the centrum.
neurocranium The braincase bones.
notochord A flexible rod running the length of the back of chordates, precursor of the spinal column in more derived forms.
nuchal Of the neck region.

occlusion   Precise meeting of the grinding surfaces of the teeth.

orbit   Eye socket.

ossified   Turned into bone.

otic capsule   The bones enclosing the inner ear region of the braincase.

outgroup   In cladistic analysis, the organisms with which one compares the organisms of interest in order to determine synapomorphies.

paedomorphosis   The maturation of an organism while retaining juvenile characters of the body.

paraphyletic   A group that arose from a single ancestor, but does not include all of the descendants of that ancestor.

pectoral   Of the shoulder region.

pelvic   Of the hip region.

phylogeny   An evolutionary tree that indicates closeness of relationships.

pineal opening   An opening in the midline of the skull roof, usually between the parietal bones, that lies close to the pineal organ of the brain (the 'third eye').

pleurodont   Teeth set in a groove.

pneumatic (of bones)   Hollow, with spaces for air sacs.

poikilotherm   An animal with varying body temperature.

polarity   The direction of change of a character, from primitive to derived.

polyphyletic   A group that arose from several ancestors.

postcranial   Those parts of the skeleton lying behind the head.

presacral   In front of the hip (sacral) region.

process   A projection on a bone.

protostomes   Those animals in which the opening of the gastrula stage in development becomes the mouth (includes everything except the deuterostomes).

radials   Bony rods within the fins of a fish.

recapitulation   A 'throwback', in which a juvenile resembles the adult stage of an ancestral form.

retraction   Pulling back.

retroarticular process   A process on the lower jaw that extends behind the glenoid articulation.

sacral   Of the hip region.

sagittal   Running along the midline of the head.

sister group   In cladistic analysis, the most closely related pair of outgroups.

spiracle   Remnant of an anterior gill slit seen in sharks and some extinct amphibians.

sprawling gait   Mode of locomotion in which the arms and legs are held out sideways from the body, with the elbows and knees bent.

streptostylic joint A joint in the skull in which the quadrate is mobile.

synapomorphy A shared derived character, characteristic of a mono-phyletic group.

temporal Of the cheek region, at the back of the side view of the skull.

tesserae Small bone plates, often forming a kind of chain mail in primitive fishes.

tetrapod A vertebrate with four legs (i.e. amphibians, reptiles, birds, and mammals).

thecodont Teeth set in sockets.

thoracic Of the chest region.

trochanter Major processes on the femur; insertion points of major muscles.

tympanum The ear drum.

wear facets Zones of the occlusal surfaces of teeth where enamel and dentine have been worn away by wear on the opposite teeth or on foodstuffs.

zygapophysis A process in front of or behind the neural arch of a vertebra, which takes part in linking the vertebrae to each other.

zygomatic arch The bony arch beneath the orbit and temporal fenestra formed from the jugal and squamosal in advanced mammal-like reptiles and mammals.

References

Akersten, W. A. 1985. Canine function in *Smilodon* (Mammalia; Felidae; Machairodontinae). *Contr. Sci (Los Angeles)* **356**, 1–22.

Aldridge, R. J., D. E. G. Briggs, E. N. K. Clarkson & M. P. Smith 1986. The affinities of conodonts – new evidence from the Carboniferous of Edinburgh, Scotland. *Lethaia* **19**, 279–91.

Alexander, R. McN. 1967. The functions and mechanisms of the protrusible upper jaws of some acanthopterygian fish. *J. Zool., Lond.* **151**, 43–64.

Alexander, R. McN. 1975. *The chordates*. Cambridge University Press.

Alexander, R. McN. 1976. Estimates of speed of dinosaurs. *Nature, Lond.* **261**, 129–30.

Allin, E. F. 1975. Evolution of the mammalian middle ear. *J. Morph.* **147**, 403–38.

Alvarez, L. W. 1987. Mass extinctions caused by large bolide impacts. *Physics Today* July 1987, 24–33.

Andrews, P. 1978. A revision of the Miocene Hominoidea of East Africa. *Bull. Br. Mus. Nat. Hist. (Geol.)* **30**, 85–224.

Andrews, P. 1985. Improved timing of hominoid evolution with a DNA clock. *Nature, Lond.* **314**, 498–9.

Andrews P. 1988. A phylogenetic analysis of the Primates. In *The phylogeny and classification of the tetrapods, volume 2: mammals*, M. J. Benton (ed.), *Syst. Ass. Spec. Vol.* **35B**, 143–75. Oxford: Clarendon Press.

Andrews, P. & L. Martin. 1987. Cladistic relationships of extant and fossil hominoids. *J. Human Evol.* **16**, 101–18.

Andrews, S. M. & T. S. Westoll 1970a. The postcranial skeleton of *Eusthenopteron foordi* Whiteaves. *Trans. R. Soc. Edinburgh* **68**, 207–329.

Andrews, S. M. & T. S. Westoll 1970b. The postcranial skeleton of rhipidistian fishes excluding *Eusthenopteron*. *Trans. R. Soc. Edinburgh* **68**, 391–489.

Archer, M. (ed.) 1982. *Possums and opossums*. Sydney: Royal Zoological Society of N.S.W.

Archer, M., T. F. Flannery, A. Ritchie & R. E. Molnar 1985. First Mesozoic mammal from Australia – an Early Cretaceous monotreme. *Nature, Lond.* **318**, 363–6.

Ax, P. 1985. *The phylogenetic system*. Chichester and New York: Wiley.

Baird, D. 1964. The aïstopod amphibians surveyed. *Breviora* **206**, 1–17.

Bakker, R. T. 1972. Anatomical and ecological evidence of endothermy in dinosaurs. *Nature, Lond.* **238**, 81–5.

Bakker, R. T. 1975. Dinosaur renaissance. *Scient. Am.* **232**(4), 58–78.

Bakker, R. T. 1980. Dinosaur heresy – dinosaur renaissance: why we need endothermic archosaurs for a comprehensive theory of bioenergetic evolution. In *A cold look at the warm-blooded dinosaurs*, R. D. K. Thomas & E. C. Olson (eds), 351–462. Boulder, Colo.: Westview Press.

Bakker, R. T. 1986. *The dinosaur heresies*. New York: William Morrow.

Barghusen, H. R. 1975. A review of fighting adaptations in dinocephalians. *Paleobiology* **1**, 295–311.

Barnes, L. G. 1984. Whales, dolphins and porpoises: origin and evolution of the Cetacea. In *Mammals, notes for a short course*, P. D. Gingerich & C. E. Badgley (eds), *Univ. Tennessee Dept. Geol. Sci. Stud. Geol.* **8**, 139–154.

Beaumont, E. H. 1977. Cranial morphology of the Loxommatidae (Amphibia: Labyrinthodontia). *Phil. Trans. R. Soc. B* **280**, 29–101.

Beerbower, J. R. 1963. Morphology, paleoecology and phylogeny of the Permo-Pennsylvanian amphibian *Diploceraspis*. *Bull. Mus. Comp. Zool.* **130**, 31–108.

Belles-Isles, M. 1987. La nage et l'hydrodynamique de deux agnathes du Paléozoïque: *Alaspis macrotuberculata* et *Pteraspis rostrata*. *N. Jb. Geol. Paläont. Abh.* **175**, 347–76.

Bemis, W. E., W. W. Burggren & N. E. Kemp (eds) 1986. *The biology and evolution of lungfishes*. New York: Alan R. Liss.

Benton, M. J. 1979. Ectothermy and the success of the dinosaurs. *Evolution* **33**, 983–97.

Benton, M. J. 1983a. Dinosaur success in the Triassic: a noncompetitive ecological model. *Q. Rev. Biol.* **58**, 29–55.

Benton, M. J. 1983b. The Triassic reptile *Hyperodapedon* from Elgin: functional morphology and relationships. *Phil. Trans. R. Soc. B* **302**, 605–717.

Benton, M. J. 1985a. Classification and phylogeny of the diapsid reptiles. *Zool. J. Linn. Soc.* **84**, 97–164.

Benton, M. J. 1985b. Mass extinction among non-marine tetrapods. *Nature, Lond.* **316**, 811–4.

Benton, M. J. 1986a. The Late Triassic tetrapod extinction events. In *The beginning of the age of dinosaurs*, K. Padian (ed.), 303–20. Cambridge University Press.

Benton, M. J. 1987a. Progress and competition in macroevolution. *Biol. Rev.* **62**, 305–38.

Benton, M. J. 1987b. Mass extinctions among families of non-marine tetrapods: the data. *Mém. Soc. Géol. Fr.* **150**, 21–32.

Benton, M. J. 1988. Mass extinctions in the fossil record of reptiles: paraphyly, patchiness, and periodicity (?). In *Extinction and survival in the fossil record*, G. Larwood (ed.). *Syst. Ass. Spec. Vol.* **34**, 269–94. Oxford: Clarendon Press.

Benton, M. J. 1989a. Patterns of evolution and extinction in vertebrates. In *Evolution and the fossil record*, K. Allen & D. E. G. Briggs (eds), 218–41. London: Belhaven.

Benton, M. J. 1989b. *On the trail of the dinosaurs*. London: Kingfisher.

Benton, M. J. 1990a. Origin and interrelationships of dinosaurs. In *The Dinosauria*, D. B. Weishampel, H. Osmólska & P. Dodson (eds), Berkeley: University of California Press.

Benton, M. J. 1990b. Scientific methodologies in collision: the history of the study of the extinction of the dinosaurs. *Evol. Biol.* **24**, 371–400.

Benton, M. J. & J. Clark 1988. Archosaur phylogeny and the relationships of the Crocodylia. In *The phylogeny and classification of the tetrapods*, M. J. Benton (ed.) *Syst. Ass. Spec. Vol.* **35A**, 295–338. Oxford: Clarendon Press.

Benton, M. J. & A. D. Walker 1985. Palaeoecology, taphonomy and dating of Permo-Triassic reptiles from Elgin, north-east Scotland. *Palaeontology* **28**, 207–34.

Blieck, A. 1984. *Les héterostracés pteraspidiformes, agnathes du Silurien-Dévonien du continent Nord Atlantique et des blocs avoisements: révision systematique, phylogénie, biostratigraphie, biogéographie*. Paris: Editions du C.N.R.S.

Blows, W. T. 1987. The armoured dinosaur *Polacanthus foxi*, from the Lower Cretaceous of the Isle of Wight. *Palaeontology* **30**, 557–80.

Bolt, J. R. 1977. Dissorophid relationships and ontogeny, and the origin of the Lissamphibia. *J. Paleont.* **51**, 235–49.

Bonaparte, J. F. 1976. *Pisanosaurus mertii* Casamiquela and the origin of the Ornithischia. *J. Paleont.* **50**, 808–20.

Bonaparte, J. F. 1981. Descripcion de '*Fasolasuchus tenax*' y su significado en la sistematica y evolucion de los Thecodontia. *Rev. Mus. Argent. Cienc. Nat., Palaeont.* **3**, 55–101.

Bonaparte, J. F. 1986. A new and unusual late Cretaceous mammal from Patagonia. *J. Vertebr. Paleont.* **6**, 264–70.

Bonis, L. de, J.-J. Jaeger, B. Coiffait & P.-E. Coiffait 1988. Découverte du plus ancien primate catarrhinien connu dans l'Éocène supérieur d'Afrique du nord. *C. r. Acad. Sci. Paris* **306**(2), 929–34.

Boy, J. A. 1972. Die Branchiosaurier (Amphibia) des saarpfaelischen Rotliegenden (Perm, SW-Deutschland). *Abh. Hess. Landesamt, Bodenforsch.* **65**, 1–137.

Briggs, D. E. G., E. N. K. Clarkson & R. J. Aldridge 1983. The conodont animal. *Lethaia* **16**, 1–14.

Brower, J. C. 1983. The aerodynamics of *Pteranodon* and *Nyctosaurus*, two large pterosaurs from the Upper Cretaceous of Kansas. *J. Vertebr. Paleont.* **3**, 84–124.

Brown, D. S. 1981. The English Upper Jurassic Plesiosauroidea (Reptilia), and a review of the phylogeny and classification of the Plesiosauria. *Bull. Br. Mus. Nat. Hist. (Geol.)* **35**, 253–347.

Buffetaut, E. 1982. Radiation évolutive, paléoecologie et biogéographie des crocodiliens mésosuchiens. *Mém. Soc. Géol. Fr.* **142**, 1–88.

Butler, P. M. 1988. Phylogeny of the insectivores. In *The phylogeny and classification of the tetrapods, volume 2: mammals*, M. J. Benton (ed.), *Syst. Ass. Spec. Vol.* **35B**, 117–41. Oxford: Clarendon Press.

Caple, G., R. P. Balda & W. R. Willis 1983. The physics of leaping animals and the evolution of preflight. *Am. Nat.* **121**, 455–67.

Carroll, R. L. 1964. The earliest reptiles. *J. Linn. Soc., Zool.* **45**, 61–83.

Carroll, R. L. 1969a. A Middle Pennsylvanian captorhinomorph and the interrelationships of primitive reptiles. *J. Paleont.* **43**, 151–70.

Carroll, R. L. 1969b. Origin of reptiles. In *Biology of the Reptilia*, C. Gans (ed.), **1**, 1–44. London: Academic.

Carroll, R. L. 1970. The earliest known reptiles. *Yale Scient. Mag.* Oct. 1970, 16–23.

Carroll, R. L. 1977. The origin of lizards. In *Problems in vertebrate evolution*, S. M. Andrews, R. S. Miles & A. D. Walker (eds), 359–96. London: Academic.

Carroll, R. L. 1978. Permo-Triassic 'lizards' from the Karoo System. Part II. A gliding reptile from the Upper Permian of Madagascar. *Palaeont. Afr.* **21**, 142–59.

Carroll, R. L. 1980. The hyomandibular as a supporting element in the skull of primitive tetrapods. In *The terrestrial environment and the origin of land vertebrates*, A. L. Panchen (ed.), *Syst. Ass. Spec. Vol.* **15**, 293–317. London: Academic.

Carroll, R. L. 1982. Early evolution of reptiles. *Ann. Rev. Ecol. Syst.* **13**, 87–109.

Carroll, R. L. 1987. *Vertebrate paleontology and evolution*. San Francisco: W. H. Freeman.

Carroll, R. L. & D. Baird 1972. Carboniferous stem-reptiles of the family Romeriidae. *Bull. Mus. Comp. Zool.* **143**, 321–63.

Carroll, R. L. & P. Gaskill 1978. The Order Microsauria. *Mem. Am. Phil. Soc.* **126**, 1–211.

Carroll, R. L. & P. Gaskill 1985. The nothosaur *Pachypleurosaurus* and the origin of the plesiosaurs. *Phil. Trans. R. Soc. B* **309**, 343–93.

Carroll, R. L. & W. Lindsay 1985. The cranial anatomy of the primitive reptile *Procolophon. Can. J. Earth Sci.* **22**, 1571–87.

Charig, A. J. 1984. *A new look at the dinosaurs*. London: Heinemann.

Charig, A. J. & A. W. Crompton 1974. The alleged synonymy of *Lycorhinus* and *Heterodontosaurus. Ann. S. Afr. Mus.* **64**, 167–89.

REFERENCES

Chatterjee, S. 1978. A primitive parasuchid (phytosaur) reptile from the Upper Triassic Maleri Formation of India. *Palaeontology* **21**, 83–127.

Chatterjee, S. 1982. Phylogeny and classification and thecodontian reptiles. *Nature, Lond.* **295**, 317–20.

Chatterjee, S. 1988. Functional significance of the semilunate carpal in archosaurs and birds [abstr.]. *J. Vertebr. Paleont.* **8** (suppl.), 11A.

Cifelli, R. L. 1981. Patterns of evolution among the Artiodactyla and Perissodactyla (Mammalia). *Evolution* **35**, 433–40.

Cifelli, R. L. 1983. The origin and affinities of the South American Condylarthra and early Tertiary Litopterna (Mammalia). *Am. Mus. Novitates* **2772**, 1–49.

Ciochon, R. L. & A. B. Chiarelli (eds) 1980. *Evolutionary biology of the New World monkeys and continental drift.* New York: Plenum.

Clark, J. & R. L. Carroll 1973. Romeriid reptiles from the Lower Permian. *Bull. Mus. Comp. Zool.* **147**, 353–407.

Clemens, W. A. 1979. Marsupialia. In *Mesozoic mammals*, J. A. Lillegraven, Z. Kielan-Jaworowska & W. A. Clemens (eds), 192–220. Berkeley: University of California Press.

Clemens, W. A. & Z. Kielan-Jaworowska 1979. Multituberculata. In *Mesozoic mammals*, J. A. Lillegraven, Z. Kielan-Jaworowska & W. A. Clemens (eds), 99–149. Berkeley: University of California Press.

Cluver, M. A. & N. Hotton 1981. The genera *Dicynodon* and *Diictodon* and their bearing on the classification of the Dicynodontia (Reptilia, Therapsida). *Ann. S. Afr. Mus.* **83**, 99–146.

Colbert, E. H. 1946. *Sebecus*, representative of a peculiar suborder of fossil Crocodilia from Patagonia. *Bull. Am. Mus. Nat. Hist.* **87**, 217–70.

Colbert, E. H. & C. C. Mook 1951. The ancestral crocodilian, *Protosuchus*. *Bull. Am. Mus. Nat. Hist.* **97**, 147–82.

Compagno, L. J. V. 1977. Phyletic relationships of living sharks and rays. *Am. Zool.* **17**, 303–22.

Coombs, W. P. 1978. The families of the ornithischian dinosaur order Ankylosauria. *Palaeontology* **21**, 143–70.

Coombs, W. P. 1982. Juvenile specimens of the ornithischian dinosaur *Psittacosaurus*. *Palaeontology* **25**, 89–107.

Coppens, Y., V. J. Maglio, C. T. Madden & M. Beden 1978. Proboscidea. In *Evolution of African mammals*, V. J. Maglio & H. B. S. Cooke (eds), 336–67. Cambridge: Harvard University Press.

Cracraft, J. 1976. The species of moas (Aves: Dinornithidae). *Smiths. Contr. Paleobiol.* **27**, 189–205.

Cracraft, J. 1986. The origin and early diversification of birds. *Paleobiology* **12**, 383–99.

Cracraft, J. 1988. The major clades of birds. In *The phylogeny and classification of the tetrapods, volume 1: amphibians, reptiles, birds*, M. J. Benton (ed.), *Syst. Ass. Spec. Vol.* **35A**, 339–61. Oxford: Clarendon Press.

Crompton, A. W. & J. Attridge 1986. Masticatory apparatus of the larger herbivores during Late Triassic and Early Jurassic times. In *The beginning of the age of the dinosaurs*, K. Padian (ed.), 223–236. Cambridge University Press.

Crompton, A. W. & N. Hotton 1967. Functional morphology of the masticatory apparatus of two dicynodonts (Reptilia: Therapsida). *Postilla* **109**, 1–51.

Crompton, A. W. & W. L. Hylander 1986. Changes in the mandibular function following the acquisition of a dentary-squamosal articulation. In *The ecology and biology of mammal-like reptiles*, N. Hotton, P. D. Maclean, J. J. Roth & F. C. Roth (eds), 263–282. Washington: Smithsonian Institution Press.

Crompton, A. W., C. R. Taylor & J. A. Jagger 1978. Evolution of homeothermy in mammals. *Nature, Lond.* **272**, 333–6.

Cruickshank, A. R. I. 1972. The proterosuchian thecodonts. In *Studies in vertebrate evolution*, K. A. Joysey & T. S. Kemp (eds), 89–119. Edinburgh: Oliver & Boyd.

Cruickshank, A. R. I. & B. W. Skews 1980. The functional significance of nectridean tabular horns (Amphibia: Lepospondyli). *Proc. R. Soc. Lond. B* **209**, 513–37.

Crush, P. J. 1984. A late Triassic sphenosuchid crocodilian from Wales. *Palaeontology* **27**, 131–57.

Denison, R. H. 1978. Placoderms. *Handb. Paleoichthyol.* **2**, 1–128. Stuttgart: Gustav Fischer.

Denison, R. H. 1979. Acanthodii. *Handb. Paleoichthyol.* **5**, 1–62. Stuttgart: Gustav Fischer.

Dick, J. R. F., M. I. Coates & W. D. I. Rolfe 1986. Fossil sharks. *Geol. Today* **2**, 82–5.

Domning, D. P. 1978. Sirenian evolution in the North Pacific Ocean. *Univ. Calif. Publ. Geol. Sci.* **118**, 1–178.

Eaton, C. F. 1910. Osteology of *Pteranodon. Mem. Conn. Acad. Arts Sci.* **2**, 1–38.

Elliott, D. K. 1987. A reassessment of *Astraspis desiderata*, the oldest North American vertebrate. *Science* **237**, 190–2.

Estes, R. 1981. Gymnophiona, Caudata. *Handb. Paläoherpetol.* **2**, 1–115. Stuttgart: Gustav Fischer.

Estes, R. 1983. Sauria terrestria, Amphisbaenia. *Handb Paläoherpetol.* **10A**, 1–249. Stuttgart: Gustav Fischer.

Estes, R. & O. A. Reig 1973. The early fossil record of frogs: a review of the evidence. In *Evolutionary biology of the anurans*, J. Vial (ed.), 11–63. Columbia: University of Missouri Press.

Evans, S. E. 1984. The classification of the Lepidosauria. *Zool. J. Linn. Soc.* **82**, 87–100.

Evans, S. E. 1988. The early history and relationships of the Diapsida. In *The phylogeny and classification of the tetrapods, volume 1: amphibians, reptiles, birds*, M. J. Benton (ed.), *Syst. Ass. Spec. Vol.* **35A**, 221–60. Oxford: Clarendon Press.

Evans, S. E. & H. Haubold 1987. A review of the Upper Permian genera *Coelurosauravus, Weigeltisaurus* and *Gracilisaurus* (Reptilia: Diapsida). *Zool. J. Linn. Soc.* **90**, 275–303.

Ewer, R. F. 1965. The anatomy of the thecodont reptile *Euparkeria capensis* Broom. *Phil. Trans. R. Soc. B* **248**, 379–435.

Farlow, J. O., C. V. Thompson & D. E. Rosner 1976. Plates of the dinosaur *Stegosaurus*: forced convection heat loss fins? *Science* **192**, 1123–5.

Feduccia, A. 1980. *The age of birds*. Cambridge: Harvard University Press.

Feduccia, A. & H. B. Tordoff 1979. Feathers of *Archaeopteryx*: asymmetric vanes indicate aerodynamic function. *Science* **203**, 1021–2.

Fink, S. V. & W. L. Fink 1981. Interrelationships of the ostariophysan teleost fishes. *Zool. J. Linn. Soc.* **72**, 297–353.

Fleagle, J. G. 1988. *Primate adaptation and evolution*. New York: Academic.

Flynn, J. J., N. A. Neff & R. H. Tedford 1988. Phylogeny of the Carnivora. In *The phylogeny and classification of the tetrapods, volume 2: mammals*, M. J. Benton (ed.), *Syst. Ass. Spec. Vol.* **35B**, 73–116. Oxford: Clarendon Press.

Forey, P. L. 1988. Golden jubilee for the coelacanth *Latimeria chalumnae. Nature, Lond.* **336**, 727–32.

Franzen, J. L. 1990. The Eocene Lake Messel and its early horses. *Palaeontology* **33**, in press.

Fraser, N. C. & G. M. Walkden 1984. The postcranial skeleton of the Upper

REFERENCES

Triassic sphenodontid *Planocephalosaurus robinsonae*. *Palaeontology* **27**, 575–95.

Frazzetta, T. H. 1986. The origin of amphikinesis in lizards. A problem in functional morphology and the evolution of adaptive systems. *Evol. Biol.* **20**, 419–61.

Frey, E. 1984. Aspects of the biomechanics of crocodilian terrestrial locomotion. In *Third symposium on Mesozoic terrestrial ecosystems, Tübingen 1984, short papers*, W.-E. Reif & F. Westphal (eds), 93–7. Tübingen: Attempto.

Gaffney, E. S. 1979. Tetrapod monophyly: a phylogenetic analysis. *Bull. Carnegie Mus. Nat. Hist.* **13**, 92–105.

Gaffney, E. S. & L. J. Meeker 1983. Skull morphology of the oldest turtles: a preliminary description of *Proganochelys quenstedti*. *J. Vertebr. Paleont.* **3**, 25–8.

Gaffney, E. S. & P. A. Meylan 1988. A phylogeny of turtles. In *The phylogeny and classification of the tetrapods, volume 1: amphibians, reptiles, birds*, M. J. Benton (ed.), *Syst. Ass. Spec. Vol.* **35A**, 157–219. Oxford: Clarendon Press.

Galton, P. M. 1969. The pelvic musculature of the dinosaur *Hypsilophodon* (Reptilia: Ornithischia). *Postilla* **131**, 1–64.

Galton, P. M. 1970a. The posture of hadrosaurian dinosaurs. *J. Paleont.* **44**, 464–73.

Galton, P. M. 1970b. Pachycephalosaurids – dinosaurian battering rams. *Discovery, New Haven* **6**(1), 23–32.

Galton, P. M. 1974. The ornithischian dinosaur *Hypsilophodon* from the Wealden of the Isle of Wight. *Bull. Br. Mus. Nat. Hist., Geol.* **25**, 1–152.

Galton, P. M. 1978. Fabrosauridae, the basal family of ornithischian dinosaurs. *Paläont. Z.* **52**, 138–59.

Galton, P. M. 1984. Cranial anatomy of the prosauropod dinosaur *Plateosaurus* from the Knollenmergel (Middle Keuper, Upper Triassic) of Germany. I. Two complete skulls from Trossingen, Wurtt. with comments on the diet. *Geol. Paleont.* **18**, 139–71.

Galton, P. M. 1985. Diet of prosauropod dinosaurs from the late Triassic and early Jurassic. *Lethaia* **18**, 105–23.

Gardiner, B. G. 1982. Tetrapod classification. *Zool. J. Linn. Soc.* **74**, 207–32.

Gardiner, B. G. 1984a. The relationships of placoderms. *J. Vertebr. Paleont.* **4**, 379–95.

Gardiner, B. G. 1984b. The relationships of the palaeoniscid fishes, a review based on new specimens of *Mimia* and *Moythomasia* from the Upper Devonian of Western Australia. *Bull. Br. Mus. Nat. Hist., Geol.* **37**, 173–428.

Garstang, W. 1928. The morphology of Tunicata and its bearing on the phylogeny of the Chordata. *Q. J. Microsc. Soc.* **72**, 51–187.

Gauthier, J. A. 1986. Saurischian monophyly and the origin of birds. *Mem. Calif. Acad. Sci* **8**, 1–56.

Gauthier, J. A., A. G. Kluge & T. Rowe 1988a. The early evolution of the Amniota. In *The phylogeny and classification of the tetrapods, volume 1: amphibians, reptiles, birds*, M. J. Benton (ed.), *Syst. Ass. Spec. Vol.* **35A**, 103–55. Oxford: Clarendon Press.

Gauthier, J. A., A. G. Kluge & T. Rowe 1988b. Amniote phylogeny and the importance of fossils. *Cladistics* **4**, 105–209.

Gauthier, J. A., R. Estes & K. de Queiroz 1988c. A phylogenetic analysis of Lepidosauromorpha. In *Phylogenetic relationships of the lizard families*, R. Estes & G. Pregill (eds), 15–98. Stanford University Press.

Gazin, C. L. 1953. The Tillodontia: an early Tertiary order of mammals. *Smiths. Misc. Coll.* **121**(10), 1–110.

Gentry, A. W. & Hooker, J. J. 1988. The phylogeny of the Artiodactyla. In *The phylogeny and classification of the tetrapods, volume 2: mammals*, M. J. Benton (ed.), *Syst. Ass. Spec. Vol.* **35B**, 235–72. Oxford: Clarendon Press.

345

Gillette, D. D. & C. E. Ray 1981. Glyptodonts of North America. *Smiths. Contr. Paleobiol.* **40**, 1–255.

Gingerich, P. D. 1984. Primate evolution. In *Mammals. Notes for a short course*, P. D. Gingerich & C. E. Badgley (eds), *Univ. Tennessee Stud. Geol.* **8**, 167–81.

Gingerich, P. D. & D. E. Russell 1981. *Pakicetus inachus*, a new archaeocete (Mammalia, Cetacea) from the early middle Eocene Kuldana Formation of Kohat (Pakistan). *Univ. Mich. Contr. Mus. Paleont.* **25**, 235–46.

Godfrey, S. J. 1984. Plesiosaur subaqueous locomotion: a reappraisal. *N. Jb. Geol. Paläont., Mh.* **11**, 661–72.

Godfrey, S. J., A. R. Fiorillo & R. L. Carroll 1987. A newly discovered skull of the temnospondyl amphibian *Dendrerpeton acadianum* Owen. *Can. J. Earth Sci.* **24**, 796–805.

Goujet, D. 1984. Placoderm interrelationships: a new interpretation with a short review of placoderm classification. *Proc. Linn. Soc. N.S.W.* **107**, 211–43.

Gow, C. E. 1975. The morphology and relationships of *Youngina capensis* Broom and *Prolacerta broomi* Parrington. *Palaeontol. Afr.* **18**, 89–131.

Grande, L. 1984. Paleontology of the Green River Formation, with a review of the fish fauna. 2nd edn. *Bull. Geol. Surv. Wyo.* **63**, 1–333.

Grande, L. 1985. Recent and fossil clupeomorph fishes, with materials for revision of the subgroups of clupeoids. *Bull. Am. Mus. Nat. Hist.* **181**, 231–372.

Grande, L. 1989. The Eocene Green River Lake system, Fossil Lake, and the history of the North American fish fauna. In *Mesozoic/Cenozoic vertebrate paleontology: classic localities, contemporary approaches*, J. Flynn (ed.), 18–28. Washington: American Geophysical Union.

Greenwood, P. H., D. E. Rosen, S. H. Weitzman & G. S. Myers 1966. Phyletic studies of teleostean fishes, with a provisional classification of living forms. *Bull. Am. Mus. Nat. Hist.* **131**, 339–456.

Gregory, J. T. 1945. Osteology and relationships of *Trilophosaurus*. *Univ. Texas Publ.* **4401**, 273–359.

Gregory, W. K. 1951/1957. *Evolution emerging*, vol. 1, 2. New York: Macmillan.

Haack, S. C. 1986. A thermal model of the sailback pelycosaur. *Paleobiology* **12**, 459–73.

Hallam, A. 1985. A review of Mesozoic climates. *J. Geol. Soc. Lond.* **142**, 433–45.

Hallam, A. 1987. End-Cretaceous mass extinction event: argument for terrestrial causation. *Science* **238**, 1237–42.

Halstead, L. B. 1973a. The heterostracan fishes. *Biol. Rev.* **48**, 279–332.

Harris, J. M. 1978. Deinotherioidea and Barytherioidea. In *Evolution of African mammals*, V. J. Maglio & H. B. S. Cooke (eds), 315–32. Cambridge: Harvard University Press.

Harrison, J. A. 1985. Giant camels from the Cenozoic of North America. *Smiths. Contr. Paleobiol.* **57**, 1–29.

Heaton, M. J. 1980. The Cotylosauria. In *The terrestrial environment and the origin of land vertebrates*, A. L. Panchen (ed.), *Syst. Ass. Spec. Vol.* **15**, 497–551. London: Academic.

Heaton, M. J. & R. R. Reisz 1986. Phylogenetic relationships of captorhinomorph reptiles. *Can. J. Earth Sci.* **23**, 402–18.

Hecht, M. K., J. H. Ostrom, G. Viohl & P. Wellnhofer (eds) 1985. *The beginnings of birds*. Eichstätt: Freunde des Jura-Museums.

Hoffstetter, R. 1967. Coup d'oeil sur les sauriens (lacertiliens) des couches de Purbeck (Jurassique supérieur d'Angleterre: résumé d'un Mémoire). *Coll. Intern. C.N.R.S.* **163**, 349–71.

Holmes, R. 1977. The osteology and musculature of the pectoral limb of small captorhinids. *J. Morphol.* **152**, 101–40.

Holmes, R. 1984. The Carboniferous amphibian *Proterogyrinus scheelei* Romer, and the early evolution of tetrapods. *Phil. Trans. R. Soc. B* **306**, 431–527.

Hopson, J. A. 1975. The evolution of cranial display structures in hadrosaurian dinosaurs. *Paleobiology* **1**, 21–43.

Hopson, J. A. 1977. Relative brain size and behavior in archosaurian reptiles. *Ann. Rev. Ecol. Syst.* **8**, 429–48.

Hopson, J. A. & H. R. Barghusen 1986. An analysis of therapsid relationships. In *The ecology and biology of mammal-like reptiles*, N. Hotton, P. D. MacLean, J. J. Roth & E. C. Roth (eds), 83–106. Washington: Smithsonian Institution Press.

Horner, J. R. 1982. Evidence for colonial nesting and 'site fidelity' among ornithischian dinosaurs. *Nature, Lond.* **297**, 675–6.

Horner, J. R. 1984. The nesting behaviour of dinosaurs. *Scient. Am.* **250**(4), 92–9.

Horner, J. R. & R. Makela 1979. Nest of juveniles provides evidence of family structure among dinosaurs. *Nature, Lond.* **282**, 296–8.

Horner, J. R. & D. B. Weishampel 1988. A comparative embryological study of two ornithischian dinosaurs. *Nature, Lond.* **332**, 256–7.

Iordansky, N. N. 1973. The skull of the Crocodilia. In *The biology of the Reptilia*, C. Gans & T. S. Parsons (eds), **4**, 201–62. London: Academic Press.

Janis, C. M. 1976. The evolutionary strategy of the Equidae and the origins of rumen and cecal digestion. *Evolution* **30**, 757–74.

Janis, C. M. 1986. Evolution of horns and related structures in hoofed mammals. *Discovery* **19**, 8–17.

Janis, C. M. & K. M. Scott 1987. The interrelationships of higher ruminant families, with special emphasis on the members of the Cervoidea. *Am. Mus. Novitates* **2893**, 1–85.

Janvier, P. 1981. The phylogeny of the Craniata, with particular reference to the significance of fossil 'agnathans'. *J. Vertebr. Paleont.* **1**, 121–59.

Janvier, P. 1984. The relationships of the Osteostraci and Galeaspida. *J. Vertebr. Paleont.* **4**, 344–58.

Janvier, P. 1985. Ces étranges bêtes du Montana. *La Recherche* **16**, 98–100.

Janvier, P. 1986. Les nouvelles conceptions de la phylogénie et de la classification des 'agnathes' et des sarcopterygiens. *Océanis* **12**, 123–38.

Jarvik, E. 1955. The oldest tetrapods and their forerunners. *Sci. Monthly* **80**, 141–54.

Jarvik, E. 1980. *Basic structure and evolution of vertebrates*. London: Academic.

Jefferies, R. P. S. 1986. *The ancestry of the vertebrates*. London: British Museum (Natural History).

Jenkins, F. A., Jr. 1971a. The postcranial skeleton of African cynodonts. *Bull. Peabody Mus. Nat. Hist.* **36**, 1–216.

Jenkins, F. A., Jr. & F. R. Parrington 1976. The postcranial skeletons of the Triassic mammals *Eozostrodon*, *Megazostrodon* and *Erythrotherium*. *Phil. Trans. R. Soc. B* **173**, 387–431.

Jenkins, F. A., Jr. & C. R. Schaff 1988. The Early Cretaceous mammal *Gobiconodon* (Mammalia, Triconodonta) from the Cloverly Formation in Montana. *J. Vertebr. Paleont.* **8**, 1–24.

Jepsen, G. L. 1970. *Biology of bats. Vol. 1. Bat origins and evolution*. New York: Academic.

Johanson, D. C. & T. D. White 1979. A systematic assessment of early African hominids. *Science* **203**, 321–30.

Kemp, T. S. 1969. On the functional morphology of the gorgonopsid skull. *Phil. Trans. R. Soc. B* **256**, 1–83.

Kemp, T. S. 1979. The primitive cynodont *Procynosuchus*: functional anatomy of the skull and relationships. *Phil. Trans. R. Soc. B* **285**, 73–122.

Kemp, T. S. 1982. *Mammal-like reptiles and the origin of mammals*. London: Academic.

Kemp, T. S. 1983. The relationships of mammals. *Zool. J. Linn. Soc.* **77**, 353–84.

Kemp. T. S. 1988a. Interrelationships of the Synapsida. In *The phylogeny and classification of the tetrapods, volume 2: mammals*, M. J. Benton (ed.), *Syst. Ass. Spec. Vol.* **35B**, 1–22. Oxford: Clarendon Press.

Kemp, T. S. 1988b. Haemothermia or Archosauria?: the interrelationships of mammals, birds and crocodiles. *Zool. J. Linn. Soc.* **92**, 67–104.

Kemp, T. S. 1988c. A note on the Mesozoic mammals, and the origin of therians. In *The phylogeny and classification of the tetrapods, volume 2: mammals*, M. J. Benton (ed.), *Syst. Ass. Spec. Vol.* **35B**, 23–9. Oxford: Clarendon Press.

Kermack, D. A. 1984. New prosauropod material from South Wales. *Zool. J. Linn. Soc.* **82**, 101–17.

Kermack, K. A., F. Mussett & H. W. Rigney 1973. The lower jaw of *Morganucodon*. *Zool. J. Linn. Soc.* **53**, 87–115.

Kermack, K. A., F. Mussett & H. W. Rigney 1981. The skull of *Morganucodon*. *Zool. J. Linn. Soc.* **71**, 1–158.

Kielan-Jaworowska, Z. 1984. Evolution of the therian mammals in the Late Cretaceous of Asia. Part V. Skull structure in Zalambdalestidae. *Palaeont. Pol.* **46**, 107–17.

Kielan-Jaworowska, Z., Bown, T. M. & J. A. Lillegraven 1979. Eutheria. In *Mesozoic mammals*, J. A. Lillegraven, Z. Kielan-Jaworowska & W. A. Clemens (eds), 221–58. Berkeley: University of California Press.

Kielan-Jaworowska, Z. & L. A. Nessov 1990. On the metatherian nature of the deltatheroids, a sister group of the Marsupialia. *Lethaia* in press.

King, G. M. 1988. Anomodontia. *Handb. Paläoherpetol.* **17C**, 1–174.

Kirsch, K. F. 1979. The oldest vertebrate egg? *J. Paleont.* **53**, 1068–84.

Krause, D. W. 1984. Mammalian evolution in the Paleocene: beginning of an era. In *Mammals. Notes for a short course*, P. D. Gingerich & C. E. Badgley (eds), *Univ. Tennessee Dept. Geol. Sci. Stud. Geol.* **8**, 87–109.

Krause, D. W. & F. A. Jenkins, Jr. 1983. The postcranial skeleton of North American multituberculates. *Bull. Mus. Comp. Zool.* **150**, 199–246.

Kühne, W. G. 1956. *The Liassic therapsid Oligokyphus*. London: British Museum (Natural History).

Langston, W. 1981. Pterosaurs. *Scient. Am.* **245**(2), 122–36.

Lauder, G. V. & K. F. Liem 1983. The evolution and interrelationships of the actinopterygian fishes. *Bull. Mus. Comp. Zool.* **150**, 95–197.

Leakey, R. E. 1981. *The making of mankind*. London: Michael Joseph.

Lewin, R. 1989. *Human evolution, an illustrated introduction*. 2nd edn. Oxford: Blackwell Scientific.

Lillegraven, J. A., Z. Kielan-Jaworowska & W. A. Clemens (eds) 1979. *Mesozoic mammals*. Berkeley: University of California Press.

Luckett, W. P. & J.-L. Hartenberger (eds) 1985. *Evolutionary relationshps among rodents*. New York: Plenum.

Lund, R. 1985. The morphology of *Falcatus falcatus* (St John and Worthen), a Mississippian stethacanthid chondrichthyan from the Bear Gulch Limestone of Montana. *J. Vertebr. Paleont.* **5**, 1–19.

Madsen, J. H. 1976. *Allosaurus fragilis*: a revised osteology. *Bull. Utah. Geol. Mineral. Surv.* **109**, 1–163.

Maisey, J. G. 1982. The anatomy and relationships of Mesozoic hybodont sharks. *Am. Mus. Novitates* **2724**, 1–48.

Maisey, J. G. 1984. Higher elasmobranch phylogeny and biostratigraphy. *Zool. J. Linn. Soc.* **82**, 33–54.

Maisey, J. G. 1986. Heads and tails: a chordate phylogeny. *Cladistics* **2**, 201–56.

Marshall, L. G. 1982. Evolution of South American Marsupialia. In *Mammalian biology in South America*, M. A. Mares & H. H. Genoways (eds), *Pymatuning Lab. Ecol., Univ. Pittsburg, Spec. Publ. Ser.* **6**, 251–72.

Marshall, L. G. 1988. Land mammals and the Great American Interchange. *Am. Scient.* **76**, 380–8.

Marshall, L. G. & C. de Muizon 1988. The dawn of the age of mammals in South America. *Natn. Geogr. Res.* **4**, 23–55.

Marshall, L. G., S. D. Webb, J. J. Sepkoski, Jr. & D. M. Raup 1982. Mammalian evolution and the Great American Interchange. *Science* **215**, 1351–7.

Märss, T. 1986. Squamation of the thelodont agnathan *Phlebolepis. J. Vertebr. Paleont.* **6**, 1–11.

Martin, L. D. 1980. Functional morphology and the evolution of cats. *Trans. Nebraska Acad. Sci.* **8**, 141–54.

Martin, L. D. 1985. The relationship of *Archaeopteryx* to other birds. In *The beginnings of birds*, M. K. Hecht, J. H. Ostrom, G. Viohl & P. Wellnhofer (eds), 177–183. Eichstätt: Freunde des Jura-Museums.

Martin, L. D. & D. K. Bennett 1977. The burrows of the Miocene beaver *Palaeocastor*, western Nebraska, U.S.A. *Palaeogeogr., Palaeoclimat., Palaeoecol.* **22**, 173–93.

Martin, L. D. & Tate, Jr. 1976. The skeleton of *Baptornis advenus* (Aves: Hesperornithiformes). *Smiths. Contr. Paleobiol.* **27**, 35–66.

Martin, P. S. & R. G. Klein (eds) 1984. *Quaternary extinctions, a prehistoric revolution*. Tucson: University of Arizona Press.

Maryánska, T. & H. Osmólska 1974. Pachycephalosauria, a new suborder of ornithischian dinosaurs. *Palaeont. Pol.* **30**, 45–102.

Massare, J. A. 1988. Swimming capabilities of Mesozoic marine reptiles: implications for method of predation. *Paleobiology* **14**, 187–205.

Mateer, N. J. 1982. Osteology of the Jurassic lizard *Ardeosaurus brevipes* (Meyer). *Palaeontology* **25**, 461–9.

Maxwell, W. D. 1989. The end-Permian mass extinction. In *Mass extinctions, processes and evidence*, S. K. Donovan (ed.), 152–173. London: Belhaven.

Mazin, J.-M. 1981. *Grippia longirostris* Wiman 1929, un Ichthyopterygia primitif du Trias inférieur du Spitsberg. *Bull. Mus. Natn. Hist. Nat. C.* (4)3, 317–40.

Mazin, J.-M. 1982. Affinités et phylogénie des Ichthyopterygia. *Géobios, Mém. Spéc.* **6**, 85–98.

McGowan, C. 1985. Tarsal development in birds: evidence for homology with the theropod condition. *J. Zool. (A)* **206**, 53–67.

McKenna, M. C. 1975. Toward a phylogenetic classification of the Mammalia. In *Phylogeny of the primates*, W. P. Luckett & F. S. Szalay (eds), 21–36. New York: Plenum.

Miles, R. S. 1969. Features of placoderm classification and the evolution of the arthrodire feeding mechanism. *Trans. R. Soc. Edinb.* **68**, 123–70.

Miles, R. S. & T. S. Westoll 1968. The placoderm fish *Coccosteus cuspidatus* Miller ex Agassiz from the Middle Old Red Sandstone of Scotland. Part 1. Descriptive morphology. *Trans. R. Soc. Edinb.* **67**, 373–476.

Milner, A. C. 1980. A review of the Nectridea (Amphibia). In *The terrestrial environment and the origin of land vertebrates*, A. L. Panchen (ed.), *Syst. Ass. Spec. Vol.* **15**, 377–405. London: Academic.

Milner, A. R. 1980. The tetrapod assemblage from Nýřany, Czechoslovakia. In *The terrestrial environment and the origin of land vertebrates*, A. L. Panchen (ed.), *Syst. Ass. Spec. Vol.* **15**, 439–96. London: Academic.

Milner, A. R. 1982. Small temnospondyl amphibians from the Middle Penn-sylvanian of Illinois. *Palaeontology* **25**, 635–64.

Milner, A. R. 1988. The relationships and origin of living amphibians. In *The phylogeny and classification of the tetrapods, volume 1: amphibians, reptiles, birds*, M. J. Benton (ed.), *Syst. Ass. Spec. Vol.* **35A**, 59–102. Oxford: Clarendon Press.

Mitchell, E. D. 1975. Parallelism and convergence in the evolution of Otariidae and Phocidae. *Cons. Intern. Explor. Rev. Rapp. Proc.-Verb. Réun.* **169**, 12–26.

Miyamoto, M. M. & M. Goodman 1986. Biomolecular systematics of eutherian mammals: phylogenetic patterns and classification. *Syst. Zool.* **35**, 230–40.

Miyamoto, M. M., B. F. Koop, J. L. Slightom, M. Goodman & M. R. Tennant 1988. Molecular systematics of higher primates: genealogical relations and classification. *Proc. Natn. Acad. Sci. U.S.A.* **85**, 7627–31.

Młynarski, M. 1976. Testudines. *Handb. Paläoherpetol.* **7**, 1–129.

Moy-Thomas, J. A. & R. S. Miles 1971. *Palaeozoic fishes*, 2nd edn. London: Chapman & Hall.

Nelson, J. S. 1984. *Fishes of the world*, 2nd edn. New York: Wiley.

Norman, D. B. 1980. On the ornithischian dinosaur *Iguanodon bernissartensis* from the Lower Cretaceous of Bernissart (Belgium). *Mém. Inst. R. Sci. Nat. Belg.* **178**, 1–105.

Norman, D. B. 1984. On the cranial morphology and evolution of ornithopod dinosaurs. *Symp. Zool. Soc. Lond.* **52**, 521–47.

Norman, D. B. 1986a. *Illustrated encyclopedia of dinosaurs*. London: Salamander.

Norman, D. B. 1986b. On the anatomy of *Iguanodon atherfieldensis* (Ornithischia: Ornithopoda). *Bull. Inst. R. Sci. Nat. Belg.* **56**, 281–372.

Norman, D. B. & D. B. Weishampel 1985. Ornithopod feeding mechanisms: their bearing on the evolution of herbivory. *Am. Nat.* **126**, 151–64.

Novacek, M. J. 1986. The skull of leptictid insectivorans and the higher-level classification of eutherian mammals. *Bull. Am. Mus. Nat. Hist.* **183**, 1–112.

Novacek, M. J., A. R. Wyss & M. C. McKenna 1988. The major groups of eutherian mammals. In *The phylogeny and classification of the tetrapods, volume 2: mammals*, M. J. Benton (ed.), *Syst. Ass. Spec. Vol.* **35B**, 31–71. Oxford: Clarendon Press.

Olson, E. C. 1951. *Diplocaulus*; a study in growth and variation. *Fieldiana* Geol. **11**, 55–154.

Olson, S. L. 1985. The fossil record of birds. In *Avian biology 8*, W. S. Farner, J. R. King & K. C. Parkes (eds), 80–128. New York: Academic.

Ostrom, J. H. 1961. Cranial morphology of the hadrosaurian dinosaurs of North America. *Bull. Am. Mus. Nat. Hist.* **122**, 33–186.

Ostrom, J. H. 1966. Functional morphology and evolution of the ceratopsian dinosaurs. *Evolution* **20**, 290–308.

Ostrom, J. H. 1969. Osteology of *Deinonychus antirrhopus*, an unusual theropod from the Lower Cretaceous of Montana. *Bull. Peabody Mus. Nat. Hist.* **30**, 1–165.

Ostrom, J. H. 1976. *Archaeopteryx* and the origin of birds. *Biol. J. Linn. Soc.* **8**, 91–182.

Ostrom, J. H. 1985. Introduction to *Archaeopteryx*. In *The beginnings of birds*, M. K. Hecht, J. H. Ostrom, G. Viohl & P. Wellnhofer (eds), 9–20. Eichstätt: Freunde des Jura-Museums.

Padian, K. 1984. A functional analysis of flying and walking in pterosaurs. *Paleobiology* **9**, 218–39.

Padian, K. 1986. *The beginning of the age of dinosaurs*. Cambridge University Press.

Panchen, A. L. (ed.) 1980. *The terrestrial environment and the origin of land vertebrates*. London: Academic.

Panchen, A. L. 1985. On the amphibian *Crassigyrinus scoticus* Watson from the Carboniferous of Scotland. *Phil. Trans. R. Soc. B* **309**, 505–68.

Panchen, A. L. & T. R. Smithson 1987. Character diagnosis, fossils and the origin of tetrapods. *Biol. Rev.* **62**, 341–438.

Panchen, A. L. & T. R. Smithson 1988. The relationships of the earliest tetrapods. In *The phylogeny and classification of the tetrapods, volume 1: amphibians, reptiles, birds*, M. J. Benton (ed.), *Syst. Ass. Spec. Vol.* **35A**, 1–32. Oxford: Clarendon Press.

Parsons, T. & E. Williams 1963. The relationships of modern Amphibia: a re-examination. *Q. Rev. Biol.* **38**, 26–53.

Patterson, C. 1965. The phylogeny of the chimaeroids. *Phil. Trans. R. Soc. B* **249**, 101–209.

Patterson, C. 1973. Interrelationships of holosteans. In *Interrelationships of fishes*, P. H. Greenwood, R. S. Miles & C. Patterson (eds), *Zool. J. Linn. Soc.* **53**, Suppl. 1, 233–305. London: Academic.

Patterson, C. 1975. The braincase of pholidophorid and leptolepid fishes, with a review of the actinopterygian braincase. *Phil. Trans. R. Soc. B* **269**, 275–579.

Patterson, C. 1982. Cladistics and classification. *New Scient.* **94**, 303–6.

Patterson, C. & D. E. Rosen 1977. Review of ichthyodectiform and other Mesozoic teleost fishes and the theory and practice of classifying fossils. *Bull. Am. Mus. Nat. Hist.* **158**, 81–172.

Pearson, D. M. & T. S. Westoll 1979. The Devonian actinopterygian *Cheirolepis* Agassiz. *Trans. R. Soc. Edinb.* **70**, 337–99.

Pilbeam, D. R. 1984. The descent of hominoids and hominids. *Scient. Am.* **250**, 84–96.

Pough, F. H., J. B. Heiser & W. N. McFarland 1989. *Vertebrate life*, 3rd ed. New York: Macmillan.

Prothero, D. R., E. M. Manning & M. Fischer 1988. The phylogeny of the ungulates. In *The phylogeny and classification of the tetrapods, volume 2: mammals*, M. J. Benton (ed.), *Syst. Ass. Spec. Vol.* **35B**, 201–34. Oxford: Clarendon Press.

Prothero, D. R., E. M. Manning & C. B. Hanson 1986. The phylogeny of the Rhinocerotoidea. *Zool. J. Linn. Soc.* **87**, 341–66.

Radinsky, L. B. 1965. Evolution of the tapiroid skeleton from *Heptodon* to *Tapirus*. *Bull. Mus. Comp. Zool.* **134**, 69–106.

Rage, J.-C. 1984. Serpentes. *Handb. Paläoherpetol.* **11**, 1–80.

Rage, J.-C. & P. Janvier 1982. Le problème de la monophylie des amphibiens actuels, a la lumière des nouvelles données sur les affinités des tétrapodes. *Géobios, Mém. Spéc.* **6**, 65–83.

Randall, J. E. 1973. Size of the great white shark *(Carcharodon)*. *Science* **181**, 169–70.

Rayner, J. M. V. 1988. The evolution of vertebrate flight. *Biol. J. Linn Soc.* **34**, 269–87.

Reader, J. 1988. *Missing links, the hunt for earliest man*. London: Penguin.

Reid, R. E. H. 1984. The histology of dinosaurian bone, and its possible bearing on dinosaurian physiology. *Symp. Zool. Soc. Lond.* **52**, 629–63.

Reig, O. A., J. A. W. Kirsch & L. G. Marshall 1987. Systematic relationships of the living and Neocenozoic opossum-like marsupials (Suborder Didelphimorpha) with comments on the classification of these and of the Cretaceous and Paleogene New World and European metatherians. In *Possums and opossums*, M. Archer (ed.), 1–89. Sydney: Royal Zoological Society of N.S.W.

Reisz, R. R. 1981. A diapsid reptile from the Pennsylvanian of Kansas. *Spec. Publ. Mus. Nat. Hist., Univ. Kansas* **7**, 1–74.

Reisz, R. R. 1986. Pelycosauria. *Handb. Paläoherpetol.* **17A**, 1–102.

351

Repenning, C. A. & R. H. Tedford 1977. Otarioid seals of the Neogene. *Prof. Pap. U.S. Geol. Surv.* **992**, 1–93.

Repetski, J. E. 1978. A fish from the Upper Cambrian of North America. *Science* **200**, 529–31.

Ricqlès, A. de 1980. Tissue structure of dinosaur bone. In *A cold look at the warm-blooded dinosaurs*, R. D. K. Thomas & E. C. Olsen (eds), 103–39. Boulder, Colo.: Westview Press.

Ricqlès, A. de & J. R. Bolt 1983. Jaw growth and tooth replacement in *Captorhinus aguti* (Reptilia: Captorhinomorpha): a morphological and histological analysis. *J. Vertebr. Paleont.* **3**, 7–24.

Rieppel, O. 1988. The classification of the Squamata. In *The phylogeny and classification of the tetrapods, volume 1: amphibians, reptiles, birds*, M. J. Benton (ed.), *Syst. Ass. Spec. Vol.* **35A**, 261–93. Oxford: Clarendon Press.

Ritchie, A. 1968. *Phlebolepis elegans* Pander, an Upper Silurian thelodont from Oesel, with remarks on the morphology of thelodonts. In *Current problems of lower vertebrate phylogeny*, T. Ørvig (ed.), *4th Nobel Symp.* 81–8. Stockholm: Almquist & Wiksell.

Ritchie, A. & J. Gilbert-Tomlinson 1977. First Ordovician vertebrates from the southern hemisphere. *Alcheringa* **1**, 351–68.

Robinson, J. A. 1975. The locomotion of plesiosaurs. *N. Jb. Geol. Paläont. Abh.* **149**, 286–332.

Romer, A. S. 1966. *Vertebrate paleontology*, 3rd edn. University of Chicago Press.

Romer, A. S. & L. I. Price 1940. Review of the Pelycosauria. *Spec. Pap. Geol. Soc. Am.* **28**, 1–538.

Rose, K. D. 1981. Composition and species diversity in Paleocene and Eocene mammal assemblages: an empirical study. *J. Vertebr. Paleont.* **1**, 367–88.

Rosen, D. E., P. Forey, B. G. Gardiner & C. Patterson 1981. Lungfishes, tetrapods, paleontology and plesiomorphy. *Bull. Am. Mus. Nat. Hist.* **167**, 163–275.

Rosen, D. E. & C. Patterson 1969. The structure and relationships of the paracanthopterygian fishes. *Bull. Am. Mus. Nat. Hist.* **141**, 357–474.

Rowe, T. 1988. Dentition, diagnosis, and origin of Mammalia. *J. Vertebr. Paleont.* **8**, 241–64.

Russell, D. A. 1967. Systematics and morphology of American mosasaurs (Reptilia, Sauria). *Bull. Peabody Mus. Nat. Hist.* **23**, 1–237.

Russell, D. A. 1969. A new specimen of *Stenonychosaurus* from the Oldman Formation (Cretaceous) of Alberta. *Can. J. Earth Sci.* **6**, 595–612.

Russell, D. A. 1970. *Tyrannosaurus* from the Late Cretaceous of western Canada. *Natn. Mus. Nat. Sci., Publ. Paleont.* **1**, 1–30.

Russell, D. A. 1972. Ostrich dinosaurs from the Late Cretaceous of western Canada. *Can. J. Earth Sci.* **9**, 375–402.

Russell, D. E. 1964. Les mammifères paléocènes d'Europe. *Mém. Mus. Natn. Hist. Nat. Paris C* **13**, 1–324.

Santa Luca, A. P. 1980. The postcranial skeleton of *Heterodontosaurus tucki* from the Stormberg of South Africa. *Ann. S. Afr. Mus.* **79**, 159–211.

Savage, D. E. & D. E. Russell 1983. *Mammalian paleofaunas of the world*. London: Addison-Wesley.

Savage, R. J. G. & M. R. Long 1986. *Mammal evolution*. London: British Museum (Natural History).

Schaal, S. & W. Ziegler (eds) 1989. *Messel – ein Schaufenster in die Geschichte der Erde und des Lebens*. Frankfurt: Waldemar Kramer.

Schaeffer, B. 1975. Comments on the origin and basic radiation of the gnathostome fishes with particular reference to the feeding mechanism. *Coll.*

Intern. C.N.R.S. **218**, 101–9.

Schaeffer, B. & D. E. Rosen 1961. Major adaptive levels in the evolution of the actinopterygian feeding mechanism. *Am. Zool.* **1**, 187–204.

Schaeffer, B. & K. S. Thomson 1980. Reflections on agnathan–gnathostome relationships. In *Aspects of vertebrate history*, L. L. Jacobs (ed.), 19–33. Flagstaff: Museum of Northern Arizona Press.

Schaeffer, B. & M. Williams 1977. Relationships of fossil and living elasmobranchs. *Am. Zool.* **17**, 293–302.

Schoch, R. M. 1986. Systematics, functional morphology and macroevolution of the extinct mammalian order Taeniodonta. *Bull. Peabody Mus. Nat. Hist.* **42**, 1–307.

Sereno, P. C. 1986. Phylogeny of the bird-hipped dinosaurs (Order Ornithischia). *Natn. Geogr. Res.* **2**, 234–56.

Sibley, C. G. & J. E. Ahlquist 1987. DNA hybridization evidence of hominoid phylogeny. *J. Mol. Evol.* **26**, 99–121.

Sibley, C. G., J. E. Ahlquist & B. C. Monroe 1988. A classification of the living birds of the world based on DNA–DNA hybridization studies. *Auk* **105**, 409–23.

Simons, E. L. 1964. The early relatives of man. *Scient. Am.* **211**(7), 50–62.

Simons, E. L. 1984. Dawn ape of the Fayum. *Nat. Hist.* **93**, 18–20.

Simpson, G. G. 1937. The Fort Union of the Crazy Mountain Field, Montana and its mammalian faunas. *Bull. U.S. Natn. Mus.* **169**, 1–287.

Simpson, G. G. 1948. The beginning of the age of mammals in South America. Part 1. *Bull. Am. Mus. Nat. Hist.* **91**, 1–232.

Simpson, G. G. 1961. *Horses*. Garden City, NY: Anchor.

Simpson, G. G. 1967. The beginning of the age of mammals in South America. Part 2. *Bull. Am. Mus. Nat. Hist.* **137**, 1–259.

Simpson, G. G. 1975. Fossil penguins. In *The biology of penguins*, B. Stonehouse (ed.), 19–41. London: Macmillan.

Simpson, G. G. 1980. *Splendid isolation, the curious history of South American mammals*. New Haven: Yale University Press.

Sloan, R. E., J. R. Rigby, Jr., L. M. Van Valen & D. Gabriel 1986. Gradual dinosaur extinction and simultaneous ungulate radiation in the Hell Creek Formation. *Science* **232**, 629–33.

Smith, K. K. 1980. Mechanical significance of streptostyly in lizards. *Nature, Lond.* **283**, 778–9.

Smithson, T. R. 1982. The cranial morphology of *Greererpeton burkemorani* Romer (Amphibia: Temnospondyli). *Zool. J. Linn. Soc.* **76**, 29–90.

Spotila, J. R., P. W. Lommen, G. S. Bakken & D. M. Gates 1973. A mathematical model for body temperature of large reptiles: implications for dinosaur ecology. *Am. Nat.* **107**, 391–404.

Stahl, B. J. 1974. *Vertebrate history: problems in evolution*. New York: McGraw Hill.

Stehli, F. G. & S. D. Webb (eds) 1985. *The great American biotic interchange*. New York: Plenum.

Stensiö, E. 1963. The brain and the cranial nerves in fossil lower craniate vertebrates. *Skrift. Norske Vidensk.-Akad. Oslo. I. Mat-Naturv. Kl.* **13**, 1–120.

Storch, G. 1978. *Eomanis waldi*, ein Schuppentier aus dem Mittel-Eozän der 'Grube Messel' bei Darmstadt (Mammalia: Pholidota). *Senck. Leth.* **59**, 503–29.

Storch, G. 1981. *Eurotamandua jorensi*, ein Myrmecophagide aus dem Eozän der 'Grube Messel' bei Darmstadt (Mammalia, Xenarthra). *Senck. Leth.* **61**, 247–89.

Storch, G. & A. M. Lister 1985. *Leptictidium nasutum*, ein Pseudorhynchocyonide aus dem Eozän der 'Grube Messel' bei Darmstadt (Mammalia, Proteutheria). *Senck. Leth.* **66**, 1–37.

Stringer, C. B. & P. Andrews 1988. Genetic and fossil evidence for the origin of modern humans. *Science* **239**, 1263–8.

Sues, H.-D. 1986. The skull and dentition of two tritylodontid synapsids from the Lower Jurassic of western North America. *Bull. Mus. Comp. Zool.* **151**, 215–66.

Sues, H.-D. 1987. On the skull of *Placodus gigas* and the relationships of the Placodontia. *J. Vertebr. Paleont.* **7**, 138–44.

Sullivan, R. M. 1987. Reassessment of reptile diversity across the Cretaceous-Tertiary boundary. *Contr. Sci. Nat. Hist. Mus. Los Angeles Co.* **391**, 1–26.

Szaley, F. S. 1976. Systematics of the Omomyidae (Tarsiiformes, Primates), taxonomy, phylogeny and adaptations. *Bull. Am. Mus. Nat. Hist.* **156**, 157–450.

Szalay, F. S. & E. Delson 1979. *Evolutionary history of the primates.* New York: Academic.

Tarsitano, S. & M. K. Hecht 1980. A reconsideration of the reptilian relationships of *Archaeopteryx*. *Zool. J. Linn. Soc.* **69**, 149–82.

Tassy, P. & J. Shoshani 1988. The Tethytheria: elephants and their relatives. In *The phylogeny and classification of the tetrapods, volume 2: mammals*, M. J. Benton, *Syst. Ass. Spec. Vol.* **35B**, 283–315. Oxford: Clarendon Press.

Taylor, M. A. 1987. How tetrapods feed in water: a functional analysis by paradigm. *Zool. J. Linn. Soc.* **91**, 171–95.

Thies, D. & W.-E. Reif 1985. Phylogeny and evolutionary ecology of Mesozoic Neoselachii. *N. Jb. Geol. Paläont. Abh.* **169**, 333–61.

Thomas, R. D. K. & E. C. Olson (eds) 1980. *A cold look at the warm-blooded dinosaurs.* Boulder, Colo.: Westview Press.

Thomson, K. S. 1969. The biology of the lobe-finned fishes. *Biol. Rev.* **44**, 91–154.

Thulborn, R. A. 1971. Tooth wear and jaw action in the Triassic ornithiscian dinosaur *Fabrosaurus*. *J. Zool., Lond.* **164**, 165–79.

Thulborn, R. A. 1982. Speeds and gaits of dinosaurs. *Palaeogeogr., Palaeoclimat., Palaeoecol.* **38**, 227–56.

Trewin, N. H. 1986. Palaeoecology and sedimentology of the Achanarras fish bed of the Middle Old Red Sandstone, Scotland. *Trans. R. Soc. Edinb.: Earth Sci.* **77**, 21–46.

Trinkhaus, E. 1986. The neanderthals and modern human origins. *Ann. Rev. Anthropol.* **15**, 193–218.

Tucker, M. E. & M. J. Benton 1982. Triassic environments, climates and reptile evolution. *Palaeogeogr., Palaeoclimat., Palaeoecol.* **40**, 361–79.

Van Valen, L. M. 1984. Catastrophes, expectations, and the evidence. *Paleobiology* **10**, 121–37.

Van Valen, L. M. 1985. Why and how do mammals evolve unusually rapidly? *Evol. Theory* **7**, 127–32.

Van Valen, L. M. 1988. Paleocene dinosaurs or Cretaceous ungulates in South America. *Evol. Monogr.* **10**, 1–79.

Viohl, G. 1985. Geology of the Solnhofen Lithographic Limestones and the habitat of *Archaeopteryx*. In *The beginnings of birds*, M. K. Hecht, J. H. Ostrom, G. Viohl & P. Wellnhofer (eds), 31–44. Eichstätt: Freunde des Jura-Museums.

Walker, A., R. E. Leakey, J. M. Harris & F. H. Brown 1986. 2.5 Myr *Australopithecus boisei* from west of Lake Turkana, Kenya. *Nature, Lond.* **322**, 517–22.

Walker, A. & M. Teaford 1989. The hunt for *Proconsul*. *Scient. Am.* **260**(1), 58–64.

Walker, A. D. 1961. Triassic reptiles from the Elgin area: *Stagonolepis, Dasygnathus* and their allies. *Phil. Trans. R. Soc. B* **244**, 103–204.

Walker, A. D. 1964. Triassic reptiles from the Elgin area: *Ornithosuchus* and the origin of carnosaurs. *Phil. Trans. R. Soc. B* **248**, 53–134.

Webb, S. D. 1986. On the interrelationships of tree sloths and ground sloths. In *The evolution and ecology and armadillos, sloths, and vermilinguas*, G. G. Montgomery (ed.), 105–12. Washington: Smithsonian Institution Press.

Weishampel, D. B. 1981. Acoustic analysis of potential vocalization in lambeosaurine dinosaurs (Reptilia: Ornithischia). *Paleobiology* **7**, 252–61.

Weishampel, D. B. 1984a. Trossingen: E. Fraas, F. von Huene, R. Seemann and the 'Schwabische Lindwurm' *Plateosaurus*. In *Third symposium on Mesozoic terrestrial ecosystems, Tübingen 1984, short papers*, W.-E. Reif & F. Westphal (eds), 249–53. Tübingen: Attempto.

Weishampel, D. B. 1984b. Evolution of jaw mechanisms in ornithopod dinosaurs. *Adv. Anat. Embryol. Cell Biol.* **87**, 1–110.

Weishampel, D. B., P. Dodson & H. Osmólska (eds) 1990. *The Dinosauria*. Berkeley: University of California Press.

Welles, S. P. 1984. *Dilophosaurus wetherilli* (Dinosauria, Theropoda), osteology and comparison. *Palaeontographica A* **185**, 85–180.

Wellstead, C. F. 1982. A Lower Carboniferous aïstopod amphibian from Scotland. *Palaeontology* **25**, 193–208.

Wellnhofer, P. 1974. Das fünfte Skelettexemplar von *Archaeopteryx*. *Palaeontographica A* **147**, 169–216.

Wellnhofer, P. 1978. Pterosauria. *Handb. Paläoherpetol.* **19**, 1–82.

Wellnhofer, P. 1987. Die Flughaut von *Pterodactylus* (Reptilia, Pterosauria) am Beispiel des Weiner Exemplares von *Pterodactylus kochi* (Wagner). *Ann. Naturhist. Mus. Wien A* **88**, 149–62.

Wellnhofer, P. 1988a. Terrestrial locomotion in pterosaurs. *Hist. Biol.* **1**, 3–16.

Wellnhofer, P. 1988b. Ein neues Exemplar von *Archaeopteryx*. *Archaeopteryx* **6**, 1–30.

Wild, R. 1973. Die Triasfauna der Tessiner Kalkalpen. XXIII. *Tanystropheus longobardicus* (Bassani) (Neue Ergebnisse). *Schweiz. Paläont. Abh.* **95**, 1–162.

Wild, R. 1978. Die Flugsaurier (Reptilia, Pterosauria) aus der oberen Trias von Cene bei Bergamo, Italien. *Boll. Soc. Palaeont. Ital.* **17**, 176–256.

Wood, A. E. 1962. The early Tertiary rodents of the family Paramyidae. *Trans. Am. Phil. Soc.* **52**, 1–261.

Wood, B., L. Martin & P. Andrews (eds) 1986. *Major topics in primate and human evolution*. Cambridge University Press.

Wyss, A. R. 1988. Evidence from flipper structure for a single origin of pinnipeds. *Nature, Lond.* **334**, 427–8.

Young, G. C. 1986. The relationships of placoderm fishes. *Zool. J. Linn. Soc.* **88**, 1–57.

Young, J. Z. 1981. *The life of vertebrates*. Oxford: Clarendon Press.

Zangerl. R. 1981. Chondrichthyes. I. Paleozoic elasmobranchs. *Handb. Paleoichthyol.* **3A**, 1–115.

Zangerl, R. & G. R. Case 1973. Iniopterygia, a new order of chondrichthyan fishes from the Pennysylvanian of North America. *Fieldiana, Geol. Mem.* **6**, 1–67.

Zapfe, H. 1979. *Chalicotherium grande* (Blainv.) aus der miozänen Spaltenfüllung von Neudorf an der March (Devinska Nova Ves) Tschechoslowakei. *Neue Denkschr. Naturhist. Mus. Wien* **2**, 1–282.

Additional references used in Figures

Andersson, K. A. 1907. Die Pterobranchier der schwedischen Südpolarexpedition 1901–1903, nebst Bemerkungen über *Rhabdopleura normani* Allman. *Wiss. Ergebn.*

schwed. Südpolarexped. **5**(1) Zoologie, 1–122.

Andrews, C. W. 1896. On the skull of *Orycteropus gaudryi* from Samos. *Proc. Zool. Soc. Lond.* **1896**, 196–9.

Andrews, C. W. 1901. On the extinct birds of Patagonia. I. The skull and skeleton of *Phororhacos inflatus* Ameghino. *Trans. Zool. Soc. Lond.* **15**, 55–86.

Andrews, C. W. 1906. *A descriptive catalogue of the Tertiary Vertebrata of the Fayûm, Egypt.* London: British Museum (Natural History).

Andrews, C. W. 1910. *Descriptive catalogue of the marine reptiles of the Oxford Clay, Part 1.* London: British Museum (Natural History).

Andrews, S. M. 1973. Interrelationships of crossopterygians. In *Interrelationships of fishes*, R. S. Miles & C. Patterson (eds), 137–77. London: Academic.

Arratia, G. 1981. *Varasichthys ariasi* n. gen. et sp. from the Upper Jurassic of Chile (Pisces, Teleostei, Varasichthyidae n. fam.). *Palaeontographica* **A175**, 107–95.

Barrington, E. J. 1965. *The biology of the Hemichordata and Protochordata.* Edinburgh: Oliver & Boyd.

Bartram, A. W. H. 1977. The Macrosemiidae, a Mesozoic family of holostean fishes. *Bull. Br. Mus (Nat. Hist.), Geol. Ser.* **29**, 137–234.

Beer, G. R. de & W. A. Fell 1937. On development of the Monotremata. Part III. The development of the skull of *Ornithorhynchus. Trans. Zool. Soc. Lond.* **23**, 1–42.

Benton, M. J. 1986b. *The history of life on earth.* London: Kingfisher.

Black, D. 1934. On the discovery, morphology, and environment of *Sinanthropus pekinensis. Phil. Trans. R. Soc.* **B223**, 57–120.

Blainville, H. M. de 1839. *Ostéographie des mammifères.* Paris.

Bonaparte, J. F. 1978. El Mesozoico de America del Sur y sus tetrapodos. *Opera Lilloana* **26**, 1–596.

Bone, Q. 1958. Observations upon the living larva of amphioxus. *Pubbl. Staz. Zool. Napoli* **48**, 236–68.

Bown, T. M. & D. W. Krause 1979. Origin of the tribosphenic molar and metatherian and eutherian dental formulae. In *Mesozoic mammals*, J. A. Lillegraven, Z. Kielan-Jaworowska & W. A. Clemens (eds), 172–81. Berkeley: University of California Press.

Boy, J. A. 1974. Die Larven der rhachitomen Amphibien (Amphibia: Temnospondyli; Karbon-Trias). *Paläont. Z.* **48**, 236–68.

Brien, P. 1948. Embranchement des Tuniciers. Morphologie et reproduction. In *Traité de Zoologie*, 2, P. Grassé (ed.), 553–930. Paris: Masson.

Brink, A. S. 1956. On *Aneugomphius ictidoceps* Broom and Robinson. *Palaeont. Afr.* **4**, 97–115.

Broom, R. 1932. *The mammal-like reptiles of South Africa.* London: Witherby.

Brown, B. 1917. A complete skeleton of the horned dinosaur *Monoclonius*, and description of a second skeleton showing skin impressions. *Bull. Am. Mus. Nat. Hist.* **37**, 281–306.

Bulman, O. M. B. & W. F. Whittard 1926. On *Branchiosaurus* and allied genera (Amphibia). *Proc. Zool. Soc. Lond.* **1926**, 533–80.

Burmeister, H. 1874. Monografía de los glyptodontes en el Museo Público de Buenos Aires. *An. Mus. Nacn. Buenos Aires* **2**, 1–412.

Butler, P. M. 1981. The giant erinaceid insectivore *Deinogalerix* Freudenthal, from the Upper Miocene of Gargano, Italy. *Scripta Geol.* **57**, 1–72.

Carpenter, K. 1982. Skeletal and dermal armor reconstruction of *Euoplocephalus tutus* (Ornithischia: Ankylosauridae) from the Late Cretaceous Oldman Formation of Alberta. *Canad. J. Earth Sci.* **19**, 689–97.

Carroll, R. L. & P. J. Currie 1975. Microsaurs as possible apodan ancestors. *Zool. J.*

Linn. Soc. **59**, 229–47.

Cassiliano, M. L. & W. A. Clemens 1979. Symmetrodonta. In *Mesozoic mammals*, J. A. Lillegraven, Z. Kielan-Jaworowska & W. A. Clemens (eds), 150–61. Berkeley: University of California Press.

Clemens, W. A. 1966. Fossil mammals of the type Lance Formation, Wyoming. Part II. Marsupialia. *Univ. Calif. Publ. Geol. Sci.* **62**, 1–122.

Cocude-Michel, M. 1963. Les rhynchocephales et les sauriens des calcaires lithographiques (Jurassique supérieur) d'Europe occidentale. *Nouv. Arch. Mus. Nat. Hist. Lyon* **7**, 1–187.

Coombs, W. P., Jr. 1975. Sauropod habits and habitats. *Palaeogeogr. Palaeoclimatol., Palaeoecol.* **17**, 1–33.

Crompton, A. W. 1968. The enigma of the evolution of mammals. *Optima* **18**, 137–51.

Crompton, A. W. 1972. Postcanine occlusion in cynodonts and tritylodonts. *Bull. Am. Mus. (Nat. Hist.), Geol. Ser.* **21**, 29–71.

Crompton, A. W. & F. R. Jenkins 1979. Origin of mammals. In *Mesozoic mammals*, J. A. Lillegraven, Z. Kielan-Jaworowska & W. A. Clemens (eds), 59–73. Berkeley: University of California Press.

Crowther, P. & J. Martin 1976. *The Rutland dinosaur Cetiosaurus*. Leicester: Leicestershire Museums Service.

Cruickshank, A. R. I. & M. J. Benton 1985. Archosaur ankles and the relationships of the thecodontian and dinosaurian reptiles. *Nature* **317**, 715–17.

Day, M. H., R. E. F. Leakey, A. C. Walker & B. A. Wood 1974. New hominids from East Turkana, Kenya *Am. J. Phys. Anthrop.* **42**, 461–73.

Dean, B. 1985. *Fishes living and fossil*. New York: Macmillan.

Delson, E. 1985. Palaeobiology and age of African *Homo erectus*. *Nature* **316**, 762–3.

Dong, Z. & Tang Z. 1984. Note on a Mid-Jurassic sauropod (*Datousaurus bashanensis* gen. et sp. nov.) from Sichuan Basin, China. *Vertebr. Palasiat.* **22**, 69–75.

Evans, F. G. 1942. The osteology and relationships of the elephant shrews (Macroscelididae). *Bull. Am. Mus. Nat. Hist.* **80**, 85–125.

Finch, M. E. & L. Freedman 1982. *Carnivorous marsupials*. R. Zool. Soc. New South Wales.

Flannery, T. F. 1982. Hindlimb structure and evolution in the kangaroos (Marsupialia: Macropodoidea). In *The fossil vertebrate record of Australia*, P. V. Rich & E. M. Thompson (eds), 507–24. Clayton: Monash University Press.

Flower, W. H. & R. Lydekker 1891. *An introduction to the study of mammals, living and extinct*. London: Evans.

Galton, P. M. 1971. The prosauropod *Ammosaurus*, the crocodile *Protosuchus*, and their bearing on the age of the Navajo sandstone of northeastern Arizona. *J. Paleont.* **45**, 781–95.

Galton, P. M. 1977. On *Staurikosaurus pricei*, an early saurischian dinosaur from the Triassic of Brazil, with notes on the Herrerasauridae and Poposauridae. *Paläont. Z.* **51**, 234–45.

Gidley, J. W. 1907. A new horned rodent from the Miocene of Kansas. *Proc. U.S. Natn. Mus.* **32**, 627–36.

Gilmore, J. W. 1914. Osteology of the armored Dinosauria in the United States National Museum, with special reference to the genus *Stegosaurus*. *Bull. U.S. Natn. Mus.* **89**, 1–143.

Goode, G. B. & T. H. Bean 1895. *Oceanic ichthyology*. Washington: Smithsonian Institution.

Goody, P. C. 1969. The relationships of certain Upper Cretaceous teleosts with special reference to the myctophids. *Bull. Br. Mus. (Nat. Hist.), Geol. Ser., Suppl.* **7**, 1–255.

Grande, L. 1988. A well preserved paracanthopterygian fish (Teleostei) from freshwater lower Paleocene deposits of Montana. *J. Vertebr. Paleont.* **8**, 117–30.

Grande, L., J. T. Eastman & T. M. Cavender 1982. *Amyzon gosiutensis*, a new catostomid fish from the Green River Formation. *Copeia* 1982, 523–32.

Grande, L. & J. G. Lundberg 1988. Revision and redescription of the genus *Astephus* (Siluriformes: Ictaluridae) with a discussion of its phylogenetic relationships. *J. Vertebr. Paleont.* **8**, 139–71.

Gregory, J. T. 1948. A new limbless vertebrate from the Pennsylvanian of Mazon Creek, Illinois. *Am. J. Sci.* **246**, 636–63.

Gregory, W. K. 1926. The skeleton of *Moschops capensis* Broom, a dinocephalian reptile from the Permian of South Africa. *Bull. Am. Mus. Nat. Hist.* **56**, 179–251.

Gregory, W. K. 1929. *Our face from fish to man*. New York: G. P. Putnam's Sons.

Gregory, W. K. 1933. Fish skulls: a study of the evolution of natural mechanisms. *Trans. Am. Phil. Soc.* **23**, 75–481.

Gregory, W. K. 1935. Further observations on the pectoral girdle and fin of *Sauripterus taylori* Hall, A crossopterygian fish from the Upper Devonian of Pennsylvania, with special reference to the origin of the pentadactyle extremities of Tetrapoda. *Proc. Am. Phil. Soc.* **75**, 673–90.

Gregory, W. K. & M. Hellman 1929. Paleontology of the human dentition. *Intern. J. Orthod.* **15**, 642–52.

Halstead, L. B. 1973b. *Ecology and evolution of the mammals*. London: Peter Lowe.

Halstead, L. B. 1985. The vertebrate invasion of fresh water. *Phil. Trans. R. Soc.* **B309**, 243–58.

Hatcher, J. B. 1901. *Diplodocus* (Marsh), its osteology, taxonomy and probable habits, with a restoration of the skeleton. *Mem. Carnegie Mus.* **1**, 1–63.

Hemmings, S. K. 1978. The Old Red Sandstone antiarchs of Scotland: *Pterichthyodes* and *Microbrachius*. *Monogr. Palaeontogr. Soc.* **131**(551), 1–64.

Heyler, D. 1975. Sur les 'Branchiosaurus' et autres petits amphibiens apparentés de la Sarre et du Bassin d'Autun. *Bull. Soc. Hist. Nat. Autun* **75**, 15–27.

Hildebrand, M. 1974. *Analysis of vertebrate structure*. New York: Wiley.

Hopson, J. A. 1966. The origin of the mammalian middle ear. *Am. Zool.* **6**, 437–50.

Huene, F. von 1929. Los saurisquios y ornitisquios del Cretáceo Argentino. *An. Mus. La Plata* (2) **3**, 1–196.

Ivakhnenko, K. F. 1978. Urodelans from the Triassic and Jurassic of Soviet Central Asia. *Palaeont. J.* **12**, 362–8.

Jaekel, O. 1915. Die Wirbeltierfunde aus dem Keuper von Halbertadt. *Paläont. Z.* **2**, 88–214.

Jefferies, R. P. S. 1969. *Ceratocystis perneri* – a Middle Cambrian chordate with echinoderm affinities. *Palaeontology* **12**, 494–535.

Jenkins, F. A., Jr. 1971b. Limb posture and locomotion in the Virginia opossum (*Didelphis marsupialis*) and in other non-cursorial mammals. *J. Zool.* **165**, 303–15.

Jensen, D. 1966. The lampreys and hagfishes. *Scient. Am.* **214**(2), 82–90.

Keast, A. 1972. Australian mammals: zoogeography and evolution. *Q. Rev. Biol.* **43**, 373–408.

REFERENCES

Kellogg, R. M. 1936. A review of the Archaeoceti. *Publ. Carnegie Inst. Washington* **482**, 1–366.

Kielan-Jaworowska, Z. 1968. Preliminary data on the Upper Cretaceous eutherian mammals from Bayn Dzak, Gobi Desert. *Palaeont. Polonica* **19**, 171–91.

Kielan-Jaworowska, Z. 1971. Skull structure and affinities of the Multituberculata. *Palaeont. Polonica* **25**, 5–41.

Kielan-Jaworowska, Z. 1975. Preliminary description of two new eutherian genera from the Late Creteceous of Mongolia. *Palaeont. Polonica* **33**, 3–16.

Kielan-Jaworowska, Z. 1978. Evolution of the therian mammals in the Late Cretaceous of Asia. Part III. Postcranial skeleton of the Zalambdalestidae. *Palaeont. Polonica* **38**, 3–41.

Kielan-Jaworowska, Z., A. W. Crompton & F. A. Jenkins, Jr. 1987. The origin of egg-laying mammals. *Nature* **326**, 871–3.

Krebs, B. 1971. Evolution of the mandible and lower jaw dentition in dryolestids (Pantotheria, Mammalia). *Zool. J. Linn. Soc., Suppl.* **50**(1), 89–102.

Kron, D. G. 1979. Docodonta. In *Mesozoic mammals*. J. A. Lillegraven, Z. Kielan-Jaworowska & W. A. Clemens (eds), 91–8. Berkeley: University of Chicago Press.

Kuhn, O. 1969. Cotylosauria. *Handb. Paläoherpetologie* **6**, 1–89.

Kuhn-Schnyder, E. 1963. *I sauri del Monte San Giorgio*. Archivo Storico Ticinese.

Lapparent, A. F. de & R. Lavocat 1955. Dinosauriens. In *Traité de paléontologie*, J. Piveteau (ed.), **3**, 93–104. Paris: Masson.

Lawlor, T. E. 1979. *Handbook to the orders and families of living mammals*. Eureka, Ca.: Mad River Press.

Loomis, F. B. 1914. *The Deseado Formation of Patagonia*. Amherst, Ma.: University of Massuchusetts Press.

Marsh, O. C. 1880. *Odontornithes: a monograph on the extinct toothed birds of North America*. Washington: Government Printing Office.

Marsh, O. C. 1884. Principal characters of the American Jurassic dinosaurs. Part VIII. The order of Theropoda. *Am. J. Sci.* (3) **27**, 329–40.

Marsh, O. C. 1885. Dinocerata. A monograph of an extinct order of gigantic mammals. *Monogr. U.S. Geol. Surv.* **10**, 1–237.

Marshall, L. G. 1980a. Marsupial paleobiogeography. In *Aspects of vertebrate history*, L. L. Jacobs (ed.), 345–86. Flagstaff: Museum of Northern Arizona Press.

Marshall, L. G. 1980b. Systematics of the South American marsupial family Caenolestidae. *Fieldiana, Geol.* **5**, 1–145.

Matthew, W. D. 1909. The Carnivora and Insectivora of the Bridger Basin, Middle Eocene. *Mem. Am. Mus. Nat. Hist.* **9**, 291–567.

Matthew, W. D. 1918. Edentata. *Bull. Am. Mus. Nat. Hist.* **38**, 565–657.

McDowell, S. B. 1958. The Greater Antillean insectivores. *Bull. Am. Mus. Nat. Hist.* **115**, 113–214.

McNab, B. K. 1978. The evolution of endothermy in the phylogeny of mammals. *Am. Nat.* **112**, 1–21.

McNab, B. K. & W. Auffenberg 1976. The effect of large body size on the temperature regulation of the Komodo dragon, *Varanus komodoensis*. *Comp. Biochem. Physiol.* **55A**, 345–50.

Milner, A. R., T. R. Smithson, A. C. Milner, M. I. Coates & W. D. I. Rolfe 1986. The search for early tetrapods. *Modern Geol.* **10**, 1–28.

Moss, S. A. 1972. The feeding mechanism of sharks of the family Carcharhinidae. *J. Zool. London* **167**, 423–36.

Napier, J. R. 1962. The evolution of the hand. *Scient. Am.* **207**(12), 56–62.

Nash, D. S. 1975. The morphology and relationships of a crocodilian, *Orthosuchus stormbergi*, from the Upper Triassic of Lesotho. *Ann. S. Afr. Mus.* **67**, 227–329.

Newman, B. H. 1970. Stance and gait in the flesh-eating dinosaur *Tyrannosaurus*. *Biol. J. Linn. Soc.* **2**, 119–23.

Olsen, P. E. 1984. The skull and pectoral girdle of the parasemionotid fish *Watsonulus eugnathoides* from the Early Triassic Sakamena Group of Madagascar, with comments on the relationships of the holostean fishes. *J. Vertebr. Paleont.* **4**, 481–99.

Olson, S. L. & A. Feduccia 1980. *Presbyornis* and the origin of the Anseriformes (Aves: Charadriomorphae). *Smiths. Contr. Zool.* **323**, 1–24.

Orlov, J. A. 1958. The carnivorous dinocephalians of the Isheevo fauna (Titanosuchia). *Trudy Paleont. Inst. Akad. Nauk. Sci. USSR.* **72**, 1–14 [In Russian.]

Osborn, H. F. 1895. Fossil mammals of the Uinta basin. *Bull. Am. Mus. Nat. Hist.* **7**, 71–105.

Osborn, H. F. 1904. The great Cretaceous fish *Portheus molossus* Cope. *Bull. Am. Mus. Nat. Hist.* **20**, 377–81.

Osborn, H. F. 1910. *The age of mammals*. New York: Columbia University Press.

Osborn, H. F. 1916. Skeletal adaptations of *Ornitholestes, Struthiomimus, Tyrannosaurus*. *Bull. Am. Mus. Nat. Hist.* **35**, 733–71.

Osborn, H. F. & C. C. Mook 1921. *Camarasaurus, Amphicoelias* and other sauropods of Cope. *Mem. Am. Mus. Nat. Hist.* **3**, 245–387.

Ostrom, J. H. & J. S. McIntosh 1966. *Marsh's dinosaurs*. New Haven: Yale University Press.

Parrish, J. M. 1986. Locomotor adaptations in the hindlimb and pelvis of the Thecodontia. *Hunteria* **1**(2), 1–35.

Patterson, B. & R. Pascual 1968. The fossil mammal fauna of South America. *Q. Rev. Biol.* **43**, 409–51.

Patterson, C. 1964. A review of Mesozoic acanthopterygian fishes, with special reference to those of the English chalk. *Phil. Trans. R. Soc. B* **247**, 213–482.

Patterson, C. 1970. Two Upper Cretaceous salmoniform fishes from the Lebanon. *Bull. Br. Mus. (Nat. Hist.), Geol. Ser.* **19**, 205–96.

Pearson, H. S. 1924. A dicynodont reptile reconstructed. *Proc. Zool. Soc. Lond.* **1924**, 827–55.

Peterson, O. A. 1909. A revision of the Entelodontidae. *Mem. Carnegie Mus.* **4**, 42–158.

Peterson, O. A. 1934. List of species and description of new material from the Duchesne River, Oligocene, Uinta Basin, Utah. *Ann. Carnegie Mus.* **23**, 373–89.

Peyer, B. 1950. *Geschichte der Tierwelt*. Zurich: Büchergilde Gutenberg.

Peyer, B. & E. Kuhn-Schnyder 1955. Placodontia. In *Traité de Paléontologie*, J. Piveteau (ed.), **5**, 458–86.

Price, L. I. 1959. Sôbre um crocodilídeo notossúquio do Cretácico brasileiro. *Bol. Serv. Min. Geol. Brasil* **188**, 1–55.

Riggs, E. S. 1934. A new marsupial saber-tooth from the Pliocene of Argentina and its relationships to other South American predacious marsupials. *Trans. Am. Phil. Soc.* **24**, 1–32.

Riggs, E. S. 1935. A skeleton of *Astrapotherium*. *Geol. Ser. Field. Mus. Nat. Hist.* **6**, 167–77.

Romer, A. S. 1933. *Vertebrate paleontology*. University of Chicago Press.

Romer, A. S. 1944. The Permian cotylosaur *Diadectes tenuitectus*. *Am. J. Sci.* **242**, 139–44.

Romer, A. S. 1952. Fossil vertebrates of the tri-state area, 2. Late Pennsylvanian and Early Permian vertebrates of the Pittsburgh – West Virginia region. *Ann. Carnegie Mus.* **33**, 47–112.

Romer, A. S. 1956. *Osteology of the reptiles.* University of Chicago Press.

Romer, A. S. & T. S. Parsons 1970. *The vertebrate body.* Philadelphia: W. B. Saunders.

Schaeffer, B. & D. H. Dunkle 1950. A semionotid fish from the Chinle Formation, with consideration of its relationships. *Am. Mus. Novit.* **1457**, 1–29.

Schoch, R. M. 1982. Phylogeny, classification and paleobiology of the Taeniodonta (Mammalia: Eutheria). *Proc. 3rd. N. Am. Paleont. Conv. Montreal* **2**, 465–70.

Scott, W. B. 1888. On some new and little-known creodonts. *J. Acad. Nat. Sci. Philad.* **9**, 155–85.

Scott, W. B. 1910. Litopterna of the Santa Cruz beds. *Rep. Princeton Univ. Exped. Patagonia* **6**, 287–300.

Scott, W. B. 1940. The mammalian fauna of the White River Oligocene. Part IV. Artiodactyla. *Trans. Am. Phil. Soc.* **28**, 363–746.

Scott, W. B. 1941. The mammalian fauna of the White River Oligocene. Part V. Perissodactyla. *Trans. Am. Phil. Soc.* **28**, 1–153.

Scott, W. B. & G. L. Jepsen 1936. The mammalian fauna of the White River Oligocene. Part I. Insectivora and Carnivora. *Trans. Am. Phil. Soc.* **28**, 1–153.

Scott, W. B. & H. F. Osborn 1887. Preliminary account of the fossil mammals from the White River formation, contained in the Museum of Comparative Zoology. *Bull. Mus. Comp. Zool.* **13**, 151–71.

Sigogneau, D. & P. K. Chudinov 1972. Reflections on some Russian eotheriodonts (Reptilia, Synapsida, Therapsida). *Palaeovertebrata* **5**, 79–109.

Simons, E. L. 1960. The Paleocene Pantodonta. *Trans. Am. Phil. Soc.* **50**(6), 1–80.

Simons, E. L. 1967. The earliest apes. *Scient. Am.* **217**(12), 28–35.

Simons, E. L. & S. R. K. Chopra 1969. *Gigantopithecus* (Pongidae, Hominoidea), a new species from north India. *Postilla* **138**, 1–18.

Simpson, G. G. 1970. The Argyrolagidae, extinct South American marsupials. *Bull. Mus. Comp. Zool.* **139**, 1–86.

Sinclair, W. J. 1906. Mammalia of the Santa Cruz beds: Marsupialia. *Rep. Princeton Univ. Exped. Patagonia* **4**, 333–460.

Steel, R. 1973. Crocodylia. *Handb. Paläoherpetol.* **16**, 1–116.

Stehlin, H. G. & S. Schaub 1951. Die Trigonodontie der simplicidentaten Nager. *Schweiz. Paläont. Abh.* **67**, 1–385.

Stensiö, E. 1969. Les cyclostomes fossiles ou ostracodermes. In *Traité de Paléontologie*, J. Piveteau (ed.), **4**(2), 71–692.

Sternberg, C. M. 1932. Two new theropod dinosaurs from the Belly River Formation of Alberta. *Canad. Field-Nat.* **46**, 99–105.

Storer, T. I. & R. L. Usinger 1965. *General zoology*, 4th edn. New York: McGraw-Hill.

Stromer, E. von 1912. *Lehrbuch der Paläozoologie.* Leipzig: Treuber.

Tattersall, I. 1970. *Man's ancestors.* London: John Murray.

Taylor, M. A. 1986. Lifestyle of plesiosaurs. *Nature* **319**, 179.

Tedford, R. H. 1966. A review of the macropodid genus *Sthenurus*. *Univ. Calif. Publ. Geol. Sci.* **57**, 1–72.

Thenius, E. 1969. Ueber einige Probleme der Stammesgeschichte der Säugetiere. *Z. Zool. Syst. Evolforsch.* **7**, 157–79.

Tobias, P. V. 1967. *Olduvai Gorge*, vol. 2. Cambridge University Press.

Trewin, N. H. 1985. Mass mortalities of Devonian fish – the Achanarras Fish Bed, Caithness. *Geol. Today* **2**, 45–9.

Turner, S. 1973. Siluro-Devonian thelodonts from the Welsh Borderland. *J. Geol. Soc.* **129**, 557–84.

Van Tyne, J. & A. J. Berger 1976. *Fundamentals of ornithology*. New York: Wiley.
Vaughan, T. A. 1972. *Mammalogy*. Philadelphia: W. B. Saunders.

Walker, A., D. Falk, R. Smith & M. Pickford 1983. The skull of *Proconsul africanus*: reconstruction and cranial capacity. *Nature* **305**, 525–7.
Ward, S. C. & D. R. Pilbeam 1983. Maxillofacial morphology of Miocene hominoids from Africa and Indo-Pakistan. In *New interpretations of ape and human ancestry*, R. L. Ciochon & R. S. Corruccini (eds), 211–38. New York: Plenum.
Wellnhofer, P. 1986. Remarks on the digit and pubis problems of *Archaeopteryx*. In *The beginnings of birds*, M. Hecht *et al.* (eds), 113–22. Eichstätt: Freunde des Jura-Museums.
Westoll, T. S. 1949. On the evolution of the Dipnoi. In *Genetics, paleontology and evolution*, G. L. Jepsen, G. G. Simpson & E. Mayr (eds), 121–84. Princeton University Press.
White, T. E. 1939. Osteology of *Seymouria baylorensis* Broili. *Bull. Mus. Comp. Zool.* **85**, 325–409.
Wieland, G. R. 1909. Revision of the Protostegidae. *Am. J. Sci.* (4) **27**, 101–30.
Wilson, J. A. 1971. Early Tertiary vertebrate faunas, Vieja Group, Trans Pecos-Texas: Agriochoeridae and Merycoidodontidae. *Bull. Texas Mem. Mus.* **18**, 1–83.
Wood, A. E. 1957. What, if anything, is a rabbit? *Evolution* **11**, 417–25.
Woodward, A. S. 1891–1901. *Catalogue of the fossil fishes in the British Museum*. London: British Museum.
Woodward, A. S. 1916. The fossil fishes of the English Wealden and Purbeck Formations. Part II. *Monogr. Palaeontogr. Soc.* **70**, 49–104.

Yalden, D. 1984. What size was *Archaeopteryx? Zool. J. Linn. Soc.* **82**, 177–88.

Zangerl, R. & M. E. Williams 1975. New evidence on the nature of the jaw suspension in Palaeozoic anacanthous sharks. *Palaeontology* **18**, 333–41.

Systematic Index

Illustrations are indicated in italics

363

Subject Index